T0291800

CAMBRIDGE LIBRARY COLLECTION

Books of enduring scholarly value

Life Sciences

Until the nineteenth century, the various subjects now known as the life sciences were regarded either as arcane studies which had little impact on ordinary daily life, or as a genteel hobby for the leisured classes. The increasing academic rigour and systematisation brought to the study of botany, zoology and other disciplines, and their adoption in university curricula, are reflected in the books reissued in this series.

English Naturalists from Neckam to Ray

C.E. Raven (1885–1964) was an academic theologian elected Regius Professor of Divinity at the University of Cambridge in 1932, who developed an interest in natural history and the history of scientific thought. First published in 1947, this volume demonstrates how changing attitudes to the natural world reflected and influenced the transformations in scientific thought between the medieval period and the eighteenth century. Raven's focus on the field of 'natural history' reveals how the scientific ideas behind modern biological studies developed from the richly illustrated and often fantastical bestiaries of the medieval world. The subjects of this volume are grouped chronologically into Pioneers, Explorers and Popularisers, with biographical details woven together with discussions of their academic work. The book provided a wealth of new information concerning the founders of natural history and remains a valuable contribution to this subject.

Cambridge University Press has long been a pioneer in the reissuing of out-of-print titles from its own backlist, producing digital reprints of books that are still sought after by scholars and students but could not be reprinted economically using traditional technology. The Cambridge Library Collection extends this activity to a wider range of books which are still of importance to researchers and professionals, either for the source material they contain, or as landmarks in the history of their academic discipline.

Drawing from the world-renowned collections in the Cambridge University Library, and guided by the advice of experts in each subject area, Cambridge University Press is using state-of-the-art scanning machines in its own Printing House to capture the content of each book selected for inclusion. The files are processed to give a consistently clear, crisp image, and the books finished to the high quality standard for which the Press is recognised around the world. The latest print-on-demand technology ensures that the books will remain available indefinitely, and that orders for single or multiple copies can quickly be supplied.

The Cambridge Library Collection will bring back to life books of enduring scholarly value (including out-of-copyright works originally issued by other publishers) across a wide range of disciplines in the humanities and social sciences and in science and technology.

English Naturalists from Neckam to Ray

A Study of the Making of the Modern World

CHARLES E. RAVEN

CAMBRIDGE UNIVERSITY PRESS

Cambridge, New York, Melbourne, Madrid, Cape Town, Singapore,
São Paolo, Delhi, Dubai, Tokyo, Mexico City

Published in the United States of America by Cambridge University Press, New York

www.cambridge.org
Information on this title: www.cambridge.org/9781108016346

This edition first published 1947
This digitally printed version 2010

ISBN 978-1-108-01634-6 Paperback

ENGLISH NATURALISTS FROM NECKAM TO RAY

ENGLISH NATURALISTS FROM NECKAM TO RAY

A Study of the Making of the Modern World

BY

CHARLES E. RAVEN, D.D.

Master of Christ's College and Regius Professor of Divinity in the University of Cambridge

CAMBRIDGE
AT THE UNIVERSITY PRESS
1947

Printed in Great Britain at the University Press, Cambridge
(Brooke Crutchley, University Printer)
and published by the Cambridge University Press
(Cambridge, and Bentley House, London)

Agents for U.S.A., Canada, and India: Macmillan

FOR MY WIFE—ALWAYS

ζηλῶ φθιμένους, κείνων ἔραμαι,
κεῖν' ἐπιθυμῶ δώματα ναίειν.

Contents

INDEXES

Preface

This book owes its existence to the kindness with which my study of John Ray was received by men and women whose opinion carries weight, to the suggestions put forward by many of them that a similar study of other early naturalists would be welcome, and to my own conviction of the interest and importance of the subject. It began as a series of biographies; and this is still its obvious characteristic. But very soon it became clear that the succession of 'lives' not only formed a very definite pattern but that this illustrated, and for me at least illuminated, the change in Western civilisation from the medieval to the modern world.

That a history of man's attitude to nature and especially to the flora and fauna of his environment would be important for the student of social development and of religious and speculative thought, has been frequently emphasised—and not least by one of the most eminent of recent philosophical theologians, Dr John Oman. But it has been something of a surprise to discover how very obviously this attitude reflected and by its alteration influenced the momentous changes taking place in the period under review. Such surprise is no doubt due partly to my own ignorance; if I had known these centuries better I might have realised the significance of their naturalists. But, indeed, the subject has never (to my knowledge) been fully treated either by the historians of science, who almost without exception pay little heed to botany and zoology, or by the students of literature and philosophy. That it richly repays investigation will I hope be evident even from so limited a survey as I have here undertaken.

My obligations are very numerous. Before expressing them in detail I may perhaps be allowed to say how deeply I have been touched by the continuous kindness and immediate help given to me by the botanists, zoologists, historians and librarians to whom I have appealed. Although in some sense an interloper and, very certainly, a nuisance to busy men and women, I have been welcomed and assisted everywhere. My recent illness and bereavement have only enlarged the circle of this generosity.

Of particular debts the following are some of the largest. To Mrs Agnes Arber whose criticism of *John Ray* in *Isis* set me upon a fresh investigation of his forerunners and whose encouragement is a continual inspiration; to Dr Julian Huxley and his colleagues who suggested to me the writing of *A History of Natural History in Britain* for which this book is a preliminary; to Mr H. Gilbert-Carter and other members of the Cambridge Botany School whose help in the identification of plants recorded by Penny and his contemporaries has been invaluable; to Dr W. H. Thorpe of the Zoological Department for similar help in other directions; to

Mr H. L. Pink and other members of the staff of the Cambridge University Library; to Mr S. Savage of the Linnean Society, Mr C. T. Onions of Magdalen College, Oxford, and to many other College Librarians; to Air-Vice-Marshal Geoffrey Keynes, especially for help over Sir Thomas Browne; to Professor A. B. Cook for the lines under the dedication; and to my own colleagues, especially Mr B. W. Downs, Dr F. H. A. Marshall and Dr C. P. Snow. In particular, I owe special gratitude to the Vice-Master of our College, Mr S. W. Grose, and to the Bursar, Mr T. C. Wyatt, who in these very difficult days have carried the chief responsibility for its welfare, freeing me from all anxiety on its account and making it possible for me to carry through a long and absorbing research.

C.E.R.

1945

A. THE PRELUDE

CHAPTER I. NATURE AND MEDIEVAL SCIENCE

Mankind has always been interested in nature, in the flora and fauna of his environment, and that not only because he lived by them and on them. When the palaeolithic artist graved and coloured the reindeer at Font-de-Gaume or the bison at Altamira; when the Egyptian painter decorated the fourth-dynasty tomb at Mêdûm with a picture of the red-breasted goose;[1] when Homer watched the wild fowl 'exulting in their wings'—'geese and cranes and long-necked swans'—over the meadow that has given its name to a continent;[2] or when the medieval craftsman filled the bosses of Southwell[3] or Ely with intricate and identifiable foliage; we have clear proof of a love of wild life for its own sake. That from the first the gathering of herbs and fruits with fishing and hunting was man's means of life, and that very long ago he had learnt to cultivate crops and to tame beasts and birds, certainly deepened and rewarded his interest, and encouraged the folk-lore and fertility festivals which are the germs of literature and of religious observance.

But though all human activities are from the first shot through with the love of nature, the student of natural history need merely notice that in most if not all high types of civilisation there has appeared what can properly be called a scientific attitude towards vegetable and animal life—an attitude, that is, which is concerned not only with the utilitarian or even aesthetic value of nature but with the objective study of its phenomena. Having noted this we can pass on to the point at which this attitude began to manifest itself when the modern Western world took shape after the Dark Ages.

In Britain, though we cannot claim any high place among the pioneers—for Italy, Germany, France and the Low Countries started in front of us—we are yet fortunate in having good if not eminent examples among our own people of all the phases through which the young sciences of botany and zoology passed before they came to adolescence. And just because this country had no such number of students as made the movement on the Continent it is the more easy to grasp and interpret the development from the frankly medieval 'bestiaries' to the 'pandects' of the Renaissance and the 'histories' of the seventeenth century.

1 *Branta ruficollis*—an unmistakable likeness of a species now only accidental in Egypt. The picture is now in the Museum in Cairo.

2 *Iliad*, II, 459–63.

3 Cf. 'The Foliage Flowers and Fruit of Southwell Chapter House', by A. C. Seward in *Cambridge Antiquarian Soc. Comm.* XXXV, pp. 1–32.

Of the general character of this development it is necessary to say something if we are to study sympathetically the individuals in whom it is represented. Even when they are men of originality they are inevitably conditioned by the outlook and prevailing concepts of their day; and we can understand neither their presuppositions nor their rebellions unless the contemporary background has been at least in outline depicted.

Particularly is this the case with the writers of our first period, that amazing thirteenth century in which the Middle Ages rose to their zenith, the century which begins with St Francis and St Dominic and includes St Albert and St Thomas Aquinas, and Roger Bacon.

Medievalism has an attitude towards the world of nature and of history which is at once universally shared by its own folk and is exceedingly difficult for the modern student to appreciate. It is not other-worldly in the sense that the natural order is regarded as illusory. A few of the more extreme ascetics, and to some extent the mystics who followed the practices of 'Dionysius the Areopagite', might regard the world as a place to be spurned and denied. But for the Catholicism of the West other-worldliness consists only in insisting that this world derives from and depends upon another; that it is the symbol and instrument of that other; and that it is to be studied and interpreted, if at all, not for its own sake but in order to disclose its supernatural meaning.

The sources of this outlook are familiar. The Church had from its early days inherited a method of allegorical interpretation from the scholars of Alexandria, a method applied not only to the interpretation of ancient literature like the *Iliad* or the Old Testament, but to the animal creation as supplying types and fables for the entertainment or instruction of mankind. At its best such an outlook enabled a coherent and Christian philosophy whereby the universe is regarded as the scene and in some sense the means of the divine self-revelation which had its full representative for man in the person of Jesus. The special incarnation in him was regarded as typical of a similar though obviously incomplete manifestation in the whole creation. At its best the period from the fourth to the seventeenth century would have echoed the words with which in 1635 Francis Quarles prefaces his *Emblemes*: 'Before the knowledge of letters God was known by Hieroglyphics. And indeed what are the heavens, the earth, nay every creature, but Hieroglyphics and emblems of his glory?'

The danger of such a view is of course that it may not only deny all worth to the natural order except in so far as its symbolic value is concerned, but may distort the interpretation of it in the interest of its supposed meaning. Nature and history, in fact, came to be regarded as the mere husk in which ideas supplied by the Church from its own tradition could find expression. This easily degenerated into a belief that the symbol itself was valueless and even that its meaning as a symbol did not depend upon any real correspondence between itself and the thing symbolised, but could be arbitrarily fixed by traditional or ecclesiastical authority. This inevitably involved the corollary that the study of the symbol was in itself useless; and therefore that the interpretation of nature consisted not in reading out of it the lesson of its true character but in reading into it an elaborate arbitrary and artificial significance. It ceased to be a means for discovering the character of reality, and became irrelevant and negligible except in so far as it was the source of imagery and anecdotes for the moralist and the preacher.

This divorce between the natural order and the supernatural had been made almost absolute by the influence of the *De Civitate Dei* of St Augustine and the consequent conviction that the secular world was totally corrupt and worthless. That the great Scholastics never assented to such a conviction, and that imperial and royal sovereignty never became wholly deprived of spiritual meaning and religious function, prevented the lapse of Christendom into complete dualism or other-worldliness: but despite St Albert the whole attitude towards nature was emblematic; men did not study flowers or birds for their own sakes or in order to learn from them new insights into reality; they went to them solely for illustrations of moral or metaphysical dogmas, accepted on authority and believed to be divinely ordained; they sought in nature not knowledge but edification, not enlightenment but the exemplification of preconceived ideas.

It is sufficiently remarkable that Christian doctrine should have so dominated the literary and intellectual world as to be able to use not merely the Old Testament but the Greek and Latin classics for its own interpretation. We know how completely the Jewish scriptures had been transformed into a prophecy, sometimes manifest, more often enigmatic and emblematic, of the Messiah and his Church. But we are less familiar with the fact that Vergil and Ovid had been similarly translated into subtle revelations of Christian truth. Vergil ever since the time of Fulgentius in the sixth Christian century, and originally on the strength of his 'Messianic' 4th *Eclogue*, had been regarded as a prophet of the Incarnation and a preacher of the Way of Christ. The *Aeneid* had become a sort of forerunner of the *Pilgrim's Progress*, an allegory of the life of man and the triumph of wisdom and virtue over folly and passion. Legends like the story of St Paul weeping over the poet's tomb[1] had gathered round him, and led almost to his canonisation. Dante was fully justified by contemporary feeling when he put Vergil forward as the supreme mystagogue of the pre-Christian world.

Ovid's case is more surprising; for even Isidore of Seville had condemned him as an evil influence. But allegorising, which had turned the Song of Songs into a rhapsody of Christ and his Church, could 'spiritualise'

1 Cf. Sandys, *Short History of Classical Scholarship*, p. 147.

even the *Art of Love* for the use of nunneries, and find rich material for fables out of the *Fasti* and the *Metamorphoses*. Lending itself admirably to quotation, and freely imitated in the elegiacs of the time, Ovid's work was probably more familiar to the scholars of the Middle Ages than that of any other classical poet. Many of the types and anecdotes most commonly used by them are drawn directly or indirectly from him.

In literature such interpretation of the work of the past is not improper: for, indeed, great poetry has a permanent significance and is in some sense inevitably representative and symbolic. The habit is more peculiar when applied to the interpretation of nature. Here we must obviously speak in greater detail.

If we are to begin at the beginning it will be with the strange volume called *Physiologus*, a collection of tales of the habits of animals put together originally, perhaps in the fourth century, perhaps in Egypt, in Greek and then in a Latin translation. Amplified in the twelfth century it became the Bestiary, and as such appears in almost every European language.[1] It has, so M. R. James declares,[2] 'no scientific or literary merits whatever', but it was furnished with pictures and as such became a usual ornament of the homes of noblemen and prelates. Thus it supplied not only the traditional fables but the heraldic beasts. Griffin, Wivern, Yale and Unicorn, Phoenix, Martlet, the familiar crests and supporters with the stories appropriate to them are derived directly from the Bestiaries; and their presence in our ancient cathedrals[3] is as obvious as it is in our modern advertisements. From these come not only the traditional ascription of certain traits to certain beasts, royalty to the Lion, cunning to the Fox, but the groundwork of legend and anecdote which colours so much of our literature and of our early natural history. That there is not a single statement in the Bestiaries which can be accepted as fact may be an exaggeration. But it is plain that it was not with fact but with doctrinal significance and moral implication that the authors were concerned. To look at such a collection is to realise how violently unscientific is its character. Natural history does not begin here.

A better starting point for the student, Alexander Neckam, or Nequam as his name is spelled by those who wish to Latinise and pun upon it, is the obvious and eminently suitable candidate.[4] His chief works, the *De Naturis Rerum* and the poem *De Laudibus Divinae Sapientiae*, a version of it in Latin elegiac verse, have been adequately edited and expounded

1 Cf. e.g. *The Bestiary of Guillaume le Clerc* (1210) translated by G. C. Druce.

2 Cf. his lecture on 'The Bestiary' in *History*, April 1931, pp. 1–11.

3 Cf. G. C. Druce in *Archaeological Journal*, LXVIII, no. 271, 2nd ser. XVIII, no. 3, pp. 173–99 (Yale); LXV, no. 264, 2nd ser. XVI, no. 4, pp. 311–38 (Crocodile), etc.

4 Adelard of Bath, for whom see C. H. Haskins, *Studies in the History of Mediaeval Science*, pp. 20–42, and whose *Quaestiones naturales* dates from early in the twelfth century, was probably used by Neckam and certainly quoted by Roger Bacon. He

by Thomas Wright in the Rolls Series (London, 1863); and recently he has been the subject of a lecture on 'Natural Science in England at the end of the twelfth century', by the late Sir Stephen Gaselee.[1] Born in 1157 at St Albans on the same night as the future King Richard Cœur-de-lion, he was the prince's foster-brother.[2] Educated, as he tells us in his poem, at the Abbey in his birthplace, he was sent from it to be master of its school at Dunstable; but left his work there to go to the great University of Paris, then probably the centre of learning in Europe. He studied under Adam du Petit-Pont, afterwards a bishop in Wales, and went through the normal course of trivium and quadrivium, proceeding on to medicine, law and theology: it was perhaps here that he gained knowledge of the mariners' compass which he is one of the first Europeans to mention.[3] Returning to England about 1186 he seems to have resumed his mastership at Dunstable, and at some time to have lectured at Oxford;[4] and then to have applied for admission to monastic orders at St Albans. Perhaps on account of the famous rebuff 'Si bonus es, venias; si nequam, nequaquam',[5] he went to the Augustinians at Cirencester before 1203, became their abbot in 1213, and lived there for the rest of his life. He seems to have died in 1217 and to have been buried in Worcester Cathedral.

For our purpose his importance is confined to the exposition of the natural philosophy and history of his day set out in the *De Naturis Rerum*. This is a typical and orderly statement of medieval cosmology—the deity, the angelic bodies, the heavens with the earth at their centre, the sun and moon, stars and planets, and so to the four elements fire, air, water and earth. These determine his arrangement of living creatures which fill Chapters 21–80 of Book I and Chapters 22–117 of Book II. The Chamaeleon as the only creature that lives on air is the first in order—it is an 'aerial animal' even more definitely than the birds and flying insects which follow it. Fishes belong to the water and so come earlier than minerals, plants, and animals.

Neckam's natural history, to which Gaselee gives the very misleading name of science, introduces us to a strange world. To study it is to discover once again why the modern man feels comparatively at home in the Athens of the fourth century B.C. or still more so (if he is honest)

deals briefly with plants, animals and man as well as with physical geography and astronomy, professes only to recount what he has learnt from the Arabs and is not himself a naturalist.

1 To the Royal Institution on 4 December 1936.

2 So the Chronicler, but cf. F. M. Powicke in *Essays presented to R. L. Poole* (ed. H. W. C. Davis), p. 247.

3 *De Nat. Rer.* II, 98: cf. R. Beazley, 'Early History of the Compass' in *History*, II (1913), pp. 46–7.

4 Cf. J. C. Russell, in *English Historical Review*, XLVII, pp. 261–2.

5 Cf. Bale, *Index Script. Brit.* (ed. R. L. Poole), p. 24.

in the Pompeii of A.D. 70, but is a conscious alien to the life of the Middle Ages. A very large amount of Neckam's material is indeed copied, often exactly, from Caius Julius Solinus, the Roman compiler of the third century A.D. who condensed in his *Polyhistor* the work of Pliny the Elder and others, and who, because he was not a Christian, was supposed by Neckam and his contemporaries to have been a contemporary of Julius Caesar. Magnus Aurelius Cassiodorus, a later and Christian polymath who produced similar anecdotes in the sixth century in his *Variarum libri*, is another principal source. The poets Martial, Vergil and Ovid are also in evidence; and there is a long passage from Claudian on the Phoenix. Wright even states that he drew from Aristotle and Pliny; if so, this is only done indirectly, at second or third hand, and in a somewhat distorted form. But, in fact, Aristotle seems merely the name of a remote and legendary sage, just as Vergil is evidently the mythical wizard and necromancer. And this is characteristic; for in science, as in philosophy and religion, the achievements of Greek and of Classical Roman civilisation had by this time been transmuted into a traditional lore in which a recognisable nucleus of original authority could scarcely be disinterred from the mass of glosses, accretions, syncretisms, and moralisings which pious imagination and fear had imposed upon it. The same influences which produced the legends of the saints produced also these tales of eagles and wrens, of foxes and bears and lions: by these influences Neckam's whole attitude towards nature is dominated and distorted.

Gaselee, who recognises that 'more than three-quarters of the facts, or supposed facts, which he relates' come from his predecessors, nevertheless affirms that he 'had the merit of being an observer and of recording facts which he had heard from eyewitnesses', and he has quoted such of his records as he regards as due to observation. These consist of the story of the Goshawk and the Eagle which may fairly be taken as typical of Neckam at his best, and of the Knight and his Parrot. The former is as follows: 'There was once a king in Great Britain who was very fond of hawking and particularly admired the skill and speed of one of his Goshawks. One day this Goshawk was chased by an Eagle, and to save himself flew or crept into a wattled enclosure made for sheep. The Eagle went peeking round this enclosure, trying to find a way in, and while so doing put his head through one of the wattles and got his neck stuck there; the Goshawk took advantage of his predicament and killed him. By this time the courtiers had come up and began to extol the Goshawk for his victory over a much stronger enemy; but the king would have none of this, deeming that the Goshawk was guilty of treason and *lèse-majesté* for killing the king of his own tribe, and ordered the unfortunate victor to be hung!'[1] It is a pleasant tale, with a moral admirably suited to feudal ideas; and it is no doubt due either to Neckam or to a contemporary.

1 *Nat. Science in England*, p. 9: *De Nat. Rer.* I, 24.

But it is not based upon observation or any sort of eyewitness; anyone who knows anything about real eagles will realise that this is as fanciful as the fable of the Fox and the Crow (l.c. I, 126). Of the Parrot, about which Gaselee says that he speaks 'of his own observation', it is sufficient to note that though the remarks about the bird's strength of beak and perhaps of its affection for its own image in a mirror may be first-hand the story quoted (*De Nat. Rer.* I, 37) is plain fiction—not least because the wild Parrot is said to have been seen on Mount Gilboa, where according to Neckam large numbers of them nested.[1] Of his bird-lore the only point that might be based upon observation is the story that there is a great river in Wales and that the Nightingale sings on the bank nearer to England, but not on the other side (I, 51; cf. *De Laud.* v, ll. 767–70). For it is a fact that Nightingales do not, now at least, cross the Severn.

With the fishes, which include the Mermaid (II, 25) and the Hippopotamus (II, 30), the minerals, coal, lime, gold, iron and quicksilver (II, 50–5), the plants which are few and traditional, and even in the single chapter devoted to the garden (II, 166) are merely listed without description or sign of interest,[2] and the reptiles and mammals, there is even less evidence of first-hand acquaintance than with the birds. With some few, especially those used for food, he may have had immediate acquaintance; in nature generally he is certainly interested: but there is (I think) no sign at all of any concern for the creatures in and for themselves or of any objective study of them. His business is to indicate correspondences so as to point a warning or give an example; to elaborate similitudes so as to illustrate his lessons parabolically; to disclose moral principles for the purpose of edification. Beyond this he is interested in the miraculous or abnormal wherever he can find it; he appreciates and can tell a good tale; and if he does not invent very much, he records most of the quaint beliefs which lingered and to some extent still linger on the confines of natural history.

Some of his records are purely legendary like the 'Grips', a bird that digs for gold (I, 31), or the Phoenix (I, 35), the Amphisbaena or two-headed snake (II, 118), or the Basilisk which corrupts trees and air (II, 120) and perhaps the Dragon which makes war on the Elephant (II, 145). Others deal with real creatures—the Crane which sets a sentry carrying a stone under its wing so that when it is disturbed this may fall and alarm the flock (I, 46); and the 'Aurifrisius', a name which Neckam derives from 'Auram frigidam sequens', but Gesner regards as a corruption of Ossifrage, a bird which 'has one foot armed with hooked claws, the other

1 Owing to the curse in 2 Samuel i, 21, Gilboa was assumed to be rainless and therefore the home of parrots, for whom rain was supposed to be fatal: cf. Gesner, *H.A.* III, p. 655 (ed. Frankfort, 1617), quoting Albert.

2 Roses, lilies, heliotrope ('solsequium'), violet, mandrake and perhaps peony, iris and gladiolus are the only ones in a long catalogue of herbs, fruit-trees and exotics (ginger, cinnamon, nard, etc.) which have any claim to beauty. Gardens were still utilitarian; and this list seems purely conventional.

suitably webbed for swimming' (I, 58)—an Osprey as pictured by tradition. Among them is an interesting reference to the 'Bernekke'[1] (I, 48), a bird which 'takes its origin from pine logs sodden with sea water'—'there comes a sticky slime from the surface of the wood which in time takes the shape of a little feathered bird so that the bird seems to hang from the wood by its beak'. So, too, he can produce queer examples of medieval physiology, as when he expounds the comb of a cock: 'Cocks have a very liquid brain: in their brain there are certain bones on the top of it, very insecurely joined together. A gross vapour rising from the liquor comes out through the cracks, and because it is gross is enclosed in the upper part of the head and forms the comb. It is not easy to assign an origin to the red whiskers which are commonly called wattles' (I, 76). But his real delight is in derivations, as Wright demonstrates by quoting his interpretation of *cadaver*, 'a corpse', as composed of the first syllables of *caro data vermibus*, 'flesh given to worms'. Two similar if less elaborate examples occur in his notes on plants and animals. Of Juniper he says: 'it is called Juniper from pir which means fire—either because like fire its shape tapers, or because once set on fire it smoulders so that if plums are buried in its ash they last a full year' (II, 82): of the Viper, 'it has got the name Viper because it bears its young violently (*vi pariat*). For when its belly is full for a birth, the young do not wait for a natural delivery but gnaw their mother's flanks and force their way out, thus causing her death' (II, 105).

Neckam later in life, after he had gone to Gloucestershire, and apparently in the year 1211,[2] embodied his account of plants in Books VII and VIII and of animals in Book IX of his long poem *De Laudibus Divinae Sapientiae*. The ingenuity with which he has put into 552 elegiacs the pharmacology and herbalism of his day is in itself admirable: but it must be confessed that the result is neither science nor art. The 424 lines of the Bestiary are less like a catalogue and are at least equally ingenious. For the benefit of Latinists four of the eight lines on the 'Crocodrillus' may be quoted—alike for their sentiments and their scansion:

> Vescitur humano nonnunquam corpore, sed tunc
> Tanquam compatiens fletibus ora rigat.
> Hic comedendo molam nunquam movet inferiorem,[3]
> Hoc tanquam proprium vendicat ipse sibi.

So 'Crocodile's tears' were familiar even in Plantagenet times.[4]

1 The fullest collection of references to this legend is in E. Heron-Allen, *Barnacles in Nature and Myth* (Oxford, 1928).

2 Fixed by the mention of the siege of Toulouse, V, ll. 442–58: cf. M. Esposito in *Eng. Hist. Review*, XXX, p. 452.

3 For this persistent error cf. F. J. Cole, *History of Comparative Anatomy*, pp. 38, 57, 408.

4 Cf. John Manderville, *Travels* (ed. A. W. Pollard), p. 190: 'Cockodrills...slay men and they eat them weeping; and when they eat they move the over jaw and not the nether jaw.'

Nevertheless, absurd as it is to speak of Neckam as a naturalist, he marks a real advance upon the Bestiary. Some at least of the tradition he rejects: some he modifies and criticises and for all his credulity and love of edification he knows that there is a school of medicine at Salerno[1] and of herbalism at Montpellier (cf. *De Nat. Rer.* II, 123), and he believes himself to be using reliable authorities for his stories.

To the generation younger than Neckam belong the three great Franciscans whose works contributed more to the reputation of Englishmen than those of any others of our early scholars—Robert Grosseteste, Bartholomew the Englishman, and Roger Bacon.

Grosseteste, the oldest and most famous, seems to have been born in 1175 in Suffolk. Educated in Paris and at Oxford he became Rector of the Franciscans there in 1224, and began his work of research into Greek and Hebrew writings, into philosophy and theology. Consecrated Bishop of Lincoln in 1235 he spent the rest of his life in practical affairs, exerting a large influence upon the king and the Church in England and visiting with tireless and sometimes pugnacious energy the monasteries and parishes of his huge diocese. Roger Bacon's tributes to his learning, 'One man alone our Lord Robert knew the sciences' (*Compendium Studii*, 8 in *Op. Inedita*, Rolls Series, p. 472; cf. p. 91, etc.), were well deserved. He had the vigorous intellect of a pioneer who anticipated the scholars of the Renaissance in his quest for ancient manuscripts and his zeal for their translation; and he had something at least of the enthusiasm which would not confine itself to the traditional learning, but undertook for himself and urged upon others tasks of research in physical and biological studies. There is not much ground for regarding him as a naturalist: but his influence upon education, especially at Oxford, makes it proper to pay tribute to his greatness.

Grosseteste died in 1253. His work was carried on by the youngest of the three great Franciscans, Roger Bacon. He was born at Ilchester in Somerset about 1214 of a well-to-do family; went to Oxford in Grosseteste's time there, and thence to Paris till about 1250. A return to Oxford was ended in 1257 by a summons from his Franciscan superiors to Paris where he was confined by them for some ten years. In 1266 the pope, Clement IV, instructed him to send to him a statement of his teaching; this gave occasion for the rapid composition of his three works, the *Opus Majus*, the *Opus Minus* and the *Opus Tertium*—each being a variant upon the same general theme. After 1267 he was in comparative freedom, returned to England, and in 1271 wrote the *Compendium Philosophiae*.[2] This for its outspoken attack upon the ignorance of the clergy seems to have been condemned by Nicholas IV and to have led to a

1 For this and its famous poem the *Regimen Sanitatis Salerni*, cf. C. and D. Singer in *History*, Oct. 1925, pp. 242–6 and below, p. 60.

2 For the fixing of the date, cf. J. S. Brewer, *Opera inedita R. Bacon*, pp. liv–v.

further long imprisonment. He seems to have died at liberty and in Oxford in 1292. His extant books, though the earlier of them deal with Alchemy, cover a very wide range. Much of them, as with Grosseteste, is confined to a survey of the defects and difficulties of scholarship, thanks to the lack of Greek and Hebrew, and of original manuscripts of the Classics and Scriptures. But much deals also with the need for observation and exact knowledge of things in themselves, with the elements of science, with the plea for experimental studies, with the results of his own researches. In Bacon we have an anticipation of both the future movements which between them created the scientific outlook—the accurate study of past records such as characterised the work of the early naturalists from Ermolao Barbaro to Gesner and Aldrovandi; and the insistence upon first-hand knowledge of the data and the inductive method of formulating and testing hypotheses which was advocated by his namesake Francis Bacon in the early seventeenth century and became the basic principle of the 'New Philosophy'. Though Bacon did not devote much time to natural history he has something to say about its study and particularly about the business of exact identification with which, as his great successor William Turner realised, such study begins. It is worth while looking at his work here in spite of its technicality; for it illustrates not only his intellectual quality but the sort of problems which were the prelude to science.

Thus in *Opus Minus* (*Op. ined.* pp. 353–4) and at greater length in the *Compendium Philosophiae*, VIII (*Op. ined.* pp. 483–8), he discusses the identity of certain animals and birds named in Scripture. First the 'Chirogrillus' of Leviticus xi, 5 and Deuteronomy xiv, 7: this he says is a five-syllable word in the old manuscripts (it is in fact 'Chirogresillus' in Codex Lugdunensis and 'Choerogyllius' in Vulgate Deut. l.c.) and the ignorant spell it 'cirogrillus'. He quotes Jerome 'On correction of the Psalter' as rendering it 'artomys'; the Gloss[1] as 'ericius' or 'erinacius', a word which Jerome had used for it in his version of Psalm civ, 18; and Proverbs xxx, 26 as 'lepusculus'. Of these he dismisses 'artomys', 'bear-mouse', of which he evidently knew nothing; criticises 'erinacius' as meaning a spiny animal whether Hedgehog or the larger 'strix' or Porcupine 'which when angered ejects its quills and shoots at the man who attacks it' (*Op. ined.* p. 484); and rejects Hare on the ground that it makes its home not in rocks but in grass. Then quoting Pliny, *Natural History*, VIII (ch. 81), he says that Hare is used to include 'cuniculus' or Rabbit which is the true meaning of 'chirogrillus'.[2] It is, of course,

1 For its authorship and character, cf. B. Smalley, *The Study of the Bible in the Middle Ages*, pp. 31–45.

2 This became the accepted identification; hence the A.V. translation where 'conies'=rabbits: cf. also the rabbits in the rocks carved on Merton College gate, possibly in Bacon's time.

actually the *Hyrax* or true 'Coney' (*Procavia syriaca*), a creature which Bacon could not have known.

Next he takes 'nycticorax' which he states is a greater difficulty. 'Brito mendax [William Briton or Guillaume le Breton, author of *Summa Vocabulorum Bibliae*, to whom he regularly refers with this epithet and who is said by Bale, *Script. Brit. Cat.* v, 89,[1] and by *D.N.B.* VI, p. 359 to have died in 1356—a century too late[2]] confirms his own opinion by quoting Isidore [*Origines*, XII, 7, p. 172] that it is noctua [the Owl].' So, too, Jerome in correction of Psalm cii, 6 has 'nycticorax in domicilio as have Aquila, Theodotion, the Septuagint and Edition v; Symmachus has upupa; Edition VI noctua. But Deuteronomy xiv, 15 prohibits both noctua and nycticorax, so that they cannot be synonyms. So does Leviticus xi, 16 in the Greek; and there the Latin has for nycticorax bubo. Moreover, Pliny in the tenth book where he discusses notable birds never mentions nycticorax but separates bubo from noctua.' After full debate he concludes that 'perhaps nycticorax is a general term for all night-birds: this would reconcile Isidore, Jerome, the Law and the Psalter: but I believe my former view [that it equals bubo] is better'.

Similarly about 'pellicanus', 'porphyrio' and 'onocrotalus', there is a difficulty which can only be decided by knowledge of Hebrew and Greek. 'Brito mendax' says that in Deuteronomy xiv, 17, where we have 'porphyrio', the Septuagint has 'pellicanus': the glosses certainly say this and lie. The Greek scholar can see that there 'pellicanus' and 'porphyrio' are both mentioned and cannot therefore be identical. 'Moreover here where the Septuagint has pellicanus our version has onocrotalus.' He then quotes Pliny (*N.H.* x, 66) to support the identity of 'onocrotalus' with the Pelican and describes from him how it has 'a sort of second stomach in its throat and regurgitates its food into it like a ruminant animal' (*Op. ined.* p. 488). Finally, he tells the story of the Pelican which 'kills its young, mourns for them three days and then brings them to life again by sprinkling them with its own blood'. 'This', he adds, 'is often carved [the "Pelican in its Piety"], is not contradicted, nor need be rejected, though it is not an opinion that must be held: Pliny says nothing about it and Isidore [*Orig.* XII, 7, Fertur, si verum est...] speaks of it as a popular tale.'

So, too, in *Opus Tertium*, XXV[3] (*Op. ined.* pp. 91–2) he illustrates the same problem by reference to a plant from the 'De Vegetabilibus of

1 So, too, his *Index Script. Brit.*, p. 117, has 'claruit A.D. 1350'.

2 The fullest discussion of Brito is that by B. Hauréau in *Histoire Littéraire de la France*, XXIX (1885), pp. 584–606. He gives a description of the *Vocabulary*, mentions a number of MSS. of it, discusses its date, drawing attention to the impossibility of 1356 and fixing upon a time shortly after 1248, and argues that its author was a Breton, not an Englishman.

3 Cf. for the same *Opus Majus* (ed. Bridges), I, p. 67, and *Comp. Phil.* p. 467.

Aristotle' (the short and spurious *De Plantis*, I, 7, 2—he is presumably referring to the Latin version from the Arabic made by Alfred of Sareshel the Englishman) where the plant has the Spanish name 'belenum'.[1] Of this Bacon declares that he could find no one at Paris or in England to tell him what plant was meant until someone, apparently Hermann the German,[2] suggested 'jusquiamus[3] or seed of cassilago'—presumably *Hyoscyamus* or Henbane. But this, in his opinion, is almost certainly wrong. He decides that the plant if not mythical seems to have been the *Persea*, which may or may not be the Peach tree.[4]

Scattered throughout his writings are similar notes of the identity of plants or animals, or allusions to their habits. Thus in the letter *De Nullitate Magiae*, that very remarkable indictment of all magic as unworthy of the serious philosopher, he cites from the same pseudo-Aristotelian *De Plantis*, I, 6 the statement that 'the fruit of female Palms grow mature by reason of the scent of the male trees'—the famous first clue to the function of pollen (cf. *Op. ined.* p. 529)—and in *Communia Naturalium*, p. 279, again from *De Plantis*, I, 2, a remarkable comparison of the structure and physiology of vegetables with those of animals; and in *Comm. Nat.* p. 308 quotes Averroes on the generation of living organisms from putrefaction in the attempt to provide a reasonable basis for this ancient belief. But the treatises recently printed do not, in this field, add very much. It seems certain that his projected writings on the generation of living things, vegetables, animals and man (*Opus Tertium*, XII) remained unwritten: in any case they are not discoverable. If he had attacked this theme with the vigour and insight and fearlessness that characterise his literary criticism, he might well have shaken the traditional mythologies and moralisings of the Bestiaries and initiated the study of the living flora and fauna two hundred years before our first herbalists.

But he can never have been an easy person to work with[5]—too keen on his results to suffer fools gladly, too outspoken in his comments to win their support, too impatient to seek for assistance and too poor to obtain it. He was a genius—perhaps one of the greatest we have had—but the time was not ready for him; and the medieval tradition and the

1 Beleño (*Hyoscyamus niger*). This passage, quoted also by Albert, *De Veg.* I, § 191, is used by A. Jourdain, *Recherches d'Aristote*, p. 188, to prove that our Greek text is a translation of this Arab-Latin version. For the whole question cf. S. D. Wingate, *Mediaeval Versions of Arist.* pp. 55–72.

2 A contemporary Arabic scholar not to be confused with his predecessors, Hermann of Reichenau, eleventh century, or Hermann of Carinthia, twelfth century.

3 Turner, *Names*, p. A. vii, gives this as the 'Poticaries' name for Henbane.

4 For this point cf. Bodaeus in *Theophrastus*, *H.P.* p. 296 (Amsterdam, 1644).

5 Cf. his diatribe against Alexander of Hales and Albert the Great in *Opus Minus* (*Op. Ined.* pp. 327–8), E. Charles, *Roger Bacon, Sa Vie*, etc., pp. 353–7, J. H. Bridges, *Roger Bacon*, p. 34, and B. Smalley, *The Study of the Bible in the Middle Ages*, pp. 242–5.

scholastic method were too strong. He was a forerunner of the scientific movement: but contributed less to its actual coming than would a man of smaller and more contemporary gifts. His writings remained always the treasured inspiration of a few: neither then nor later was their influence widespread. Even if we say with François Picavet that 'if the Church had followed the road along which he wished to guide her, there would have been no room for a Renaissance often hostile to Christianity, nor for a Reformation wholly separated from Catholicism, nor for an open struggle and total rupture between theology, philosophy and science',[1] we must add that even the splendid tribute to his achievements recently paid by the Oxford edition of his *Opera* does not make his failure to lead more difficult to understand.

The best known of all medieval treatises upon nature by an Englishman is of course that of Grosseteste's younger Franciscan contemporary, Bartholomaeus Anglicus, which was not only widely circulated and freely translated all over Europe before the end of the fourteenth century but was still probably 'the standard authority on Natural History in Shakespeare's youth'.[2] Its authorship has been obscured by the confident statement of Stephen Batman its editor in 1582 that 'Bartholomew Glantvyle descended of the noble familye of the Earles of Suffolk was a Franciscan Frier and wrote this worke in Edward the thirds time, about the yeare of our Lord 1366' (*Batman uppon Bartholome*, p. 436*b*). But, in fact, it is more than a century older and is perhaps quoted by Bacon, *Opus Tertium*, XXXVII (*Op. ined.* p. 118), in 1267.[3] On internal evidence it is plainly later than Albert the Great whose work in Paris before 1248 is quoted, but not much later, since it quotes none of the great men, Thomas Aquinas, Vincent de Beauvais or Roger Bacon, who were also in Paris shortly after. It seems that Bartholomew was sent by his Order to Saxony in 1231, and Salimbene (*Chronica*, p. 48) states that at some time he had been a professor of theology in the University of Paris. The earliest known MS. is in the Bodleian and was written in 1296. In any case he was clearly active about the middle of the thirteenth century and apparently in close touch with the splendid life of that great period.[4]

The book itself, entitled *De Proprietatibus Rerum*, was, as we have said, widely and generally known. But for our purpose its printed versions are more important than manuscripts. Three of these, based upon the English translation made by John Trevisa, Chaplain to 'Sir Thomas lorde of Berkeley' in 1397, are famous—that by Wynkyn de Worde in

1 *Roger Bacon Essays*, p. 87.

2 So H. W. Seager, *Natural History in Shakespeare's Time*, p. vi.

3 So R. Steele, *Medieval Lore*, p. 5. It is quoted as 'The book *De Proprietatibus Elementorum*'. This may be Bartholomew's book, but is more likely to be the pseudo-Aristotelian work translated from Arabic by Gerard of Cremona.

4 Cf. L. Delisle, *Hist. Litt. de France*, XXX (1888), pp. 352–65.

1491, by Thomas Berthelet in 1535, and by Stephen Batman with omissions and additions in 1582. We shall quote from Berthelet and at a later point refer to Batman's alterations.

Steele's book *Medieval Lore*, from which we have already drawn, gives so adequate an epitome of the *De Proprietatibus* that we need not attempt any elaborate summary.[1] God; the Angels; the Rational Soul; the Body with its four elementary qualities of heat and cold, dryness and moisture; the Members of the human body; Age, sex and status; Diseases; Heaven and earth; Times and seasons; Matter and form and the elements, fire; Air; Birds; Water and fish; Earth and the hills; Countries; Stones; Trees and herbs; Animals; Colours, smells, tastes and foods, number, weight and measurement—these form the subjects of the nineteen books, the particular headings under each (e.g. the separate birds or countries) being arranged alphabetically as chapters. The general order is clearly the same as in Neckam and is determined by that of the four elements: but Bartholomew puts man outside the sequence, treating him as on a more godlike level than the rest of the world.

It is a vast encyclopaedia,[2] obviously dominated in its scope by the desire to expound such knowledge as is found in the Bible and the Gloss, and therefore devoting chapters to every hill in Scripture from Ararat to Syon and Ziph, but including the Classical tradition so far as it had become part of common knowledge and so inserting Caucasus and Olympus and the 'hylles Riphei' or Alpes. But here, as with Neckam, it is with the books dealing with nature, XII, part of XIII, XVII and XVIII, that we are concerned; and these are, in fact, the longest and richest of them all.

For birds, as indeed for fishes and animals, there is a small nucleus of material drawn from Michael Scot's Aristotle and having some remote resemblance to fact. But much more comes from Isidore Archbishop of Seville, who died in A.D. 636 and whose *Etymologia* or *Origines* was the most authoritative of all text-books until the Renaissance. Some of the rest comes from Pliny, some professedly from St Basil's *Hexaemeron* (Lenten sermons on the Works of Creation delivered at Neocaesarea in Cappadocia about the year A.D. 370), some from St Ambrose who produced a *Hexaemeron* in imitation of Basil's in A.D. 389, from Bede on the Calendar, from Avicenna, and from the Aurora or metrical Latin version of the Bible. These were accepted not, in most cases, from the originals, but from *Physiologus* and the Bestiaries, in which the traditional anecdotes had been given a symbolic meaning. As Emile Mâle says of them in his fascinating treatise *L'Art Religieux du XIIIe siècle en France*, p. 49: 'Quant aux savants, Vincent de Beauvais, Barthélemy de Glanville,

1 It is well to note that Steele's selection emphasises the best qualities of the book: as a whole it is much less interesting and original.

2 Berthelet's edition contains 388 double pages, small folio.

non seulement ils ne dédaignent pas ces fables, mais ils les mettent au rang des vérités scientifiques.'

Of birds there is almost no sign of any first-hand knowledge except perhaps in the account of the Swan and the twining of their necks in courtship[1] (p. 169): the great majority of the stories are merely legend. In XII, Bees, Gnats, Grasshoppers, Locusts, the Phoenix and the 'Remouse' or Bat are included among birds, presumably as belonging with them to the air. Some of the legends retold by Bartholomew persisted into the scientific era, as when like Neckam he declares that the Sea Eagle 'hath one fote close and hoole as the fote of a gandar and therwith she ruleth herselfe in the water, when she cometh downe by cause of her praye. And her other fote is a clove fote with full sharp clawes with the which she taketh her pray' (p. 164); or when he declares that Vultures hunt by smell (ch. 35), an error already exposed by the Emperor Frederick II, but defended even with passion until very recent times.[2] Others, like that of the Kite which 'taketh Cockowes uppon his shoulders, leaste they fayle in space of longe wayes, and bryngeth them out of the countrees of Spaine as Isidore sayth'[3] (p. 173*b*), are less persistent. Several of the English names given by Trevisa are interesting. Thus *Coturnix*, a bird of which he tells us (p. 167*b*) that they have fixed times of coming, and when crossing the sea settle on the water, spread their wings and sail, that they dread the Goshawk and love their own kind, is translated 'Curlewe', this being the name used by Hampole and Wyclif to translate Quail in their versions of the Psalms. 'Myredromble' (usually supposed to be a local English name for the Bittern) is used to render 'Onocrocalus', of which Pliny's story that it puts its beak into the water and bellows is told without acknowledgment and which Bartholomew separates from the Pelican (p. 172*b*), and also to render *Ulula*, 'a foule that hath that name of shrychynge' (p. 175). 'Lapwynge' is *Upupa*, although the account, 'a birde most filthy and is copped on the heed, and dwelleth alway in graves' (p. 175), suggests that the Hoopoe is meant.

On fish six and a half pages contain stories from Jorath, *De Animalibus*, the 'Jorach who often lies' of Albertus, *De An.* XXII, 10*b*, who is apparently the source of the tale about a small fish fastening on to the hulls of ships and stopping them (p. 185*b*), though this also occurs in Pliny (*N.H.* IX, 41), Aelian (*H.A.* II, 17) and Neckam, and of the Whale which genders amber and is in old age so covered with earth that bushes grow on it and seamen mistake it for an island. Bartholomew does not seem to have been a fisherman.

Book XVIII deals with animals and includes all creatures that live on the land: the Bee which is treated again here, ch. 12, pp. 314*b*–16*b*,

1 This is not invariable but has been recorded by modern observers.
2 E.g. by Charles Waterton, *Essays on Natural History*, pp. 17–48 (London, 1838).
3 The statement is under Tucus or Cuculus, *Origines*, XII, ch. 7, p. 174 (ed. 1601).

although it has already appeared in XII, ch. 4, pp. 165–6*b*, the 'Malshragge' (Caterpillar) and the Flea; the Scorpion, the Worm, the Snail, the 'Tortuse', the 'Cocodrill' and the Asp; the Chamaeleon, the 'Salamandra' and the Frog; the 'Orix', the Hyena and the Ape; the 'Rinoceros' (or Unicorn), the Pigmy and the 'Mermayden'; the 'Grife' (or Gryphon), the 'Ficarius' (or Faun) and the 'Onocentaur'. It is a wonderful medley; and the legends are as strange as the list. Most of them derive from Isidore's *Origines* and thence from Pliny and the Graeco-Roman world; but some are evidently eastern in character and source; and almost all represent a heavy accretion of tradition even if embedded in it there is a nucleus of fact. In some cases, notably that of the Oryx, there is discussion as to the kind of beast intended: 'orix' in Scripture Bartholomew decides to be a Water Mouse or Glis (that is, Dormouse), although he admits that in Pliny it is an Egyptian beast that gazes worshipfully at the dogstar, and in Juvenal a bird[1] who 'blontethe the knyfe with his fatnesse'[2] (p. 341), and elsewhere (apparently when Briton's declension 'orix, origis' is found) a Mountain Goat of the sort which 'lepe downe of highe rockes and falle uppon their owne hornes'. Under Fauns and Satyrs (ch. 48) there is a mixed collection of human and semi-human monsters borrowed from *Physiologus* and the Bestiaries—'cenophali' (presumably 'cynocephali') with dogs' heads, 'ciciopes' (Cyclopes) with one eye, some with 'closed mouthes and in theyr brestes onely one hole', and in 'Sithia' some with ears so large as to cover their whole bodies. Four chapters, 42–5, are devoted to the Elephant, the first based upon Isidore and Pliny, the second on Aristotle and Avicenna, the third on Solinus, a version of the *Polyhistor* similar to that given by Neckam (*De Nat.* II, 144), and the last on *Physiologus*. There are occasional suggestions of real knowledge as in the account of the growth of the frog from the tadpole, or of the caterpillar which 'weveth certayne webbes of his owne guttes as the spynner [spider] doothe, and wrappethe hymselfe in those webbes...all the wynter longe...and chaungeth his shape...for he taketh thynne winges and brode: and fleeth uppe freely in the ayre...and suche a fleeyng worme hyghte Papilio' (pp. 132–2*b*). Perhaps the best of these is that of the Cat in ch. 76, quoted by C. H. Haskins, *The Renaissance of the Twelfth Century*, pp. 335–6—a vivid and pleasant description of the best-loved of medieval pets. But to cite the two or three passages that show any sign of first-hand knowledge as if they indicated the beginning of a scientific outlook is to exaggerate their place in the mind of its author. In Bartholomew, unlike Albert, there is not a trace of any desire to correct his fables by observation or to test his statements by experiment.

1 A curious mistake due presumably to the fact that Juvenal, *Sat.* XI, 140, places Gaetulus oryx' next to pheasants and flamingoes as an expensive luxury.

2 Sheer mistranslation!

The book dealing with Trees and Herbs, Book XVII, is much less interesting than those on birds and animals. As Steele rightly notes, 'the subject does not so readily lend itself to fables' (*Medieval Lore*, p. 87). But this does not mean that the records are more exact or objective. Isidore and his fantastic accounts of the derivation of names, 'Aristotle' and the *De Plantis*, Dioscorides, Pliny, Isaac, an Arab physician of the seventh century who wrote on foods, and Platearius, one of the famous Salernitan school of twelfth-century doctors, are his chief authorities. In ch. 169, p. 291 *b*, he quotes by name 'Alexaundre Nequam' and his poem, but only to prove that the word 'tisana' has a long middle syllable; and this reference is probably derived not from Neckam direct but from William Briton or perhaps Bacon.[1] Among the most interesting sections are those which deal with vegetable products unknown to the Graeco-Roman world and for which he only has comparatively recent authorities —'Gynger hyghte Zinziber and is the roote of an herbe and is hotte and moyste as it is sayde in Platearius' (of this he evidently knows nothing more though he expands upon its uses) and 'Sugre called Zucarum and Sucara also. And is made and yssueth out of certen Canes and Redes which groweth in lakes and pondes faste by a Ryver that is in Egypte called Nilus. And the Juys that yssueth oute of those canes or redes is called Canna mellis and of that juys is Sugre made by sethynge, as salte is made of water' (ch. 196, p. 299 *b*). So too of the 'Morus', ch. 50, p. 272, he seems to be writing independently when he says 'Leaves of highe mulbery trees be great and brode, and are greved with Malshragges and flyes, and gladly wormes eate therof, and so silke wormes ben best fed and nouryshed with such leaves'.[2] But perhaps the nearest approach to first-hand knowledge is in his account of 'Carix, Sedge, an herbe mooste harde and sharpe, and the stalke therof is three cornered, and kittethe and kervethe the honde that it holdeth, if it be hard drawen there through....Plinius saith...that the rote of a three edged russhe is of goode smelle...but I understonde this is not generalle but specyall'—almost the only occasion on which he criticises his authorities (ch. 35, p. 254 *b*) or uses the first person.

These are, however, rare exceptions. In the vast majority of the 196 chapters he is simply restating traditional accounts, stories and prescriptions. Almost all the plants dealt with are either edible or of supposed medical value. Some like the Vine occupy several chapters—the Vine has 20 columns and 14 chapters, one being on the wine-cellar. There are also many headings that deal with general subjects—meal, hay, fruit,

1 Under Tipsana (Ptisana) Briton has 'Alexander Nequam sic ait' and quotes a verse which Esposito (*Eng. Hist. Review*, XXX, p. 470) says cannot be found in Neckam's extant works. Roger Bacon also quotes and discusses this in *Comp. Phil.* pp. 454, 457.

2 Cf. also under Bombax, Bk. XVIII, ch. 18, p. 318 *b*.

buds, chaff ('quisquilie'), groves, planks, hedges, stubble, incense and suchlike. Almost none of the plants except the Lily, the Rose and the Violet are flowers of the garden.[1] Here, as in the rest of his work, there is no sign of personal interest or direct acquaintance with the objects of which he writes. It is, indeed, astonishing that anyone should have been able to produce such a mass of material without adding to it one single fact of his own observation, or been content to do so with hardly a flicker of criticism or of comment. Beyond an evident enjoyment in the retailing of some of his legends and a genuine readiness to point their moral, he remains curiously impersonal and obscure, giving no hint of his age or temper, his likes or dislikes, his adventures or experiences. Even when on occasion he speaks of a particular case of healing or a particular habit of plant or animal, it is never associated with himself, with his own life or friends. The anecdotes and reminiscences, the records of personal observation and practice, which are so delightful a part of the seventeenth-century herbals and nature-books, are wholly lacking here. From internal evidence Bartholomew is a *nudum nomen*, a mere cypher. His book draws its charm from the quaintness and variety of its subject-matter, not from the quality of its author. He is content to amuse, instruct and edify by the recital of the tradition: on occasion he improves upon it: but of concern for truth in a scientific sense, indeed of any love of nature apart from books and legends, he shows not a trace. And this is characteristic of his age and general outlook. 'Guard that which has been entrusted to you'—guard it and improve upon it—was a text faithfully followed; and its effects were far-reaching.

To understand this feature of medieval natural history it will be well to take an instance or two from his handling of stories which have an ascertainable origin. The tale of the Crocodile-bird is a good one, for it is not purely fictitious and the stages of its degradation (or enrichment!) can be easily traced. It relates to the Egyptian Plover or Courser (*Pluvianus aegypticus*) which lives on the sand-banks by the river and has been seen by A. E. Brehm to run along the backs of Crocodiles and pick up food from them; or perhaps to the larger Spur-winged Plover (*Hoplopterus spinosus*), as stated by J. M. Cook, who claims to have seen this species enter a Crocodile's mouth (cf. *Ibis*, 1893, pp. 275–7). Our first record of this comes from Herodotus, *Histories*, II, ch. 68, who writes of the Crocodile: 'since it lives in the water its mouth is all full within of leeches...whenever the crocodile comes ashore and opens its mouth the trochilus [the courser in question] goes into its mouth and eats the leeches; the crocodile is pleased by this service, and does the bird no harm.'[2] There is a plain tale which may well on occasion be true. Thence it passes

1 Like Neckam he shows neither knowledge of nor interest in gardening.
2 Translated by A. D. Godley, Loeb Series, I, p. 357.

to Aristotle, *Historia Animalium*, IX, 6: 'When the crocodile yawns the trochilus flies into his mouth and cleans his teeth. The trochilus gets food thereby, and the crocodile gets ease and comfort; he makes no attempt to injure his little friend, but, when he wants to go he shakes his neck in warning lest he should accidentally bite the bird.'[1] The story grows: Pliny helps it in *Historia Naturalis*, VIII, ch. 37: 'When the crocodile is sated with a meal of fish, a small bird which is there called the trochilus but in Italy the King of birds[2] invites him to open his mouth wide to give it food: and first it jumps in and cleans the mouth, then the teeth and inner throat also, as it yawns open as wide as possible for the pleasure of this scratching; then the ichneumon [its enemy according to Pliny] seeing him sunk in gratification and sleep darts into his throat and gnaws out his belly.' Solinus, *Polyhistor*, XXXII, 25, has a similar account with more stress on the bird's scratching. Isidore does not tell this particular story. Neckam says nothing under Crocodile, but under 'Strofilos' records that this little bird scratches the mouth of the 'cocodrill', gets into its throat, and comes out from its vitals after biting its belly (*De Nat. Rer.* I, 57). Bartholomew—or Trevisa—renders it as follows: 'The Cocadrylle eateth ryghte moche: and soo whan he is ful, he lyeth by the brynke... and bloweth for fulnes, and then there cometh a lyttell byrde, whiche hyght Cuschillus among them, and is called kynge of foules amonge the Italyens, and this byrde fleeth tofore his mouth, and somtyme he putteth the byrde of, and at the laste he openeth his mouth to the byrde, and suffreth hym entre. And this byrde claweth hym fyrste with clawes softly and maketh hym have a maner lykynge in clawing, and falleth anone a slepe, and when this byrd Cuschillos knoweth and perceyveth that this beast sleepeth, anone he descendeth into his wombe, and forthwith stycketh hym as it were with a darte and byteth hym full grevously and fulle sore' (Book XVIII, ch. 33, p. 326). The names of beast and bird have been debased; the story has been expanded from Pliny; and the parts of the bird and the ichneumon have been united so as to round off the tale and give it an edifying moral. One could hardly find a better example of the growth of a legend.

Similar, though somewhat less expanded, is the traditional story of the Cretan Dittany (*Origanum dictamnus*), of which there is a charming picture in the *Botanical Magazine* for 1795 (vol. IX, pl. 298). Aristotle in *Historia Animalium*, IX, 6, had recorded, 'Wild goats in Crete are said when wounded by arrows to go in search of Dittany (Dictamnon) which is supposed to have the property of ejecting arrows in the body'. He is at least non-committal and only repeats by hearsay; and it is a fact that Dittany abounds in Crete and that Goats are fond of it. Thence the story

1 Translated by D'Arcy Thompson, Oxford Series.

2 Pliny has mistakenly identified the trochilus with the regulus or Wren!

spread through most of the Greek writers on plants and animals down to Aelian, *Variae Historiae*, I, 10; and so came to Cicero who writes in *De Nat. Deorum*, II, ch. 50: 'The story goes...that wild goats (caprae) in Crete when they have been pierced by poisoned arrows look for a herb called Dictamnus and that when they have tasted it the arrows fall out of their body; and hinds a little before giving birth purge themselves with seseli'—this addition being from Aristotle, *H.A.* IX, 5. Vergil no doubt got from Cicero his allusion to the 'Dictamnus from Cretan Ida, a plant with woolly leaves and topped with purple blossom, herbs not unknown to wild goats when swift arrows have stuck in their back', *Aeneid*, XII, 412–15. Pliny seems also to have drawn this particular story from Cicero; for he has transformed the Goat into a Deer, no doubt misled by Cicero's juxtaposition: he writes in *Hist. Nat.* VIII, ch. 41, 'The value of the herb Dictamnus for extracting arrows was shown by stags when wounded by that weapon and ejecting it by grazing on that herb'. He has also a short account in *H.N.* XXV, 52, 53: 'Hinds have shown seseli when they are pained by birth-pangs. When wounded they have shown Dictamnus as we have already said, for the weapons fall out at once when they eat it....It grows in Crete and is much sought by goats.' Solinus condenses this, *Polyhistor*, XIX, 15: 'Stags showed Dictamnus, for feeding on it they shake out the arrows that have hit them.' So too St Ambrose, *Hexaemeron*, VI, 26: 'The wounded Roe (caprea) searches for Dictamnus and draws out arrows from her wound.' Isidore is similar, *Origines*, XVII, ch. 9: 'According to Vergil the wounded hind (cerva) searches the glades for Dictamnum: so great is its potency that it expels iron and shakes out arrows from the body.' Neckam, *De Nat. Rer.* II, 136, exactly copies Solinus but spells the plant Dictannus. Bartholomew gives a full account of it: he begins with a medical description of Diptannus professedly drawn from Dioscorides and Platearius: he refers to Pliny, 'Book XXVI, ch. 8' (*sic*): 'it is sayde that a Hynde taught firste the vertue of diptannus, for she etith this herbe that she may calve eselier and soner: if she be hurt with an arowe, she seketh this herbe and eteth it which puttith the yren out of the wôund as Basilius sayth in Exameron and Ambrosius and the expositor': he then goes on to quote Isidore both in his description of the plant and in the passage cited above: to this he adds: 'Therfore bestes smyten with arowes eate therof and dryve the yren out of the body. For this herbe hath a maner myghte of werre to drive out arowes and dartes and quarelles, as Isidore sayth' (Book XVII, ch. 49, p. 258). Thus he conflates the records both of Deer and Goat, and of Seseli and Dittany;[1] ascribes to St Basil what is in fact only in St Ambrose; and adds a generalisation which seems to be all his own. So legends grow.

1 This conflation is found as late as 1553 in a serious botanist, Amatus Lusitanus, *Enarrationes in Dioscor.* p. 318: he adds, 'to-day we know this to be true'.

We have examined these two illustrations of the medieval mind at some length because they show how remote it is from that associated with any scientific appreciation of nature or of history. Similar examples can be quoted almost without limit from the legends of the saints or of the events of earlier days. The 'plain tale' of happenings in Palestine or Rome had been enriched, distorted, transformed, until it bore almost no resemblance to the original records; saints like 'St Oracte which is a dunce-like corruption of Mount Soracte or St Amphibolus an English saint which is a dunce-like corruption of the cloak worn by their St Alban'[1] had been invented; characters like King Arthur or Saint George whose original existence was purely mythical had become the familiar heroes of romance or hagiology; fictitious events like the landing of Brutus in Britain or the coming of St Joseph to Glastonbury had been given a primary place in the traditional story of our country and faith; and a world-picture very widely removed from actuality had become characteristic of all educated men.

CHAPTER II. NATURE AND MEDIEVAL ART

That neither Bartholomew nor his readers had the slightest desire to check the accuracy of their stories or to relate them to facts such as must have come under their own notice reveals the vast difference between the medieval and the scientific outlook. In regard to nature as in regard to history legend was accepted not only without hesitation but in preference to accurate record. It is not true to say that the result was a world of chaos in which since there was no concept of verifiable cause and effect anything might happen: there was, on the contrary, a definite world-picture deduced from general principles about God, the hierarchies of being, the elements, the vegetable and animal realms, a picture universally accepted, a picture whose only defect was that it could not be maintained by empirical proof. So soon as men began to investigate, such a picture was bound to be challenged: the evidence of its universality is that for centuries the challengers could make no impression upon it. Bartholomew's book kept its popularity for three hundred years; it was repeatedly quoted and copied both at home and abroad, as, for example, in the very popular *Ortus Sanitatis*.[2] Shakespeare and the Elizabethans if they had

1 Cf. C. Reade, *The Cloister and the Hearth*, ch. lxxiv.

2 E.g. *Tractatus de Animalibus*, chs. 47, 75, 116, 144, 145: *De Avibus*, ch. 51. This book whose origin and bibliography are very complex was apparently written by Johann von Cube of Frankfort late in the fifteenth century, and printed first at

broken away from the tradition in regard to history had not yet done so in regard to nature:[1] they rejected the legends of the saints but preserved those of the beasts. It was not until the fourth century after its composition that the *Novum Organum* proclaimed the dethronement of the *De Proprietatibus*.

Nor was the change, when it came, due mainly or in the first instance to a comparison between the tradition and the facts. For us, if we were confronted with a cosmology that seemed fantastic, the first appeal would be to observation of the data. 'You say that the crocodile behaves like this; let us go and find out.' 'Does Dittany draw out iron? An experiment can easily be made.' But argument of this kind seems to have been wholly strange to the mentality of medievalism. The schools agreed that this and that was so: men might find it hard to accept: yet the schools must be right, or the whole structure of life was imperilled. And in fact sceptics seem to have been rare. We call that epoch the age of faith; and if faith means power to disregard facts of observation in obedience to statements by authority, and to perpetuate for centuries an almost purely imaginary *Weltanschauung*, it deserves the title. That towards the end of the period faith was maintained by Nominalism in philosophy and the Inquisition in practice, is hardly surprising. But it surely is a surprise that the change which shattered the old world-picture was not the discovery of the world of nature, but the recovery of the Greek and Latin classics. It was the comparison between St Paul and Catholic theology that accomplished the Reformation: it was the contrast between Aristotle and Bartholomew that initiated modern science.

Hence it happens that before the first experimentalists, Vesalius in anatomy and the herbalists in botany, had effected any serious change, there came the makers of commentaries who strove to purify the text and interpret the meaning of the botanists and zoologists of Greece and Rome, and the makers of pandects who collected and collated the utterances of the ancients. Ermolao Barbaro and Francisco Massario on Pliny, and Jean Ruel on Dioscorides, and Pierre Gilles on Aelian represent the former; Gesner and Aldrovandi are the outstanding examples of the latter. Our English representatives of them are neither eminent nor early: but they have a good claim to be our first naturalists; and their influence in encouraging a new outlook was more than their merits would lead one to suspect.

But before we deal with them there are records not in the books about nature but in the chronicles that deserve to be quoted with honour.

Mainz by Peter Schöffer in 1485, and in an enlarged form by Jacob Meydenbach in 1491. It is very similar to and largely reproduced in *Le Grand Herbier* of Paris of which *The Grete Herball*, published by Peter Treveris in London in 1526 and 1529, is a translation.

1 Cf. E. M. W. Tillyard, *The Elizabethan World Picture*, London, 1943.

Among the earliest of these are those of Reginald the monk of Durham in 1167, who chronicled in the *Miracles of Cuthbert* the tameness of his birds on the Farne Islands—birds which 'the English call "Lomes" and the Saxons of Friesland "Eires"' and which nest 'in the houses and even under the beds' and come clamorously to the hands of those who make pets of them;[1] and of his monastic contemporary Thomas of Ely who in the elaborate pages of the *Liber Eliensis* interpolates in his record of Hereward an account of the activities of the fenmen and the beasts, birds and fishes of the Isle[2] which anticipates and is in many respects identical with the famous description by Camden[3] four hundred years later. He notes the abundance of Stags, Hinds, Roe-deer and Hares; of Otters, Weasels, Stoats and Polecats; of Eels, Pike and Jacks, Perch, Roach ('roceae'), Burbots and 'lampreys which we call water-snakes'; of ? Salmon ('isicii'), the Royal Sturgeon ('rumbus'); of Geese, ? Divers ('fiscedulae'[4]), Coots, Grebes ('mergae'), Cormorants ('corvae aquaticae'), Herons and Ducks 'which I have seen caught by the hundred and three hundred in winter or when moulting their pens'. No wonder that, as he reports, 'in our isle men do not worry, the ploughman does not turn his hand from the plough, nor the hunter abandon the chase, nor the fowler rest from snaring birds'—whatever kings and nobles may demand.

A few years later is the earlier contemporary of Neckam, the famous Welshman Gerald de Barri, usually known as Giraldus Cambrensis, one of the most voluminous of early British writers and the author among many other works of the *Topographia Hibernica* in 1188 and the *Itinerarium Cambrense* in 1191. Born at Manorbier in Pembrokeshire in or about 1146, the youngest son of William de Barri and the famous Princess Nesta, educated in Paris and ordained in 1172, he visited Ireland with Prince John in 1184–5 and preached the Crusade in Wales in 1188. After years of ecclesiastical strife he died at St David's about the year 1220. Pompous and monstrously conceited, in consequence quarrelsome and vituperative; credulous and inaccurate and unable therefore to present an objective picture of nature or history; Giraldus is yet an observer interested in the ways of beasts and birds and, where he is not misled by traditional fables or edificatory symbolism, capable of seeing and recording real facts.

1 I owe knowledge of this record to the Rev. E. A. Armstrong. 'St Cuthbert's Birds' are properly Eider Ducks (*Somateria mollissima*), which still nest as described in the garths and sheds of the Farne. 'Lome' is usually the Guillemot (*Uria aalge*): cf. below, p. 45, for Leland's account of the islands.

2 *Liber Eliensis*, ed. D. J. Stewart, pp. 231–2: quoted in an English translation by J. E. Marr and A. E. Shipley, *Handbook to Nat. History of Cambs.* pp. 6, 7.

3 *Britannia* (ed. Gibson), cc. 408–9, quoted by G. M. Trevelyan, *English Social History*, pp. 147–8. Cf. also the description of the birds of the fens in Drayton, *Poly-olbion*, XXV, ll. 31–138.

4 Fiscedula is a solecism: 'ficedula' is a small bird, 'ortolan' or finch: is it a mistake for 'fuligula', a diving duck?

J. F. Dimock, who edited the *Topographia* for the Rolls Series in 1867, puts the point admirably in his Preface, p. lxxi: 'Of course Giraldus gives us plenty of the stock medieval legendary lore...and plenty of the then common symbolical adaptations; but, and I know not where to name anyone else of the sort within ages of his time, he was also an acute original observer and had the boldness to put his own observations by the side of the received traditions.' Dimock gives an admirable example of this from his *Anglia Sacra*, II, p. 431,[1] where the differences between the wild or Whooper and the tame Swan (*Cygnus cygnus* and *C. olor*) are clearly and accurately described. But the *Topographia* is full of similar instances; and there are others in the *Itinerarium*.

Thus in the chapters on Irish fauna, the two on fresh-water fishes are a straightforward list of those of the Shannon (Sinnenus), Salmon, Trout, Eels, Shads and Lampreys; then lists of those not found, Pike, Perch, Roach, Barbel, Gudgeon, Loach, Miller's Thumb and others; finally, three which he describes as peculiar, a Trout called Salar, like the Grayling but with a larger head, a second described as like the sea 'hake', and a third, an unspotted Trout, perhaps a Char. There is little of importance in the list except that it is a plain factual and very early record. Of birds he begins with Hawks and Falcons; and gives a number of facts, each followed by the lesson from it—as, for example, that in birds of prey the females are always larger 'which perhaps signifies that the female sex is far stronger than the male in every sort of evil-doing' (*Top. Hib.* p. 36). Then comes a chapter on the Eagle, a long and disconnected sermon; then the story of the Cranes and their sentinel with a stone held in its foot; of the Barnacle Geese (bernacae) bred from pine trunks in the sea: 'I have seen many times with my own eyes more than a thousand tiny bodies of this bird on the shore hanging from a log, shut up in shells, and already formed' (p. 48); and of the Osprey (aurifrisius) with one foot clawed and the other webbed, of whose mode of fishing, however, he gives an accurate and plainly authentic description. There follows an interesting chapter on hibernation, appended to the statement that Storks spend the winter in the bottom of rivers, but describing the sleep of the Dormouse as a parallel (pp. 52–3).

Of animals after mention of Deer and Boar he records the Irish Hare (*Lepus hibernicus*), 'numerous but small, like rabbits in size and in the delicacy of their fur', and found in overgrown rather than open country (p. 57). Badgers he says are found—and he tells the legend of their

1 In Rolls Series, vol. VII, pp. 75, 109 (the story being twice told). The words describing the wild swan tamed by St Hugh are 'praeter quantitatem hoc distante quod tumorem in rostro atque nigredinem more cignorum non praeferebat, quinimmo locum eundem rostri planum croceoque decenter colore, una cum capite et colli parte superiore distinctum habebat'.

method of carrying earth: Beavers are not, though they occur in Cardigan, on the Teifi (Tivy)[1] and rarely in Scotland (p. 59). There are Weasels but no Moles; Mice in vast numbers but no poisonous Snakes. Yet one Frog had lately been found at Waterford (p. 65).

As will be seen, there is here some genuine observation—not enough to banish or even restrict credulity, still less to threaten the tradition: but enough to prove that Giraldus, like many of his countrymen, had a real interest in nature, and that not merely in the creatures appropriated to sport.

In the *Itinerary* there are similar notes on many of the fauna of Wales. Thus speaking of the Usk and Wye he records the distribution of Salmon and Trout in them, and the splendid Grayling to be found in the latter river (*It.* Rolls Series, p. 33). In the chapter on the 'Teivi' (pp. 115–18) there is a long account of the habits of the Beaver, not all of it legendary. At Bangor (p. 125) he gives clear descriptions of the Woodpecker 'which the Welsh call Spec' (probably *Picus viridis*) and of the Oriole 'with its yellow colour and sweet whistle' (*Oriolus oriolus*); and notes that the Nightingale has never entered Wales.

Finally, here is a further and more striking observation: 'At the turn of the same year, at the season of fruits, certain wonderful birds never before seen in England appeared, particularly in orchards. They were a little bigger than larks and ate the pips of apples and nothing else from the apples. So they robbed the trees of their fruit very grievously. Moreover they had the parts of the beak crossed (*cancellatas*, literally "lattice-wise") and with them split the apples as if with pincers or a pocket-knife. The pieces of the apples which they left were apparently tainted with poison.'

This note is a paragraph in the concluding portion of Matthew Paris's *Chronica Majora* under the year 1251. With it is a drawing of bird and apple which confirms the description. He has told us of an occurrence which has since then been many times repeated but is always rare and noteworthy, an immigration of Crossbills.

Such records, belonging to a different realm from the traditional lore of Neckam and Bartholomew, remind us that then as now there must have been men who noticed the world around them, who watched the growth of plants and the nesting of birds and the ways of beasts with joy and interest, but without moralising or myth-making.[2] Deep

1 Drayton, *Poly-olbion*, VI, ll. 56–86, has a long description of their former presence in the Tivy and of their habits. He, like Merret in his *Pinax* (1666), notes that they had become extinct: see below, p. 313.

2 Perhaps the most remarkable observation of nature in medieval times comes from the *De Arte Venandi cum Avibus* of Frederick II written in the middle of the thirteenth century and discussed by C. H. Haskins, *Studies in the History of Mediaeval Science*, pp. 299–326. Here in the Paris Bibl. Maz. MS. 3716, p. 70 (Haskins, pp. 321–2) is a record that deserves translation in full: 'In the case of Ducks and other non-

embedded in the Englishman, deeper no doubt than the medieval world-picture, is this delight in nature, preserved more truly in the quaint dialect names of flowers and birds than in the legends of the clerks. Among the common folk there must have been many like the unknown but excellent artist, presumably a monk of the great abbey at Bury St Edmunds, who in the early years of the twelfth century painted plants in a manuscript of the *Herbarium* of Apuleius Barbarus (MS. Bodley 130), reproduced and described by R. T. Gunther (Oxford, 1925) and discussed by Dr C. Singer in an article in the *Journal of Hellenic Studies*, XLVII, pp. 40–3. Most of these are of the conventional type common to such Herbals, but among them are a few, Bramble, Danewort and Ivy for example, that are naturalistic in treatment and drawn from living specimens.

Like his prototype John Falstaff, John Bull babbles of green fields. Out of the life of the countryside, going on almost untouched by education or literature, grew up a natural lore of plants, fishes and livestock, the possession of the herb-woman, the angler and the shepherd, handed on by word of mouth, preserved with secrecy and tenacity, and in some parts of the land surviving to the present day. It is a lore free from the influence of doctrinal and ethical theories, though it has its own symbolisms and superstitions. So far as it was mixed up with white magic or with witchcraft it had points of contact with the tradition. No doubt its fancies easily passed over into the more educated world; and men like Neckam, who must have lived nearer to the land in his monastery at Cirencester than did Bartholomew at Paris or in Saxony, drew from it material which by way of sermon and story passed into their books. Springing from an authentic love of nature it has supplied much of the driving force to natural history and so to the biological sciences: it has also been characteristic of the poetry of England and is linked up with the Englishman's love of gardens, of fishing and of hunting and shooting. This deep-seated native delight in the woods and valleys, the plants and birds of the country, an aesthetic, almost a religious, rather than an intellectual experience, may well be regarded, more truly than the fables and moralisings of Bartholomew and his fellow-scholars, as the source of our scientific

predatory birds we have seen that when anyone approached their nests, acting as if diseased they pretended not to be able to fly and so moved a little way from their eggs or chicks, and deliberately flew badly so that they might be supposed to have damaged wings or legs. So they pretended to fall to the ground so that the man might follow and try to catch them.' Anyone who has seen Mallard or Ringed Plover or Snipe doing its 'decoy' performance will recognise the authenticity of this record, although Haskins couples it with one of the hunted Crane which is misinterpreted and partly fabulous. Frederick also tested the claim that Vultures find prey by smell and found it false; and sent for Barnacles and found that they were not young Geese (MS. l.c. pp. 29, 63).

achievements. But before this delight found expression in organised methods of observation and experiment there was a long period of transition.

For the England of the fourteenth century Geoffrey Chaucer is a sufficient representative. Of his delight in the countryside, in the singing of blissful birds and the springing of fresh flowers, there is perpetual evidence in his works. Of his powers of exact observation, of insight and imagination the *Canterbury Tales* are plain proof. Yet when he speaks of nature it is in terms of symbolism. The charming *Tale of the Nonne Preeste* about Chanticleer and Dame Partlet [Damoysele Pertelote] and Dan Russell the Fox, with its wealth of pleasant illustration and its vivid realism of description, is yet pure fable—an elaboration of Æsop and of Marie de France,[1] a morality in which cock and hen play their symbolic and traditional parts. Physiologus on the Mermaidens' song, Dan Burnel the Ass, Jack Straw and his 'meynee' and the killing of the Flemings,[2] these are allusions which give literary distinction, native relevance and contemporary appeal to a theme thoroughly familiar. It is a delicious example of the fable, but is poles away from the world of science.

For 'the empty fifteenth century'[3] Bartholomew maintained his fame unrivalled: but towards the end of it from the new printing-press of Wynkyn de Worde came in 1486 the treatises upon hawking and hunting which with those on fishing with an angle and on coat-armour added in 1496 form the famous *Boke of St Albans*. Traditionally ascribed to Dame Juliana Barnes or Berners, Prioress of Sopwell Nunnery,[4] a house under the great Abbey of St Albans, and said to have been edited and enlarged by an anonymous schoolmaster of the city, it was reprinted repeatedly during the whole sixteenth century, and became so far as hawking and angling are concerned the bible of the craft. That the book is composite, the work of several hands and perhaps of several times, seems clear; the evidence for attaching an author's name to any of it is very slight; the various parts of it differ widely in interest and fullness; only some of them seem to be complete and some are mere fragments. But as showing the character and intricate technicality of medieval sporting achievement, especially in hawking, the *Boke* is valuable; and the *Treatyse of Fysshing* is deservedly a classic.

Hawking is, as the term strictly denotes, confined to the short-winged hawks, Sparrowhawk and Goshawk: the only mention of other species

1 Cf. W. W. Skeat, *Works of Chaucer*, III, pp. 431–3.

2 L.c. V, p. 25.

3 A valuable article by H. S. Bennett, 'Science and Information in English Writings of the Fifteenth Century' in *Modern Language Review*, XXXIX, 1, pp. 1–8, protests against this calumny.

4 For criticism of this, cf. William Blades's introduction to his edition of the treatises on 'Hawking, Hunting and Cote Armour', London, 1881.

is in the list which assigns the Eagle to an emperor, the Jerfalcon to a king, the Peregrine ('Fawkon gentyll') to a prince, the Merlin to a lady and the Hobby to a young man. Most of the treatise is concerned with the names proper for the various actions and parts of the birds; the rest deals with their feeding and illnesses. Obviously the intention is rather to initiate the reader into the correct language of the craft and to enable him to take a proper interest in his hawks than to describe their habits or discuss their quarry.

Hunting is described first in rhymed couplets; the beasts of venery, Hart, Roe, Boar, Hare; the hunting of the Hare; a dialogue between master and man on various terms and points of the chase, ending in a long and detailed description of the 'breaking up' of a Hart: here, too, the centre of interest is the vocabulary and ritual of the sport. Then follow a few notes and proverbial sayings in prose, and the famous list of the 'Companies' of beasts and fowls which includes next to a 'flyghte of Douves' 'an unkyndnes of Ravens', 'a claterynge of Choughes' and 'a dyssymulacõn of Byrdes' and goes on to 'a pontifycalyte of Prelates', 'a bomynable syght of Monks' and finally 'a dysworshyp of Scottes'. Other lists follow; and the treatise concludes with droll verses in praise of 'myn owne purse'.

The treatise on coat-armour which begins with tracing class distinctions back to Cain the churl and Seth the gentleman, and coat-armour to the siege of Troy—although the law of arms being grounded upon the nine orders of Angels is older than the ten commandments or any law except that of nature—is largely concerned with the story of the first Knight Olybyon. It is a typical example of the medieval attitude towards history.

'Fysshynge wyth an Angle' is the best part of the *Boke*—a classic around which has gathered a considerable literature, of which it is sufficient to mention, in addition to Joseph Haslewood who edited it with full notes in 1811, R. B. Marston, *Walton and Earlier Writers on Fish*, and J. W. Hills, *History of Fly-fishing for Trout*. It is much the most veracious of all medieval books on nature in English, and is quite certainly the work of a different author from any of the other treatises issued with it. There is here a serene philosophy worthy of Isaak Walton himself: here, too, a sense of humour—thus of the huntsman 'he blowyth tyll his lyppes blyster; and whan he weneth it be an hare, full oft it is an hegge hogge'; and a *joie de vivre* appropriate to the naturalist—'yf there be nought in the water, atte the least he hath his holsom walke, a swete ayre of the swete savoure of the meede floures, that makyth hym hungry. He hereth the melodyous armony of fowles. He seeth the yonge swannes, heerons, duckes, cotes and many other foules wyth theyr brodes whyche me semyth better than alle the noyse of houndys....And yf the angler take fysshe, surely thenne is there noo man merier than he is in his spyryte.'

After the introduction come eight pages of advice about the making of 'harnays': the rod, in three pieces; the lines dyed for different waters and plaited of horse-hair; the hooks and their fastening to the lines; and the lead weights. There follows a short but admirably judicious section on the six manners of angling, on the places and times to be chosen, and on the twelve impediments. So to the several fishes and the baits appropriate to them for each month of their season—the list is worth notice as the first catalogue of any part of our fauna: Salmon, 'the moost statelye fyssh that any man may angle to in fresshe water'; Trout, 'a ryght fervente byter'; Grayling, 'by a nother name called umbre' (*Thymallus vulgaris*); Barbel (*Barbus fluviatilis*), 'a quasy meete and a peryllous for mannys body'; Carp (*Cyprinus carpio*), 'there ben but fewe in Englonde'—but even so their presence is hard to reconcile with Leonard Mascall's claim, in his book published in 1590, that he had introduced Carp and Pippins; Chub, 'Chevyn' (*Leuciscus cephalus*), 'strongly enarmyd wyth scalys on the body'; Bream (*Abramis brama*); Tench (*Tinca vulgaris*), 'heelith all manere of other fysshe that ben hurte yf they maye come to hym'—a belief whose origin seems as obscure as its persistence is remarkable;[1] Perch (*Perca fluviatilis*), 'passing holsom and a free bytyng'; Roach (*Leuciscus rutilus*), 'an easy fysshe to take'; Dace (*L. leuciscus*); 'Bleke' (*Alburnus lucidus*); Ruff (*Acerina vulgaris*), like the Perch, 'savynge the ruf is lesse'; Flounder (*Pleuronectes flesus*)—an unlikely fish to get at St Albans though it runs up the Thames at least as far as Hampton Court; Gudgeon, 'Gogen' (*Gobio fluviatilis*), 'a good fysshe of the mochenes [size]'; 'Minnow (*Phoxinus laevis*) though his body be lytyll yet he is a ravenous biter and an egre'; Eel (*Anguilla vulgaris*); and Pike (*Esox lucius*), 'devourers of the brode of fysshe'—for the Pike after recommending a dead Roach or a live Frog as bait the following is suggested: 'take the same bayte and put it in Asa fetida and cast it in the water wyth a corde and a corke...and yf ye lyst to have a goode sporte thenne tye the corde to a gose fote, and ye shall se god halynge whether the gose or the pyke shall have the better.' A few instructions for the care of bait and a list of twelve artificial flies to be used for Trout and Grayling of which eleven can be identified, and resemble those still in use;[2] then a series of drawings of eighteen hooks; and finally some general advice—'fyssch not in noo poore mannes severall water'; 'breke noo mannes hegges...ne open noo mannes gates'; 'ye shall not use this forsayd crafty dysporte for no covetysenes...but principally for your

1 It obviously arose before the *Boke of St Albans* but does not seem to be from Pliny or the Roman world. It was only exploded in the middle of last century. There is a version of it—the Tench healing the Pike opened by the fishmonger—in Camden, *Britannia* (ed. 1687), p. 312, and in Aldrovandi, *De Piscibus* (ed. 1629), p. 245, who quotes both the British practice and a record by Cardan.

2 So J. W. Hills, l.c. pp. 25–6, 146–69.

solace and to cause the helthe of your body and specyally of your soule'—bring the treatise to its end. Appended by way of a Colophon is a characteristic note: 'And for by cause that this present treatyse sholde not come to the hondes of eche ydle persone whyche wolde desire it yf it were emprүntyd allone by itself and put in a lytyll plaunflet [pamphlet], therefore I have compylyd it in a greter volume of dyverse bokys concernynge to gentyll and noble men, to the entent that the forsayd ydle persones whyche sholde have but lytyll mesure in the sayd dysporte of fysshyng sholde not by this meane utterly dystroye it.'

Whoever be the author, he clearly makes no claim to have written the rest of the *Boke*. He was evidently a good fisherman and at least something of a naturalist. As there exists an earlier manuscript version of the treatise, the Denison text printed with preface and glossary by Thomas Satchell in 1883, and pronounced by Skeat to date from earlier than 1450, and as this differs in such fashion from Wynkyn's version as to imply that they are collateral descendants of a common original, he may well have lived in the early part of the century. Certainly he wrote later than its first decade; for he quotes 'the ryght noble and full worthy prynce the Duke of Yorke late callid Mayster of Game'—Edward cousin to Henry IV whose *Master of Game*, an English version of Gaston de Foix's *Livre de la Chasse*, was certainly written about 1405. He refers in the section on the Carp to 'bokes of credence' as among his sources of information; and this, together with the whole quality of the work, involves that it has behind it a long tradition both oral and written. The fact is that fishing was in medieval times a craft of great economic and religious importance; the lack of variety in diet and the fixed fasts of the Church made the provision of fish a necessary part of domestic life. The *Treatyse* is not the work of an isolated or exceptional naturalist: it is the first printed version of a lore many centuries old, built up by long and varied experience and kept from becoming legendary and inaccurate by the repeated and practical testing of its efficiency. The fables of lions and eagles could hardly have been accepted as true if their capture had been a matter of importance.

Here then is an indication of that curious paradox with which the medieval mind confronts us alike in its religion, its science and its art. On the one side there is a complete and ordered system of deductive and explanatory lore, giving an interpretation of the universe in terms derived from a dogmatic theology and elaborated into a vast and intricate symbolism which gave significance to every phenomenon and sanction to every rule of life. It was theoretically complete, logically exact and ethically edifying. Under the combined pressure of Church and State men accepted it as irreformable, and professed to shape their thought and conduct in accordance with it. Unfortunately it bore little resemblance to reality; and its acceptance came to depend more and more upon the maintenance of ignorance and credulity by means of the distortion of evidence and the suppression of enquiry. On the other hand, there was the native and naive response of mankind to the beauty and interest of his environment, a response stimulated by the need to understand the true character of plant or beast since life and health depended upon such knowledge. Gardeners and agriculturists, fishermen and stockbreeders could not avoid first-hand acquaintance with the actual as contrasted with the emblematic values of vegetable and animal life. Tradition, however powerful, must be checked by fact; and for all its influence could only dominate realms in which facts were few and hard to understand.

Nevertheless, as the art of the time shows even more plainly than its science, the two worlds—of tradition and of actuality—normally existed side by side. This has not always been recognised. Indeed, during most of the last century Catholic apologists devoted themselves to an insistence upon the symbolic character of every ornament with which the Gothic craftsmen had enriched their cathedrals. 'In these majestic shrines', said the Abbot Auber, 'not a detail, not a carved head, not a leaf on a capital but represents a thought or speaks a language that all can apprehend.'[1] Symbolism was the clue to it all—as it obviously is the clue to many of the traditional images on carved misericords or bench-ends. But it is not less obvious that alongside the symbolic and moralising ornaments there is another and a more luxuriant artistry at work in the roof-bosses and the spandrels, in places which only a field-glass can explore, and that here the sheer beauty of foliage and flower is the only reason for their appearance. Mâle, to whom we have already alluded, makes a strong case for the contention that when they were released from the conventional emblems proper to certain parts of the edifice, the sculptors and masons let their own creative impulse have full play. As the early painters adorned the foregrounds of their Nativities with studies of finches or primroses, so these other artists 'did not set themselves to read into the flowers of April the mystery of the Fall and the Redemption. In the early days of spring they went out into the woods where lowly plants were beginning to break from the soil. They gazed at them with the tender and impassioned curiosity which we only feel in our childhood but which the true artist keeps all his life.'[2] So the capitals of Notre Dame or the bosses of Southwell burgeoned into beauty, and in this art there is a combination of accurate observation and sensitive interpretation which is totally lacking in the heavy symbolism of the *De Proprietatibus*.

It is indeed almost certainly true that it is to artists rather than to

1 *Revue de l'Art Chrétien*, x, p. 133, quoted by E. Mâle, *L'Art religieux du XIIIe siècle*, p. 63.

2 Mâle, l.c. pp. 69–70.

scientists that we owe the beginning of modern zoological and botanical science. Leonardo or Dürer in their pictures of plant and animal life obviously lived in a different world from the writers on natural history who were their contemporaries. It may of course be argued that Dürer's Irises or Anchusa, now at Bremen, his Columbine and Celandine and Grasses, his Hare or his dead Roller at Vienna, are the most perfect studies of nature ever produced in the Western world;[1] and that such genius is wholly exceptional. But the contrast is hardly less emphatic when we compare Hans Weiditz's exquisite drawings in the *Herbarum Vivae Eicones* of 1530 with Otto von Brunfels's pedantic and tradition-ridden text. Even Leonhart Fuchs's descriptions in his *De Historia Stirpium* of 1542 based upon those of Jerome Bock are hardly worthy of the superb folio-size designs of Albrecht Meyer and Heinrich Füllmaurer his artists: when Mattioli in the Dedicatory Letter of his *Commentarii* pays tribute to Giorgio Liberale[2] and Wolfgang Meyerpeck he is discharging an obvious debt. Indeed, throughout all the period of the Herbals the pictures are almost always more useful and valuable than the letter-press. Men had a love of nature and a power of appreciation and expression which had not been crushed by the fantasies and fables of the official cosmology. This love, and the understanding to which it gave rise, could not permanently acquiesce in theories and stories which falsified fact. Sooner or later the whole edifice of medieval philosophy would have to be transformed or overthrown. But the artists, the naturalists, were inarticulate and in the main uninfluential. It was not from them that the revolt actually came, though it was their spirit that, when the tradition had been effectively challenged, inspired those who set themselves to replace it by a new and inductive philosophy. But this took time—a very long time.

Mâle's summary of his study of the treatment of nature in the thirteenth century is so clear and so relevant as to deserve quotation. 'Pour les théologiens du moyen âge, la nature était un symbole, les êtres vivants exprimaient des pensées de Dieu. Ils imposèrent parfois leur conception du monde aux artistes et firent exécuter sous leurs yeux un petit nombre d'œuvres dogmatiques, où chaque animal a la valeur d'un signe. Mais de pareilles œuvres sont rares. La plupart du temps, les sculpteurs peuplèrent à leur gré l'église de plantes et d'animaux. Ils choisirent ces formes en purs artistes, mais avec la pensée confuse que la cathédrale est un abrégé du monde, et que toutes les créatures de Dieu peuvent y entrer.'[3] It was the tragedy of medievalism that this last 'confused belief' never

1 For admirable reproductions, cf. K. Gerstenberg, *Albrecht Dürer*, *Blumen und Tiere*.

2 Presumably the artist who attached G. S. (*Giorgio sculpsit*) and a burin to certain plates of the 1562 edition: cf. Arber, *Herbals*, pp. 224–6.

3 L.c. p. 82.

received ecclesiastical endorsement—that, in spite of theological readiness to admit the 'congruity' of nature and grace, the world of plants and animals was in fact set in contrast with the world of revelation. Strictly speaking orthodoxy maintained that the universe was a sacramental and harmonious system wherein the outward and visible was the sign and instrument of the inward and spiritual: but in practice, for the custodians of the supernatural, the temptation to magnify their office, to decry the natural as mere vanity, to ignore its lessons and to impose upon it an arbitrary and dogmatic interpretation, was too strong. The two swords, the two cities, the antithesis between secular and sacred, these had become basic elements in traditional Christianity. Pre-Reformation Catholicism had emphasised their separateness: Luther and Calvin by contrasting works and faith elevated it to the status of a dogma. Hence the Reformers, even though their attack upon the tradition helped to reveal the errors of its natural history, did not in fact alter and in some respects even strengthened the belief that man was a denizen of two worlds, and must be content to live as the 'Great Amphibium'.

To us it would seem impossible that this maintenance of two worlds side by side could continue when once mankind had been awakened to the faults and insufficiencies of current belief. We do not realise how deep-seated and long-enduring has been the influence of medievalism: indeed, to judge by many of the older accounts of (say) the Elizabethan age it was assumed that when once the trumpets of the Renaissance sounded the walls of the medieval fortress fell flat, and mankind marched forward into the modern world. Few pictures could be more absurd. We have not yet fully completed the journey; and in the sixteenth century it had hardly begun.

We shall be concerned in the following chapters with the pioneers and their achievements; with men and events that emphasise the movement from the old to the new. Such a record will inevitably suggest that change was more rapid and more continuous than was in fact the case. Therefore, before we commence the task it is well to remember that during the whole period with which we are dealing the medieval outlook was still the popular *Weltanschauung*, that scientific study and the inductive method were hardly appreciated at all until the middle of the seventeenth century, and that the outlook of Bartholomew even in his most extravagant beliefs was still expounded and published not only by poets and preachers and the writers of emblem-books, but as authentic natural history.

A glance at the writing of John Maplet, *A Greene Forest or a Naturall Historie wherein may be seene first the most sufferaigne vertues in all the whole kinde of stones and mettals: next of plants, as of herbes, trees and shrubs: lastly of brute beastes, foules, fishes, creeping wormes and serpents*,[1] will show

1 References to it are quoted from the Hesperides Press reprint, London, 1930.

how popular and persistent was the old outlook. Maplet went up to Queens' College, Cambridge, in 1560 when (as we should suppose) the effects of Renaissance and Reformation were established. He graduated and became a fellow of Catharine Hall in 1564 and of Caius in 1566. In 1567 he published his book, dedicating it to the Earl of Sussex. It may well serve as a sample of what an educated man of some academic distinction thought and believed; and it is, despite its date and provenance, naïvely and contentedly medieval.

The book, after its dedication and preface, contains a list of 'our chiefest authors herein'. All are of the Classics except 'Avicen and Albertus Magnus, Harmolaus Barbar and Laurentius Lippius, Cardane, Lonicer and Ruellius'. Of these only Cardane and once Lonicer are quoted except on matters of nomenclature. For the Marigolde and the Saint Johnes Seale Ruellius and Manardus are mentioned as giving alternative names. Otherwise there is hardly a scrap of evidence that the author had read anything later than Bartholomew, whom he does not mention on his list but whose book contains practically everything that he writes. Here are the same quaint etymologisings: 'Melanite as you would say honeystone' (p. 31); 'Coriander seemeth to be so called. Apo tou koriou of plentie of seede' (p. 71); 'The Bee is called in Latin Apes for that it is first born without any feete' (p. 125). Here is the same dependence upon Pliny and Isidore, but rather less upon Scripture. Here are the same legends—the Dictamnus from Crete which 'the Deare or Harte stricken with anye Dart eateth as a present remedie' (p. 74) (though Maplet gives 'Tullie' (Cicero), not Pliny or Vergil, as his authority); the Crocodile and 'a certaine little small birde called of us the Wren' which 'goeth into his throte...goeth further to his heart and pecketh at it' (pp. 134–5) (though here again 'Tullie' is made responsible). The fact is that of the three parts of the book the first on stones is almost a verbal transcript from the *De Proprietatibus*; the second has hardly any material not derived from it; and the third, though less definitely copied, is in most of its contents plainly dependent upon Bartholomew. A few records seem to be from other sources—'The Quiren is a stone which is found in Ilandes and Fennes most commonly in Lapwings nestes: this is a betrayer of dreames, and of a man's secrets when as he is in sleepe' (p. 36); 'Laus tibi or white Daffadill in Greeke is called Narkissos'[1] (p. 86); 'The Carpe is a kinde of fish well knowne of us...Erasmus called it in Latin Carpa when as other called it Carpio' (p. 132); 'Of the Pearch...Lonicer sayth that when as the fish Lucius is hurte and cannot helpe hir selfe she seeketh out the Pearch which toucheth and suppleth his woundes' (p. 166) (a story similar to that which Camden tells of the Tench). There

1 This is exactly similar to Turner, *Libellus*, Bk. III; cf. below, p. 63.

are also a number of quatrains, English renderings of quotations from Latin poets. Two of these deserve reproduction.

Of the Cameleon

With earth the Moule is said to feede,
With flame the Salamander:
And water is the Herrings meate
The Cameleons the ayer.[1]

And

Of the Fesaunt

By Argolike ship I first was brought
And shewed to other landes
Before that time I knewe no place
But the iland Phasis sandes.

That a fellow of two Colleges should have produced such a book at such a date is evidence of the small extent to which Renaissance and Reformation had as yet altered the world-picture. His uncritical acceptance of legend and superstition, his bare-faced copying from Bartholomew whom he never mentions, his apparent indifference to the work that had recently been done not only by men like Gesner and Belon, but by John Caius who was at the time Master of one of his own Colleges, this is surprising enough. It is the more remarkable when we discover that there are two or three statements like that about the Narcissus already quoted which are almost certainly borrowed directly from Turner: one such is the name Crowtoe applied to 'Iacynthos' (*Greene Forest*, p. 72; Turner, *Names*, p. D. vi)—he makes a special heading of this though he includes 'Jacinct' with the same story of a dead boy (p. 84); another is the statement about the Restharrow, 'It is called in Cambridgeshire whine' (p. 67 and verbatim in *Names*, p. A. viii*b*). So, too, his notes on fishes suggest that he has used an authority like Gesner that has given him their German and French names; for example on the Barbell (p. 129), the Gylthead (p. 147), the Lamprey (p. 153), the Pearch (p. 166) and the Salmon (p. 168). But the book is too unimportant to deserve an elaborate research into its sources. Its significance for us is that a man who should have known the work done by his own senior contemporaries could find a publisher for a book which ignores them and reflects only the outlook and knowledge of three hundred years previously.

1 Translated from Bartholomew, *De Prop.* XVIII, c. 21, p. 320. For 'the camelion's dish' cf. above, p. 5, and Shakespeare, *Hamlet*, Act 3, Sc. 2.

B. THE PIONEERS

CHAPTER III. THE MEN OF THE RENAISSANCE

To trace in any detail the history of the breakdown of the medieval *Weltanschauung* or even to show how its characteristic outlook upon nature was assailed and at length transformed would be to write a history of civilisation during nearly three hundred years. From the beginning of the Renaissance, which is familiarly but not too accurately associated with the fall of Constantinople in 1453, to the middle of the seventeenth century, when the inductive method found its instrument in Britain by the foundation of the Royal Society in 1662, the change from medieval to modern was taking place. In the earlier stages England was not directly concerned: Italy, France, the Rhineland, the Low Countries all contributed and in that sequence before any serious part in it was played by Englishmen. The first phase, that of the Renaissance properly so called, the phase of the recovery, publication and intensive study of the Greek and Latin classics and culture, hardly came to us until the end of the Wars of the Roses and the last years of the fifteenth century, although Wiclif and the Lollards on one side and the increase of grammar schools on another had to some extent prepared the way. Then the University of Oxford, with William Grocyn, Thomas Linacre and William Latimer, John Colet and William Lily, began to convert the country to the new learning.

These five great men had all gained their knowledge by direct intercourse with the scholars and humanists of Italy. Perhaps the most typical of the five, and for our purpose the most important because he was both physician and scholar, is Linacre. His journey in 1488 in the train of William de Selling, his teacher and Henry VII's ambassador to the Pope, introduced him to the Renaissance leaders who were afterwards regarded as the founders of natural history. He went first to Bologna and to Florence where he worked with Angelo Poliziano and Demetrius Chalcondyles and no doubt met the remarkable scholar and Platonist Marsilio Ficino and his brilliant younger contemporary Giovanni Pico della Mirandola, who died before his dream of harmonising Christian, Platonic and Cabbalistic lore could be put into the way of fulfilment. Then he went on to Rome, and working in the Vatican Library made acquaintance with Ermolao Barbaro, the Venetian, who produced in 1492 in a concentrated burst of study amounting only to twenty months an edition of Pliny's *Natural History* with five thousand corrections of the traditional text. Going on to Venice he met

Aldus Manutius (Theobaldo Manucci), greatest of contemporary printers, who formed and expressed a high opinion of the 'English Thomas'. At Padua, where he spent some time in medical studies and gained his M.D. in August 1496,[1] he made a reputation in what Sir William Osler has described as 'then and after one of the most famous schools in Europe' (*Thomas Linacre*, p. 13). Finally, in Vicenza, where he also studied, he worked with Nicolaus Leonicenus (Niccolo da Lonigo), who published his *De Erroribus Plinii* in 1492. He was certainly back in Oxford some time before the visit of Erasmus, who came in 1499 to study Greek with him and Grocyn. In 1509 he became one of the king's physicians to Henry VIII, received a succession of preferments in the Church, and settled down in London. In 1518 he undertook his greatest public service, the foundation of the College of Physicians, modelling it upon similar institutions which he had known in Italy. He died in 1524, and just before his death founded by letters patent the three Linacre lectureships in medicine, two at Oxford, and one at Cambridge.

Like that of his contemporaries, Linacre's influence upon science and natural history was indirect. There seems little evidence that he studied medicine (or nature) with the devotion that he certainly gave to grammar or to his translations of Proclus and Galen from Greek into Latin. But in these works, as in the similar classical studies of his contemporaries, there was in fact a powerful solvent of the medieval tradition. Aristotle's *History of Animals* and other related books; Theophrastus's work on plants, which was now made familiar in the Latin version of Theodore of Gaza, the Greek scholar who had come to Italy before the fall of Constantinople, had dedicated his translations to Pope Nicholas V and had died in 1475; Pliny's *Natural History*, very much less valuable but nevertheless a useful compilation of previous knowledge; Dioscorides's authoritative work on vegetable and other drugs; Aetius, Paulus Aegineta and other late Greek physicians and pharmacists; these represented an outlook far more accurate in its records of flora and fauna, far less cumbered with myth and fable, than the works of Neckam or Bartholomew. The first attack upon medievalism came when the relatively scientific literature of classical antiquity was compared with the products of the thirteenth and fourteenth centuries. Quotation, the appeal not to observation but to authority, was the instrument of assault, and was used on occasion to an almost ludicrous extent, 'as if', said Nehemiah Grew in 1681, 'Aristotle must be brought to prove a man hath ten toes'.[2]

Renan in his *L'Avenir de la Science*, p. 145, makes the pronouncement: 'Les plus importantes révolutions de la pensée ont été amenées directement ou indirectement par des hommes qu'on doit appeler littérateurs

1 Cf. R. J. Mitchell in *Eng. Hist. Review*, L, pp. 696–8.

2 *Musaeum Reg. Soc.* Preface; he is criticising Aldrovandi.

ou philologues.'[1] No doubt he has the Renaissance chiefly in view; and of its effect upon science the saying is not less true than it is of religion. Men like Linacre discovered the ancient wisdom of Greece: they found the sources from which, albeit with slight acknowledgment, the traditional lore constituting the contemporary outlook had been originally derived: they compared the recently discovered texts of Aristotle and Pliny with the parodies of them preserved by a half-literate acquaintance with such interpreters as Isidore of Seville or the Western versions of the great Arabians: and they saw that the boasted authority of the schools was maintaining legends and fables in place of truth. The consequent reformation was neither so speedy nor so violent as that which followed upon the similar discovery in religion. But though science had no one to represent Luther, it had plenty to attempt the part of Erasmus.

How rapid had been the discovery and publication of classical texts between the years 1465 and 1515 (the date of Manucci's death) may be seen from the lists of first editions of Latin and Greek authors in Sir John Sandys's *Short History of Classical Scholarship*. These were almost entirely produced by the presses of northern Italy, and testify to the extraordinary zeal and energy of the scholars at work there. But despite Linacre and his colleagues the task of introducing the new learning and especially the Greek language into England was almost insuperably hard. Oxford, although recognising the worth of its Renaissance scholars, was deeply involved in the maintenance of the traditional language, curriculum and philosophy. Greek was an innovation and when Erasmus desired to return and continue his studies in it Latimer had to warn him of the obstacles.[2] John Clement, More's secretary who married Margaret Giggs his adopted daughter, came back to the University for one year as reader in 1519 and revived the enthusiasm for Greek: but he turned immediately to medicine, and hardly fulfilled the promise of his youth and upbringing. In fact, in spite of Wolsey's magnificent foundation of Christchurch and the transfer to it of a number of outstanding Cambridge scholars in 1525, the cultivation of Greek tended to pass from Oxford to Cambridge. There the influence of John Fisher, President of Queens' College, Chancellor of the University, and confessor and adviser to the Lady Margaret whose benefactions founded Christ's College in 1505 and St John's in 1511, was thrown into the advocacy of the new learning; and a post as the Lady Margaret's Professor of Divinity was found for Erasmus in the years 1511–14, in which he prepared his epoch-making edition of the Greek New Testament.

From this time dates the origin of the very distinguished band of Cambridge scholars, humanists and reformers. They were not indebted to

1 I owe this reference to my colleague, Mr B. W. Downs.

2 Cf. J. N. Johnson, *Life of Linacre*, pp. 200–2.

Erasmus so much personally; for as a teacher in the University he was not a great success: but his spirit, his writings and above all his New Testament gave an impetus to Greek studies and opened up to the University a new understanding of classical civilisation and the Christian religion.

Of the group at Cambridge round Thomas Bilney, Hugh Latimer, Robert Barnes and the rest, who from their interest first in Erasmus's Greek Testament and then in the work of Martin Luther were known as Little Germany, this is not the occasion to speak. With them began the Reformation in England, and from them, as we shall see, sprang the first English scientist. But for the twenty years after Erasmus's departure, it is with Sir Thomas More and his circle, with Colet's new foundation of St Paul's, and with Cuthbert Tunstall and a few other learned churchmen that the promotion of scholarship chiefly rested. More in particular, with his large family, his generous and learned hospitality at Chelsea, his wide range of interests and of friends, stood for all that was best in the Renaissance. Though not specially devoted to nature he had a great love of birds and animals, and according to his 'best friend' Erasmus[1] had an aviary and live monkeys, weasels and foxes in his gardens. Religion, art, music, law, letters, philosophy—he ranged over them all with a knowledge and humour that explain the affection which the scholarship of Europe felt for him. That he shared the habit of personal and often violent abuse of those who differed from him (a fault common to most of his contemporaries), that his actions were not always consistent with his opinions, and that in consequence he did less to shape the history of the period than several less gifted and influential contemporaries, are facts which must qualify the extravagant eulogies with which his memory has recently been injured. But if the generous spirit of his earlier days had prevailed in the Church of which he was so loyal a critic the transition from the medieval to the modern would have been easier and less destructive.

Before we pass to the second stage in this transition we must consider the type of literature to which the Renaissance and the recovery of Greek and Latin classical texts gave rise in the field of natural history. This was what is usually called the 'pandect'—a name derived from the title of the huge compendium of Roman civil law made under the orders of the Emperor Justinian in the sixth century, and consisting of extracts from famous authors and authorities collected into an ordered system. The term came to be applied to the similar compilations from writers on natural history of which Conrad Gesner's immense tomes published in the middle of the sixteenth century at Zürich, and Ulisse Aldrovandi's even more huge series half a century later, are the familiar representatives.

Of this class of literature, the natural outcome of the antiquarian

1 *Epistolae* (London, 1642), x, *Ep.* 30, c. 536, written in 1519.

interests of the new learning, we have in England only one proper representative, Edward Wotton. Born in Oxford in 1492, educated at 'the Grammar School joining to Magdalen College' (Wood, *Ath. Oxon.* I, c. 226), entering the College as a demy in 1506, he graduated in 1513, and became a fellow in 1516. He came under the influence of John Claymond, President there from 1504 to 1516 and author of four volumes of manuscript notes on Pliny's *Natural History* to which he freely refers in his own book.[1] When Claymond was persuaded by Bishop Foxe to move from Magdalen to Corpus Christi, Wotton soon afterwards, in 1520 or more probably 1523, became 'socio compar' of that College. He was then granted three years' leave of absence for travel and went to the medical school at Padua. There he took his doctorate in medicine, and on his return in 1526 was admitted to the same degree in Oxford. Soon afterwards he became a member of the College of Physicians and, according to Wood, physician to Henry VIII. He seems to have moved to London and to have devoted himself to his profession, becoming President of the College 1541–3. In 1551 William Turner wrote of him as one of the English savants who might have produced a treatise upon plants; and Gesner paying tribute to his book said that 'he teaches nothing new but gives a complete digest of previous works on the subject' (*Historia Animalium*, IV, Enum. Auct.). He died in 1555, leaving a large family.

In the dedication of his book, the *De Differentiis Animalium*, to Edward VI he gives a few particulars of its production. He describes how he had for many years been devoted to the study of all the best writers whose books could contribute anything of value to medicine, how finding it difficult to gather and to retain all the material available from so large a field he had prepared commentaries and compilations of their works, and how when he had spent a long time on this and made some progress he had seen the books first of Jean Ruel the French botanist and then of his former friend and associate Georg Bauer (Georgius Agricola the German who wrote *De Natura Fossilium* and on metals), and realised that he had been relieved of a large part of his labour. Plants and minerals being thus disposed of, there remained all the various ancient writings about animals, birds and fishes. To this task he devoted himself; and his friends welcomed and encouraged him. When it was finished, he still hesitated about its publication, and only yielded when his friend and patron, Sir John Mason, French Secretary to Henry VIII and subsequently Dean of Winchester and Chancellor of Oxford University, going as Edward's ambassador to France, carried off the script, promising that if he approved it he would

1 Gesner, who refers to these commentaries of 'Clemundus Anglus' in his list of authors prefixed to Kyber's Latin version of Tragus (Bock), seems to have assumed that they had been published.

see to its publication. This he did and it was admirably printed in 1551 at Paris by Michael Vascosanus (Michel de Vascosan), one of the first printers to give up the black letter type.

The *De Differentiis* is in ten books, the first three dealing with the parts and functions and differences of animals in general, Book IV with man, V with 'quadrupeds that bear young', VI with 'quadrupeds that lay eggs', VII with birds, VIII with fishes, IX with insects and X with squids, crustaceans and molluscs. It is an astonishing mosaic of extracts from every sort of Graeco-Roman writer: Pliny, Aristotle, Galen, Dioscorides, Oppian, Athenaeus, Varro, Ovid, Xenocrates, Aelian, Diphilus, Columella, Pausanias, Nicander, Strabo, Vergil, Cornelius, Celsus, Martial, Aristophanes and very many others—roughly in the order of frequency thus indicated. Unfortunately his method is to give a list of authors in the synopsis of each chapter prefixed to the book: these are without exact references; nor are there many notes in the text to indicate the source of any of his statements. This makes it hard to discover what if anything is original. He quotes no 'moderns' as authoritative: but at the end of each chapter gives a number of footnotes in which the text is discussed and commentator's notes upon it are cited. Thus he also quotes very freely Theodore of Gaza, the Greek scholar; Ermolao Barbaro, the great Venetian; Pierre Gilles, the translator of Aelian's book on animals; Francisco Massario, the Venetian commentator on Pliny; Jean Ruel, the French herbalist; Antonio Musa Brasavola, his Italian successor; Niccolo da Lonigo, the doctor; Aldo Manucci, the printer; Michel Angelo Biondo, the doctor whose book *De Canibus et Venatione*, 1544, he quotes on Italian bird-catching; and John Claymond, whose commentaries on Pliny he must have seen at Oxford. Occasionally in the notes there is a reference to some fact which a modern has recorded: but in the main Wotton is plainly concerned with the ancient references—to make his catena of quotations as accurate and complete as possible. There is no evidence that he specially cared for, or indeed had ever seen, any of the creatures which form the subjects of his quotations. He had obviously read widely, copied his extracts diligently, and fitted them together ingeniously. The result is a useful compendium of the traditional lore, a 'pandect' less, vastly less, in size than Gesner's huge tomes, and much more strictly confined to the ancients, but not a book with much originality or human feeling. Nevertheless as compared with Neckam or Bartholomew he is both accurate in his rendering of his authorities and restrained in his acceptance of the mythical. Thus he omits the story of Goat or Hind and Dittany; follows Pliny without addition in that of the Crocodile-bird (p. 128*c*); says nothing about the webbed foot of the Osprey; expresses scepticism about the uniqueness of the Phoenix (p. 134*f*); and concerning the Quail, instead of the legend that it settles on the sea, spreads its wings and sails, says that it returns from the south in flocks which often alight on the sails of ships (p. 116*a*).

In parts, and especially in the two last books where he deals with insects and molluscs, he shows a measure of independence. Here for example is a definition of real importance: *Bivalve appello quod gemina testa clauditur; univalve quod testa singulari continetur* (p. 207*f*), 'a bivalve is shut up in a double shell, a univalve in a single one'. In Book IX there is sufficient evidence of independent knowledge to make his reputation as a student of insects intelligible—though the extent of it cannot easily be estimated. He seems to have seen a certain number of larvae as well as the beetles or moths to which they give rise. He knows that some insects at least are produced from parents, though many including the caterpillars on cabbages he regards as spontaneously generated. A large part of the book is devoted to the medicinal uses of insects; squashed, powdered or pickled specimens especially of those parasitic upon man or living in his dwellings being regarded as potent drugs. This superstition survived for at least another two hundred years: but it can rarely have been more prevalent than with Wotton.

He is particularly interested in the problem of the generation of insects. In Book IX, chapter 200, pp. 176*d*–7*b* he discusses it, claiming that some insects like Spiders produce young by sexual procreation: others like Flies and Butterflies produce creatures of different type, Caterpillars: others like Gnats and many more are spontaneously generated, from dew on leaves, from mud or dung, from wood or plants, in furs, or in the internal organs. He illustrates those born sexually from the Butterfly, though his record is in the crucial question confused; for having said that whereas most insects give birth to worms the Butterfly produces 'something like a seed but fluid inside', he then goes on to say that 'butterflies come from caterpillars and caterpillars from plants especially from cabbage'. He gives a good account of the growth of the caterpillar, of its changing into a hard-skinned chrysalis, of the breaking of the skin and the emergence of the winged insect: but both here and in the chapter on Caterpillars (p. 193) he cannot give up his belief in spontaneous generation. In later chapters as that on Lice who 'copulate but produce only nits from which nothing more comes' (p. 196) and who are themselves born of human sweat, or in dealing with clothes-moths born of wool or papers or books, he shows how far he is from any true interpretation of metamorphosis. That the various stages, egg, larva, pupa, imago, egg are a series which leaves no room for sexless or spontaneous creations is a fact that has not yet been appreciated.

Contemporary with Wotton and like him known later to Gesner was one whose claim to a place in the history of science is much less serious, Thomas Elyot, diplomatist and man of letters. Son of Sir Richard Elyot and born about 1490 in Wiltshire he inherited properties both at Combe near Woodstock and at Weston Colvile in Cambridgeshire. Of his education there seems to be little clear evidence—both Universities claim him and neither can show convincing proof: but he seems to have learnt Latin early and well, and then to have read medicine, possibly with the great Linacre. Taken up by Wolsey in 1523; knighted and dismissed; a friend of Sir Thomas More; a follower of Thomas Cromwell; Ambassador on two occasions to the Emperor Charles V; Sheriff for Oxfordshire in 1527 and M.P. for Cambridgeshire in 1542, he had a varied and not very stable career. Evading several situations which might easily have been fatal to one more important or less pliant, he gave himself increasingly after 1531 when his first book *The Governour* was published to literature and the translation of the classics. In 1534 he published his *Castel of Helth*, though of this edition no copy seems to be known. It was a dietary and dealt not only with food and exercise but with the symptoms and treatment of disease. In consequence it provoked criticism from the doctors; for it was written by a layman and in English. To this Elyot replied in a vigorous introduction to the next edition in 1541; and the book became very popular. In 1538 and 1545 he produced the first Latin-English Dictionary to be printed, a storehouse of antiquarian and other lore.

The *Castel* is not a herbal but it contains enough about plants and birds to show its author's interest in them. It is on the English renderings of the names of flora and fauna in the *Dictionary* and the notes occasionally appended to them that his position as a naturalist depends. He was evidently interested in zoology, and although the bulk of his material is derived from Pliny, probably in most cases by way of the *Ortus Sanitatis*, a few details show signs of personal knowledge. Onocrotalus 'a bird like a swan'; Platalea 'a byrde which foloweth water foules that do take fyshes and do pecke them so on the heade that they let go theyr prey whiche Platalea taketh and lyveth therwith'—an interesting tale derived by way of Pliny from Cicero, *De Natura Deorum*, II, 124,[1] and seeming to recall an observation of the habits of the Frigate-bird (*Fregata aquila*), or of the Skuas; Phoca 'a sea-calfe it may be supposed to be a seale'; Trigla 'a fyshe I suppose it to be a soore mullet such as ar taken in Devonshyre and in Cornwal';[2] these are perhaps the most interesting of them, the two first being from the *Ortus*, the two last original. Three more have a certain importance from their connection with Turner: Caprimulgi 'byrdes lyke to gulles whiche appere not by daye, but in the nyght they

1 There appears to be no known source for this statement, though most of Cicero's material in this section of his treatise comes from Aristotle. The Skuas of the genus *Stercorarius* which have this habit occur occasionally in the Mediterranean: *Fregata* is solely tropical.

2 Maplet, *Greene Forest*, identified Trigla as Barbel.

come into gote pennes and do sucke the gootes wherby the udders of them be mortifyed'—a legend derived from Aristotle and Pliny through the *Ortus Sanitatis*, De Avibus, ch. 22, but accepted by Turner with a touch of irony; Attagen 'they are deceived that take hym for a woodcocke', with which Turner wholly agrees; Upupa 'a lapwynke or blacke plover', which Turner quotes and rejects.

As a botanist he is still more interesting. Acknowledging help from doctors and naming Ruel, Manardo and Brasavola among them, he often shows considerable research. His basis is Pliny and the tradition; under 'Dictamnus' he refers to its power to draw out weapons; under 'Pulegium' he cites Theophrastus and Pliny besides mentioning the views of Ermolao Barbaro and Manardo; under 'Rhamponticum' he cites Ruel's comment upon Pliny; under 'Iberis' Ermolao, Manardo and Aetius; under 'Polygonatum' Manardo and Johannes Agricola; under 'Pityusa' Actuarius. Ruel is mentioned many times—for example under Doronicum, Radicula and Parthenium. Moreover, he knows the apothecaries' usage and on occasion, though rarely, writes as if he had first-hand knowledge of the plant itself. As we shall see it is certain that he knew Turner's first book, *Libellus de re herbaria*, and more than likely that he had close contact with him; for the number of times in which he gives the same English rendering is very large—far too large to be accidental. But he is a real and independent student who gives many names not found in the *Libellus*, and some which Turner never mentions at all; and if the two men were acquainted neither of them hesitates to contradict or to ignore the other. So far as can be traced Elyot nowhere mentions Turner by name; but this is quite characteristic. The only contemporary of whom he speaks is a worthy London merchant named Goodman eulogised under Isatis as the man who introduced into England for the first time since the days of the Ancient Britons the growing and grinding of woad.

The evidence of the *Dictionary* is fully sufficient to prove that Elyot also, while obviously interested in natural history, was mainly concerned with the business of finding in the English tongue and country equivalents for the plants and animals of classical antiquity. And though his own efforts are not important, his connection with Turner increases their significance. It was precisely this desire not only to collect the lore of Greece and Rome, but to attach it to particular and locally recognisable species that characterised the beginnings of modern science. Elyot illustrates the transition from Wotton to Turner.

Of a rather different type is the third contemporary whom we must mention as representing the beginning of the new age—John Leland the antiquary. Born in London, but presumably of a Lancashire family, educated at St Paul's School under Lily, he graduated at Christ's College, Cambridge, in 1521–2 and then went on to All Souls, Oxford, and to

Paris. His *Itinerary* in or about the years 1537–9 anticipates the work of William Camden a generation later; and though mainly topographical and archaeological contains a number of notes on natural history. He is particularly interested in fishes and draws attention on several occasions to those that can be found in various lakes and rivers. Thus in his account of Shropshire he says: 'At the hither ende of Whitchirch is a veri faire poole having bremes, pikes, tenches, perches and daces, the wich except bremes be the commune fisches of al the pooles of Shropshire, Ches and Lancastre shire, in sum be also trouttes.'[1] Similar notes occur of the mere at Ridley Place (l.c. p. 3) and of Cumbermere (v, p. 16); and of Windermere he has a more interesting record: 'Wynermerewath wherin a straung fisch called a chare' (v, p. 47). So, too, of birds he has a note of the isles adjoining to 'Farne Isleland' (iv, p. 123): 'Certen bigge fowles caullid S. Cuthebertes Byrdes brede in them and Puffins birdes less then dukkes having grey fethers like dukkes but withoute paintid fethers and a ring about the nek be found breding there in the cliffy rokkes'—which if not very clear evidence of first-hand observation at least shows a desire to describe and identify.[2] Of beasts apart from deer he has little to say; nor is he interested in plants. But he comments upon the snake stones (ammonites) at Keynsham (v, p. 103) and the fossil cockles and oysters at Alderley in Gloucestershire (v, p. 95) as well as on fir trees buried in the peat. With him we may surely say that the exploration of Britain for its natural resources and fauna has begun.

But before we deal with this exploration and the real founder of the serious study of nature in this country, it is well to note that during the next century, quite apart from the pioneers whom we shall consider in detail, there was a two-fold result of the increased interest in flora and fauna. On the one hand pandects of the type represented by Wotton's work continued to be produced: and on the other the traditional fables and allegories maintained their place by the writing of the emblem-books.

Of the former category a good example can be seen in William Langham's volume, *The Garden of Health*, first published in 1597 (the printed date is a textual error, 1579), and reprinted in 1633. Of Langham little seems to be known; his book, though describing him as a 'practitioner in physicke' and speaking of his travel, is wholly impersonal and uninformative. It consists of an alphabetical series of plants with a vast compendium of their uses and a long list of all the ailments and diseases which each may be expected to cure. Apart from the names no attempt is made to define any species or to describe it or to state its locality: but the preparation of potions, simple or compound, from it is elaborately stated. It is in fact more a pharmacopœia than a herbal, and contains

1 Ed. L. Toulmin Smith, iv, p. 1.

2 He notes that sea-fowl nest on 'Godryve' (Godrevy) off St Ives (i, p. 317).

hardly a single original word, although no authorities are quoted or even mentioned.

Similar, but much more voluminous, are the two volumes published by Robert Lovell of Christ Church, Oxford, the *Pambotanologia* in 1659, and the *Panzoöryktologia* in 1661. These are a monument to the patient industry of their compiler, who must have ransacked and copied out a large percentage of the relevant literature ancient and modern to make the elaborate mosaic of quotations which he prints under every name. The former, a small but thick volume, closely printed of 80 unnumbered and 670 numbered pages, is a tolerably complete alphabetical catalogue of plants with a highly elaborate introduction on the principles of 'herbarism'. Under each name are lines for its locality, almost always given in general terms, 'In India and Arabia' or 'Moist meadows and fields'; for its time of flowering, and for its synonyms. Then follows a series of quotations from all sorts of authors as to its character and uses. There are no descriptions, no helps to identification and, so far as can be seen, no records which are not copied. Gerard and Parkinson are laid under frequent contribution: but Turner, Tradescant and the *Phytologia Britannica* are also listed among the authors quoted. There is an astrological classification occupying ten pages of the introduction: but this is not referred to elsewhere. The latter book, 84 pages of introduction, 520 of animals, 152 of minerals, has pages twice as large and the same close print. Dedicated to Charles II it contains nothing of importance except evidence of its author's industry, and of his self-repression; for there is hardly a word in it that gives any clue to his character, interests, knowledge or ability. Its 'Isagoge' begins with a list of every possible beast, bird, fish, serpent and insect, grouped together as 'irrational', and subdivided partly on structural and partly on geographical lines. Very many of the names, especially of quadrupeds, are South American—the 'Viviparous digitates wild' for example begin 'Lion, puma, mitzli, quamitzli, macamitzli, cuitlamitzli, tlalmitzli; pardal, theotochtli; lynx, tigre, tlacoocelotl'—creatures whose identity might provide an interesting subject for research. He classifies the Bat among the birds (p. 131), and the Puffin among 'fishes lesse used in meat' (p. 234). Man, the one 'rational' creature, has more than 200 pages including a complete 'dispensatory' of drugs for his diseases. The Minerals begin with Alanian earth useful for cleansing gold and Ampelite earth which 'dissipateth, refrigerates and honestats the eyebrowes'. The lists of authorities include for zoology Bartholomew, 'Dr Cay', 'Gyraldus', Jonston, Mouffet, Topsell and Wotton, and for minerals Roger Bacon and no other Englishman.

According to Wood, *Athenae Oxonienses* (ed. Bliss), IV, c. 296, Lovell came up to Oxford from Warwickshire in 1648, and after the publication of his books settled in Coventry, practised medicine there and died in 1690—a sufficiently uninforming record. His brother Sir Salathiel seems to have had a similar laboriousness and a similar mediocrity: but he lived long enough to gain some reward for his labours.

The other effect of the new outlook upon nature is to be found in the development at this time of the emblem-book as a characteristic expression of the period. Moralising by symbolic anecdote had been as we have seen the universal method by which writers like Neckam or Bartholomew dealt with birds and beasts: Æsop was their master even if Pliny was the ultimate source of their material. Now when interest in nature for its own sake was banishing its emblematic significance, the taste for allegory previously satisfied by the *De Proprietatibus* had to find its further satisfaction elsewhere. It was an essential part of life; and at first was given expression in the pageantry and masques, the tournaments and ceremonial of the Court and of society: the *Faerie Queene* is not the only poem of the period to carry a mystical meaning. Then emblem-books, modelled upon that first published in Italy in 1531 by the Italian lawyer Andrea Alciati, began to attract attention. French, Spanish and Dutch books of this kind were produced; and Antwerp, already famous for the woodcuts and herbals of the Maison Plantin,[1] found a new opening for its artists and publishers. England only began to produce such books in 1586 when Geffrey Whitney published *A Choice of Emblems and other Devises*, but they became numerous by the beginning of the next century, and persisted until long after the scientific age had begun. In them the traditional images, the Pelican in its Piety, the Stork or Crane keeping vigil with a stone in its claw, the Salamander in the flames, the Lily and the Palm, retain the meaning formerly given to them by medieval naturalists; and their prophetic and hortatory value was thus preserved. But in this form their persistence no longer inhibited truer concepts of plant and animal life. There was room for scientific studies free from all concern with allegorising; and for a century or more the study of nature for its own sake by observation and experiment was carried on alongside of its use for hieroglyphics, emblems, symbols, fables and moralisings. Only with John Ray was a complete separation of the two insisted upon; and then the fate of the emblem-book as a serious interpretation of life was sealed.[2]

The process of change was inevitably slow. The detailed study of its advance and of the compensatory developments which furthered but

1 The volumes of the *Correspondance de Christophe Plantin* by M. Rooses and J. Denuce show plainly how great was the popularity of Alciati and his imitators.

2 I am indebted for much information on this subject to a thesis by Miss R. Freeman, Ph.D., *The English Emblem-books*. Those wishing to see how vast was the literature of the subject should refer to the collections of emblems in the *Hieroglyphica* of Giovanni Pietro Valeriano (Lyons, 1626) or to that in the *Mundus Symbolicus* of Filippo Picinelli (Cologne, 1695).

also prolonged it must not blind us to the greatness of its scope and significance. What we are considering is a series of events which has modified, indeed in measure transformed, the whole thought and life of man. As such it is traceable, and its effects can be seen, in every field of human activity. When for example Dr G. M. Trevelyan writes: 'It was a factor in the development of the new England of adventure and competition replacing the old England of custom and settled rights',[1] he is in fact speaking of the rise of prices and its economic results in the sixteenth century which promoted what we commonly call the rise of capitalism. But his words apply as accurately to the changes by which the England of the *De Proprietatibus* was replaced by the England of the *Novum Organum*. It is, of course, not true that the new outlook on nature was created or even fostered by the new economic system: the reverse statement would be much more accurate. But the development of criticism and revolt and of exploration and experiment is common to the two spheres, and indeed to the whole life of the epoch in our Western world, and springs from a profound alteration in the whole relationship between mankind and his environment.

And at this point, so far as England is concerned, we have an outstanding though somewhat neglected pioneer who represents this alteration alike in its political and religious and in its scientific aspect.

CHAPTER IV. WILLIAM TURNER: HIS TRAINING

Wotton, the Oxford scholar and doctor, was a man of books rather than of first-hand observation; and has little claim to be called a naturalist. He belongs to the category of Salviani rather than of Belon. The case is very different with his two younger contemporaries of the University of Cambridge, William Turner and John Caius, with whom the great succession of field-naturalists in Britain may properly be said to begin. They were almost exact contemporaries: both were doctors, both studied in Italy, and both were friends of the great Zürich naturalist Conrad Gesner: they had many common associates like the eminent doctor Thomas Wendy: but they represented different religious convictions and though they must almost certainly have met there seems to be no evidence of their friendship or co-operation.

Of the two William Turner was for our subject by far the more important: indeed his work both as one of the most vigorous of the English

1 *English Social History*, p. 122.

Reformers and as the first scientific student of zoology and botany in this country deserves full-length study. He was born at Morpeth in Northumberland, presumably in or about the year 1508. His father was perhaps a tanner; the family certainly belonged to the neighbourhood; and young William was evidently educated there. From his will, printed by B. D. Jackson, it appears that he had at least one brother and one sister.

As a boy he began to notice the ways of birds and plants. After his description of a Robin's nest 'built at the roots of brambles where oak-leaves lie thick, and concealed with leaves as if with topiary work' and the note, 'There is no obvious approach to the nest; it is reached only along a single way: on the side where the entrance is placed she strews a long vestibule with leaves before the door of the nest; when she comes out to feed she closes the end of this with leaves', he adds: 'This which I now write I noticed when a boy; though I will not go so far as to deny that she may make her nest differently: if anyone has observed a different method of nesting let them publish it and earn the gratitude of all students of the subject, and mine among them: I have merely told what I have seen' (*Avium Praec.* p. 84). He records, too, the nesting of Kestrels (*Falco tinnunculus*) in the church tower at Morpeth (l.c. p. 90); describes the Dipper (*Cinclus cinclus*) 'another sort of Kingfisher' and says that 'the people of Morpeth among whom I saw it call it the Water Crow';[1] notes the cry of the Corn-crake (*Crex crex*)[2] 'among the crops and flax only in Northumberland' (p. 57); and tells how he had seen Cormorants 'nesting on sea-rocks off the mouth of the Tyne as well as in tall trees in Norfolk'[3] (l.c. p. 57, this in comment upon the fact that Aristotle described them as nesting on rocks but Pliny in trees). His clear and accurate account of the Black Cock and Grey Hen (*Lyrurus tetrix*) (l.c. p. 37) which he knows as the two sexes of a moorland bird that lives on corn and shoots of heather, and which he tries to identify with the 'Attagen' of his authorities, must reflect knowledge gained in his boyhood.

In his plant-books he has still more references to the neighbourhood. Thus he notes that 'Aconitum Pardalianches' which following Fuchs

1 L.c. p. 5: he also describes later, on p. 25, the 'Cinclus' or Water Swallow: the description is very similar, 'a little larger than the Lark with dark back and white belly, with long legs and beak by no means short: it is very noisy, and makes short and frequent flights, on river banks in spring'. By reference to Gesner, *H.A.* III, pp. 557–8, it seems that to him this Cinclus was the Grey Wagtail: perhaps to Turner it was the Sandpiper (so A. H. Evans, *Turner on Birds*, p. 57) (*Actitis hypoleucos*) or Green Sandpiper (*Tringa ochropus*).

2 'Non aliam habet vocem quam crex crex.'

3 Cf. Browne, 'Natural History of Norfolk' (*Works*, ed. G. L. Keynes, v, p. 379), 'building at Reedham upon trees from whence King Charles the first was wont to be supplyed'.

he identified with *Paris quadrifolia*[1] 'whiche we may call in englishe Libardbayne or one bery' grows 'much in a wodde beside Morpeth called Cottingwod' (*Names*, p. A. v back, cf. *Herbal*, p. B. i*b* and I, p. 19); that 'Lepidium is wel knowen in Englande and is called with a false name Dittany...it groweth in Morpeth by a water called Wanspeke [Wansbeck] in great plentie alone without any settyng or sowyng' (*Names*, p. E. i); that 'that kinde of bear-foot that goeth every yeare into the grounde [*Helleborus viridis*] groweth in the west parke besyde Morpeth a little from the river called Wanspek' (*Herbal*, II, p. 160*b*); that Orobanche 'is called about Morpeth in Northumberland, newchappell floure; because it grewe in a chappel there in a place called bottell bankes whereas the unlearned people did worshyppe the Image of Saynt Mary'[2] (*Herbal*, p. P. v); that of 'Great Pilletorie of Spayne' (*Peucedanum ostruthium*) 'I never hard that it grew wilde in Englande savinge aboute Morpeth in the North parke there' (*Herbal*, III, p. 37) and that of Beeches 'two of the greatest that ever I sawe growe at Morpeth on ii hylles right over the Castle' (*Names*, p. D. 1*b*).

He knew the country and coast pretty well and the *Herbal* refers to them not infrequently. 'I have sene Sea Wormwode in Northumberlande by holye Ilande' (I, p. 11, cf. p. A. iv*b*), 'I never sawe thys herbe [*Meum athamanticum*] in Englande savynge once at saynte Oswarldes where as the inhabiters called it Speknel' (*Names*, p. E. v, cf. *Herbal*, II, p. 57, 'in New castel in a gardin'). 'The same bent or sea rishe [J. Britten, *Turner's Names*, p. 101, identifies this as *Stipa tenacissima*: it is surely *Ammophila arundinacea*] have I sene in Northumberlande besyde Ceton Dalavale [Seaton Delaval, 10 miles south-east of Morpeth] and ther they make hattes of it' (*Herbal*, II, p. 144*b*). 'In the Bishopricke of Durram, the housbandmen of the countre that dwel by the Sea syde, use to fate their lande with Sea wrake' (*Herbal*, I, p. 142). Here are reminiscences which give something of his quality. Under the heading 'Of Byrche' (*Herbal*, p. F. v*b* or I, p. 84) he writes:[3] 'I have not red of anye vertue that it hath in Physick. Howbeit, it serveth for many good uses, and for none better than for betinge of stubborne boyes, that ether lye or will not learne. Flechers make pricke shaftes of Birche because it is heavier than Espe is.

1 'Aconitum pardalianches' was afterwards identified as *Doronicum plantagineum*: cf. Dodoens, *Pemptades*, pp. 437, 444; Mattioli in his *Commentarii in Diosc.* (ed. Bauhin), pp. 766–7, proved that Fuchs's plant ought to have been called 'Herba Paris' and Turner, *Herbal*, III, p. 35, accepted this correction. The whole question of the identity of Aconitum pard. was the subject of a treatise by Gesner (Zürich, 1577) in which he cites Turner, p. 15*b*, and of a controversy between him and Mattioli.

2 Cf. Leland, *Itinerary*, v, p. 63: 'A qwarter of a mile owt of the towne on the hithere syde of Wanspeke was Newe Minster abbay of White Monks plesaunt with water and very fayre wood about it.'

3 Quoted, as to spelling, from *Herbal*, I, p. 84.

Byrders take bowes of this tre, and lime the twigges and go a batfolinge with them. Fisherers in Northumberland pyll of the uttermoste barke, and put it in the clyft of a sticke, and set it in fyre, and holde it at the water syde, and make fishe come thether, whiche if they se, they stryke with their leysters or sammonsperes: other use of Birche tre knowe I none.' Under 'The vertues of Hiacinthus' (*Scilla nutans* of which he says that 'it is called in Englishe crowtoes and in the North partes Crawtees', *Names*, p. D. vi) he gives a more vividly personal remembrance (*Herbal*, II, p. 18): 'The boyes in Northumberlande scrape the roote of the herbe and glew theyr arrowes and bokes wyth that slyme that they scrape of'; and in his first book (*Libellus*, p. B*b*) under Githago, 'of it among my people at Morpeth boys weave garlands on the day of St John Baptist'.

So, too, of the coast where he must have acquired his interest in and much of his knowledge of fish. Of Phocaena he writes, 'English in Northumberland where I uttered my first infant wails call it a Porpess': 'Codlings are taken in quantity near Bednel [Beadnell] a town on the Northumbrian coast': 'Salar is called by the Northumbrians Burntrout': 'Candele [Sand-eels], so called in Northumberland, are dug out of the sand with toothed reaping-hooks': and most personal of all, 'When I was a boy the quantity of Whitings in Northumberland was so great that twenty full-sized fish were sold for a "nummus Henricianus" [1^{d}]'. Yet he does not mention the famous Whale stranded at Tynemouth in August 1532 and described in a letter by Polydore Vergil (cf. Gesner, *Historia Animalium*, IV, p. 212); and this is perhaps a sign that when he left the north for Cambridge he did not return.

We have pieced together these allusions to Morpeth and his boyhood which in fact represent the full extent of our knowledge. Here as for the whole story of his life we have to depend upon scraps—casual allusions in his notes upon birds or plants, a few brief biographical remarks in the dedications or introductions of his books, a few letters, a few references by contemporaries or successors.[1] Out of them it is hardly possible to construct even a chronological outline, let alone an integrated story. Certain events stand out clearly; others if their occasion or sequence cannot be precisely fixed are obviously authentic: but when we try to fit them all together the extent of our ignorance is speedily revealed, and is depressing. Even his character which was plainly strong and individual

1 References to the existence of his Commonplace Book suggest that if it should ever be recovered it might throw valuable light on his career. G. S. Boulger states that it was listed in a sale catalogue by Thomas Kerslake of Bristol in 1841 as a thick quarto bound in old stamped calf with green edges, and that an account was given of its contents (*Notes and Queries*, 12th ser., II, p. 507). J. Ardagh similarly states that 'a commonplace book' by him was sold by Puttick and Simpson in 1861 (l.c. 14th ser., CLX, p. 443).

is not easy to indicate except in broad strokes, and these may well be unfair to him. There is no extant portrait of him in any of his books or elsewhere, and no descriptions except Strype's remark that he was 'a very facetious man' and 'delivered his reproofs and counsels under witty and pleasant discourse' (*Cranmer*, p. 512, ed. Oxford, 1812), and Bale's that he was 'very handsome in person and both witty and facetious, and withal a sound and elegant scholar'.[1]

There is one possible source of evidence as to his boyhood and youth which might have been expected to be illuminating, the fact that at exactly the same period and from the same town there came another naturalist, Thomas Gibson, the herbalist and physician, the printer and reformer. J. Hodgson in his memoirs of Northumbrian naturalists has pointed out that it is curious and disappointing to find that neither of them ever seems to mention or allude to the other. He suggests that the *Grete Herball* printed by Treveris and subsequently edited and reprinted by Gibson was a cause of estrangement between the two men; for we know that Turner roundly condemned this book as 'full of cacographees' and may have resented the attempt to perpetuate it. But this seems hardly sufficient to account for the silence, especially as Gibson is explicitly mentioned by Gesner in his list of learned men from whom he had obtained help, and in the *Historia Animalium*, 1, p. 365 (cf. Topsell, *Of Foure-footed Beastes*, p. 111) is an account of the breeding habits of the Rabbit quoted from Gibson and showing good knowledge: he is said, in Gesner's list of authors prefixed to Tragus, *De Stirpium*, to have written a volume on herbs—though this probably refers only to his edition of the *Grete Herball*. Moreover the two men shared the same religious convictions and were both exiles during Mary's reign. The breach, if such it was, may well have gone back to their early years and be due to social or other differences at Morpeth: but it remains unexpected and with our present knowledge inexplicable.

Of Turner's schooling we know nothing unless perhaps his story of a local scandal suggests that he like Ridley was at school in Newcastle.[2] But whatever their circumstances his family seems to have gained the interest of Thomas first Baron Wentworth of Nettlestead in Suffolk. William Turner, who in 1538 dedicated to him his *Unio Dissidentium*, dedicating the second part of the *Herbal* in 1562 to his son the second Baron, writes, 'who hath deserved better to have my booke of herbes

1 Quoted by R. Potts, introduction, p. iv, to reprint of *Huntyng*.

2 In *The Rescuyng of the Romishe Fox* is the story of how 'one Tom Story Steelgate bewrayed hys sonnes at the sessions in Newcastel': the said Tom Story 'was a strong thefe and had iiii tall felowes to hys sonnes' (B. 8*b*). This is introduced in reference to Stephen Gardiner's attitude to 'that excellent yong man Germane Gardiner' put to death in 1544.

to be given to him than he whose father[1] with his yearly exhibition did helpe me, beyng student in Cambridge of physik and philosophy?' And it seems that the relationship was not solely financial; for under 'Tithymales' in *Herbal*, II, p. 154, he writes of the Wood Spurge (*Euphorbia amygdaloides*), 'Thys kinde have I sene in diverse places of England, fyrst in Suffock in my lorde Wentfurthis parke besyde Nettelstede'—which must imply a personal and early visit perhaps during his time at the University. The connection probably explains Turner's later attachment to Edward Seymour, Duke of Somerset, whose mother Margaret was an aunt of his patron.

From Morpeth he went up in 1526 to Pembroke Hall in Cambridge. The College had at that time a regular connection with Northumberland. Thomas Patinson, to whom Turner dedicated his first plant-book in 1538, came from the county, had graduated in 1492–3 and spent his life in the College. Nicholas Ridley, born as Turner told Foxe at Willimoteswick[2] and at school in Newcastle, entered Pembroke in 1518, and became a Fellow in 1524. Soon after Turner came into residence Ridley was abroad, studying at Paris and Louvain, but by the time that he took his B.A. in 1529–30 Ridley had returned and settled down to College work. Strype (*Eccles. Mem.* III, I, p. 386) has printed the letter which Turner sent to Foxe to supplement the account in his *Actes and Monuments*, and this is of great interest. After describing Ridley's birth from a good family and his connection with Pembroke he writes: 'He first instructed me in a further knowledge of the Greek tongue....His behaviour was very obliging and very pious, without hypocrisy or monkish austerity; for very often he would shoot in the bow and play at tennis with me.... When I had nothing to give to the poor he often supplied me that I might give too; what aid he sent out of England to us in our exile in Germany, Doctor Edmund Grindal can testify.'[3] The passage, which ends with a restrained but moving outburst against Ridley's judges, is an attractive tribute from a younger man of lower social status.

But Ridley who did not become an avowed supporter of the Reformers until 1536 was less influential in shaping the views of the young Turner than Hugh Latimer his fellow-martyr. In 1529 Latimer who had graduated in 1510 and was then a Fellow of Clare Hall preached his famous 'cards' sermon, and Turner came under his influence. Dedicating to him 'A preservative or triacle against the poyson of Pelagius' in 1551 Turner writes:

1 He died in 1551.

2 Or Willowmontswick, now called Ridley Hall, on the South Tyne. Turner (Letter to Foxe) derives it from Willowmont, the Northumbrian name of the Rock-duck or Guillemot!

3 Also printed in a Latin and an English version in *Works of Ridley* (Parker Society), pp. 487–95.

'First in Cambrydge about XX yeares ago ye toke great paynes to put men from their evil workes...we that were your disciples had much to do in Cambrydge, after your departing from us, with them that defended praying unto sayntes....' Latimer in fact until his appointment to Worcester in 1535 inspired the group which made the University the pioneer of the Reformation in Britain; and Turner was an early and prominent member of it.

Of his College life until Ridley's return we have a hint in the dedication, to Edward VI then Prince of Wales, of his book of birds published at Cologne in 1544. Here he mentions George Folbery as 'formerly my teacher, a man of distinguished learning, and an admirable craftsman in the right education of the young'. Folbery had graduated at Clare Hall in 1514, but had then been elected to a fellowship at Pembroke where he seems to have spent the rest of his life. He took his B.D. in 1524, was given a prebend in York in 1531, and became Master of the College in 1537.

It was an extraordinarily interesting time in the history of the University. Indeed Mullinger, whose three great tomes contain a mass of information about it, regards the year 1535 as the date of transition from the medieval to the modern; and though this is perhaps arbitrary, since the process lasted more than a century, the events of that year marked a definite and important stage in the change. Erasmus had left Cambridge in depression if not in disgust in 1513. But the spirit which he symbolised and to some degree inspired lived on. Greek prevailed; and twenty years later under the leadership of John Cheke, who is credited with having laid the foundations of learning in St John's, and was the first Regius Professor of Greek, and of Thomas Smith of Queens' College his friend and predecessor in the Oratorship of the University, the Humanists had driven everything before them. William Grindall and Roger Ascham, successively tutors to Elizabeth, John Aylmer tutor to Lady Jane Grey, William Bill, afterwards Master of St John's College, Walter Haddon afterwards President of Magdalen, Thomas Wilson, Robert Horne, James Pilkington and many more were among them. Turner certainly knew a number of these scholars ranging from Richard Cox, who had migrated from King's to Wolsey's new foundation in Oxford, had thence become Master of Eton, and had returned to Cambridge to sit at Smith's feet, to William Cecil then just beginning his career and already credited with brilliant parts.[1] He was himself described by John Bale as 'Cantabrigiae Latinus et Graecus, rhetor ac poeta clarissimus'.[2]

Erasmus had opened to the University the means to study the New Testament: Luther's writings showed them how revolutionary was its

1 Cf. Strype, *Cranmer*, I, p. 394.

2 *Scriptorum Illustr. Britanniae*, Basel, 1557, p. 697.

significance. But by 1530 the first group at the White Horse seemed to have lost its power: Robert Barnes was in exile; Thomas Bilney and Thomas Arthur had been forced to recant; George Joye had fled to Strasburg; George Stafford[1] had died in 1529; the other 'Germans' had mostly gone down. But Latimer carried on; and his eloquence and judgment commended the cause to many who had previously stood aloof. His preaching vividly described by Turner's young contemporary, Thomas Becon of St John's,[2] gathered a number of followers. Ridley soon openly joined them; and Turner who may well have been won over earlier stood in with them. They gave themselves with enthusiasm to the study of the Greek Testament and to the interpretation of St Paul; and what they read fired them with a passion to proclaim it and to challenge the distortions by which as they were convinced it had been perverted.

Twenty years later on the eve of his martyrdom Ridley apostrophising the College which had been his own, his 'cure and charge', could say of it 'Thou wast ever named sithens I knew thee, which is now 30 yeares agoe, to be studious, well learned and a great setter forth of Christes Gospell and of Gods true word: so I found thee, and blessed be God so I left thee indeed....In thy orcharde (the wals, buts and trees if they coulde speake woulde beare me witnes) I learned without booke almost all Paules Epistles...the sweet smell thereof I trust I shall carie with me into heaven' (Foxe, *Actes and Monuments*, pp. 1609–10). Turner, to whom he had just previously sent an affectionate message,[3] as a contemporary product of the College, is no bad evidence of the truth of Ridley's words. He, too, loved the orchard and 'Paules Epistles', and found the College 'studious and well learned'.

Another Northumbrian of exactly the same age as Turner and with whom he 'lived for many years on terms of intimacy'[4] was Rowland Taylor of Rothbury who took the degree of Bachelor of Civil Law in 1529–30, and became Principal of Burden Hostel in 1532. In the letter to Foxe already quoted for its information about Ridley, Turner, after describing his friendship with Taylor, goes on to say that he used to exhort him to embrace the evangelical doctrine and obtained for him a copy of the famous *Unio Dissidentium*. By this and Latimer's sermons he was finally brought 'into our doctrine'. He was burnt at his parish of Hadleigh in Suffolk in 1555.

From Turner's references to Edward Crome and Nicholas Shaxton, brought before Stephen Gardiner in Lent 1546 (cf. *The Huntyng of the*

1 Of whom it was said, 'When Master Stafford read and Master Latimer preached then was Cambridge blessed': T. Becon, *The Jewel of Joy*.

2 Cf. *The Jewel of Joy* (Parker Soc.), p. 425.

3 *Works of Ridley* (Parker Society), pp. 389 and 394, in letter to Grindal.

4 So his letter to Foxe, *Works of Ridley*, pp. 490, 494.

Romyshe Wolfe, p. E. v*b*), it is probable that these two who were members of the White Horse group were well known to him. And how deeply he valued the friends whom he made among the Reformers may be seen from the passion with which, in the same book, p. E. 1, he recounts the names of those who had by that time given their lives. 'The clergy of Englande killed Bilney[1] [Thomas Bilney, martyred at Norwich, 1531], Baynam [James Bainham, at Smithfield, 1532], Bayfield [Richard, at Smithfield, 1531], and Anthony Person [Peerson, at Windsor, 1543], Mekins [Richard, a boy of fifteen, in London, 1541], Lambert [John, at Smithfield, 1538] and Philips[2] with many others. But the Wolfe of Winchester [Bishop Gardiner] killed Barnes [Robert, whose sermon at St Edward's, Cambridge, had caused the stir described in Cooper, *Annals*, I, pp. 311–23], Jerom [William Jerome] and Garret [Thomas Gerard: all three were burnt together at Smithfield, 1540].[3] The citie of London, the townes of Colchester, Braintre and Chensfurth [Chelmsford] can tell how many his felowe, bloudy Boner [Edmund Bonner, Bishop of London], hath killed.'[4] But in the thirties the movement, though creating much local excitement, had not become influential, and was still academic rather than evangelistic. Henry's matrimonial affairs which brought Thomas Cranmer of Jesus College and other scholars into public life drew attention to the Cambridge group: but it was from its own sense of vocation, its own growing convictions of the urgent importance of its message, that its activities originated.

Turner's first publication vividly illustrates both the religious situation and his own reactions to it. 'A Comparison betwene the Olde learnynge and the Newe. Translated out of Latyn into Englysh by Wyliam Turner 1537. Printed in Sowthwarke by me James Nicolson' and reprinted next year, is a version of a book by Urbanus Regius,[5] a straightforward statement of the new or Roman Catholic doctrines set out, subject by subject, and answered by the old or scriptural teaching in a catena of biblical quotations. After the translation Turner has added an appeal 'To the Christen reader' which summarises the argument and sets out the grounds on which a return to the Scripture is needful. This is the only original section of the book; and it is worthy of its place. Britten, himself a Catholic,[6] and most recent writers, have charged Turner with being a

1 Thomas Allen, then a fellow of Pembroke, was with Bilney at Norwich and gave Turner a report of his death: Foxe, *Actes and Monuments* (ed. Pratt), IV, p. 651.

2 Possibly a mistake for Filmer—Henry Filmer, burnt with Peerson.

3 For Gardiner's version of his dealings in the matter cf. J. A. Muller, *Letters of S.G.* pp. 165–75.

4 In a similar but shorter passage (l.c. p. B. iii*b*) he adds the names of John Frith (at Smithfield, 1533) and John Lascelles (Smithfield, 1546).

5 This treatise in Latin is the second of those printed in his collected works, edited by his son Ernest Regius, in 1562, pp. xvii–xxx.

6 Cf. Preface to his edition of *Names*, p. vii. The ground alleged for this charge and cited in *Dictionary of Plant-names*, p. 367, does not seem very adequate.

violent (or as he would himself call it 'unmanerly', cf. *Herbal*, II, p. 128*b*) and unscrupulous controversialist. This is certainly untrue of his early books, and never really justified even when he had himself been in peril and exile and had seen his friends done to death.

For us the interesting point of this book is not so much its contents which are familiar, but the fact that it reverses our usual statement of the position. We speak of the Reformation as the 'New learning' and regard the Catholic tradition as old. Turner, quite genuinely, believed that the newly discovered Greek Testament represented the ancient learning and faith of Christendom, and instead of regarding himself and his friends as destroyers or innovators based his case upon the claim to be restoring the ancient purity of the faith. It is indeed arguable, and on many grounds probably true, that the strictly religious influences which effected the Reformation were in intention conservative rather than modernising; and that it was the humanistic and scientific studies of the time, rather than the biblical and theological, that gave the Reformers their liberal and progressive element.[1]

Turner's own conservatism is plainly stated in his letter to Foxe (*Works of Ridley*, p. 494): complaining that the martyrologist had printed the book of William Thorpe (his *Examination* in 1407) not in the old English then in use, but in a modernised version (by Tyndale), he adds: 'So great an admirer am I of antiquity that I could ill bear treasures of such antiquity to perish from amongst us. On which account I feel no great obligations to those persons who have translated Piers Plowman, Gower and Chaucer and authors of a similar stamp into a mongrel language neither true English nor pure French.'

Certainly, as Mrs Arber has justly observed,[2] he had a strong reverence for the ancients even in his botanical and zoological studies. His primary intention was to interpret the true meaning of the past: to learn what St Paul and the Gospels had to say about Christianity, or Aristotle and Pliny about ornithology, or Theophrastus and Dioscorides about herbs; and he is critical, often severely so, of those moderns who ignore or misinterpret the authorities of the classical age. His purpose was not to break away from the past or to rebel against tradition but to reaffirm and defend a yet more ancient wisdom. To do so involved scholarly handling of documents and the business of verifying their findings by precise definition of their content. This led to an approximation to a strictly scientific method: but it is an anachronism to regard that method as established in the sixteenth century. Learning was still synonymous with

1 To discuss this important and far-reaching problem would take us far from our subject and involve a consideration not only of the relations between Luther and Melanchthon, but between (say) Erasmus and Bucer.

2 *Herbals*, pp. 123–4.

the study of the proper authorities; and its process was by deduction from traditional lore to current fact. Indeed, so far as respect for the past is concerned all that had happened was the replacement of the infallibility of the tradition by that of the ancient texts.

Nevertheless, if conventional, his training was finely scholarly; and it gave him just the power to handle ancient authors and extract from them their precise meaning which was important for his life's work. In this he was taking up the task traditionally laid upon Adam[1] and, as we have seen, begun by Roger Bacon of naming the animals; and now he, unlike his predecessor, had resources sufficient for at least a measure of success. Anyone who has tried to identify from their original descriptions the plants of Theophrastus or Dioscorides or Pliny or the Arabians will know that to do so requires not only a profound knowledge of the language used both in its general significance and in the particular idiosyncrasies of each author, but a flair for fastening upon the points of importance in each record and appreciating their value as compared with similar points in allied species. Turner had the further difficulty that in the case of his Greek authorities he was dealing in general with Latin translations. Theodore of Gaza, the Greek scholar who had died in 1479, had translated Aristotle's *Historia Animalium*, *De Partibus* and *De Generatione* and Theophrastus's *Historia Plantarum*, and his version had been published at Venice in 1503–4 and again in 1513. Turner quotes him freely both for plants and birds and speaks of him as 'a learned man both in Greke and in Latin and an excellent translator'; but he certainly knew Theophrastus at least in the Greek; for in his long and interesting discussion of the Pine-tree in *Herbal*, II, p. 88*b*, he compares a passage in the original Greek with Theodore's Latin, and collates a number of places in which Theodore translates his Πίτυς. Of Dioscorides there was no such authoritative version. Turner apparently used that by Jean Ruel published by Stephanus at Paris in 1516; for he mentions it in a significant passage on the seed vessel of Oleander (*Herbal*, II, p. 65), 'which when as it openeth sheweth a wollyshe nature lyke an thystel down, as Ruellius translation hath. It seemeth that hys greke text had ἀκανθινοις παπποις. But my greke text [no doubt that by Marcello Vergilio, Cologne, 1529: it is on p. 520] hath ὑακινθινοις παπποις. And so semeth the old translator [? Peter of Padua, 1478] to have red, for he translateth thus lanam deintus habens similem hyacintho....I lyke Ruelliusses Greke text better than myne, for the down is whyte and lyke thestel down and nothynge lyke hyacinthus.' Of Aëtius, the other Greek doctor to whom Turner refers for plants and, on occasion, birds, there was a translation by Janus Cornarius whom he describes in *Herbal*, II, p. 104*b* as 'well learned in the knowledge of the Greke tong, and a very good Grammarian there in'. This was published at

1 Cf. Genesis ii. 19.

Basel in 1542, and evidently Turner got hold of it during his second exile and drew freely on it. He also refers at least once to Linacre's translation presumably of Galen: 'som translate agron into amerinam, as Theodore and our Linaker do' (*Herbal*, II, p. 165*b*).

Though he used translations of Greek authors—whose original works were still hardly obtainable except in manuscripts—he not only knew Greek well enough to discuss the variants in a passage with competence, but recognised the importance of getting a wider and fuller acquaintance with such works in Greek. Interpolated in one of his long discussions in the *Herbal* (II, pp. 100*b*, 101, 'on the vertues of the Popler') and after a complaint that we have not got Nicholas of Alexandria in the original Greek, is the remark: 'I wold wish that they that fynde any old Greke examples or copies of old authores and intend to translate them, that they shuld as well set out and cause to be printed the Greke texts as theyr own translationes: for so myght men the better examin theyr translationes, and the studiouse youthe by comparyng of them together myght profit much more in the greke tong and practicioners myght be more bold to work accordyng to it that they have translated.'

How eagerly he followed out his own exhortations and compared versions with originals can be seen in almost any one of his longer discussions. A good example is to be found a few pages further on in the same *Herbal* (II, p. 104, 'of the plum tre'), where he compares Vergil, whom he calls 'a great folower and translater of tymes [oft-times] of Theocritus', *occultant spineta lacertos*, with Eobanus Hessus,[1] the translator of Theocritus, *et virides recubant subter consepta lacerti*, and adds: 'Marke where as the translator of Theocrytus hath consepta, and Theocrytus hath hys ownself in hys Greke verse αιμασιαις, Virgil hath spinetum'[2]—all this to prove that spinus is the blackthorn! Or (l.c. p. 146) we may illustrate by a similar passage in which he proves that 'acantha in Greke signifieth a thistell' by quotation from St Luke viii, the parable of the sower, where the Latin spina, used to translate acantha, is wrongly translated thorn—'some fell among thorns' being a rendering which he condemns both on linguistic and botanical grounds.

With Pliny, 'ye noble clearke' of *Herbal*, II, p. 7, he was in a much better position to deal. For in *Herbal*, II, p. 17, when commenting on a reading of Mattioli he speaks with evident pride of 'mi Plini corrected by Erasmus and prynted by Frobenius'[3]—a reference which he amplifies on p. 89, 'my Plini corrected and set out by Erasmus after that Hermolaus Barbarus [Ermolao Barbaro, 1454–93], Nicolaus Beroaldus [surely Filippo

1 German scholar born at Hesse, 1488; translated Theocritus into Latin verse, 1530.

2 The references are Theocritus, *Id.* VII, 22 and Vergil, *Ecl.* II, 9.

3 It was Frobenius who in 1516 had published the *Novum Instrumentum*—Erasmus's epoch-making edition of the Greek New Testament.

Beroaldo who edited Pliny with notes, Parma, 1476], Guilhelmus Budaeus [Guillaume Budé, 1467–1540] and Johannes Cesareus [Jean Caesarius, 1460–1551] had done to Pliny what they could do'. He evidently possessed the superb folio headed: 'Joannes Frobenius Lectori S.D. En damus C Plinii Secundi divinum opus cui titulus, Historia Mundi, multo quam antehac umquam prodiit emaculatius: idque primum ex annotationibus eruditorum hominum, praesertim Hermolai Barbari: deinde ex collatione exemplariorum...postremo ex fide vetustissimorum codicum ...Basileae apud Jo. Frobenium, mense Martio, An. MDXXV.' Of this edition Erasmus wrote the dedicatory letter, mentioning the four scholars whom Turner names. It is a grand piece of printing and a good text, and has a magnificent index which must have greatly simplified its user's labour. In general he has a high regard for Pliny—too high as most of us would agree. But there is at least one passage in which he reveals a less than reverent attitude. In *Herbal*, II, p. 70*b*, he writes: 'It is playn that Plini had Dioscorides, howsomever like a falslying goodlesse man he pretendeth as thoughe he never saw Dioscorides, of whom he hath conveyed so much learned stuf into hys omnigatherum.'

Of other authorities he knows the famous *Regimen Sanitatis Salerni*, 'the Phisiciones of Salern' (*Herbal*, II, p. 108*b*), 'theyr booke whiche they wrote unto the Kyng of Englande' (l.c. p. 93).[1] But the most frequently quoted is the Arab doctor of the ninth century, Yahya ibn Masawaih, or his younger namesake, whom he quotes as Mesue, or on one occasion as Joannes Mesue (*Herbal*, II, p. 42*b*).[2] This man's works had been freely translated into Latin, and the 'Joannes Manardus' (Giovanni Manardo, professor at Ferrara, died 1536) of whom Turner speaks in the preface to his first *Herbal* had published 'Letters' on him at Basel in 1535. Mesue had compiled a pharmacopoeia and this, published as *De Re Medica* in 1548, is probably the work with which Turner was most familiar.

The other great Arabians, Avicenna and Averroes, he quotes occasionally but rather by allusion than by extracts from their books. Of commentators upon them he quotes in *Herbal*, III, p. 79, 'Arnolde of Newton' (Arnaldo de Villa Nova, died 1313), author of the *De Theriaca* of whom Gesner said that 'he followed both the names and the mistakes of the Arabs and of their interpreters' (*H.A.* I, *Catal. Auth.*).

But it would be wholly misleading to represent Turner as merely an antiquarian and scholar. Along with his reverence for antiquity is another and for his equipment as a scientist equally important characteristic. He was not a disciple of Latimer for nothing; and from him he learnt that concern with social conditions, that sympathy with the common folk in

1 For this cf. F. R. Packard, *The Schoole of Salerne* (Oxford, 1922).
2 Cf. G. Sarton, *Introd. to History of Science*, I, pp. 574, 728.

their ignorance and weakness, their worth and aspirations which are so plain a feature of his writings. With him as with Latimer a prophetic hatred of injustice and greed and the oppression of the poor was more powerful even than scholarship. He had seen the great Wolsey in the heyday of his pomp and ambition bestowing the high places of the Church upon his son[1] and diverting to his personal aggrandisement the riches of his see and of his victims. He saw the changes that were coming over the land, ancient abuses being overthrown only to give rise to new and not less flagrant evils. If, associating the Church with its representatives, he denounced the chicanery and superstitions of 'clerks' even more vehemently than a Chaucer or a Fisher, he was not less vigorous in his denunciations of the nobles who grew fat upon the spoils of the monasteries and of the new rich who squandered upon self-display resources that should have been devoted to the common weal. Latimer's sermons and Turner's writings are full of an indignant appreciation of the failure of Church and State alike to grasp the opportunities and fulfil the aspirations of the time. They were men observant, sensitive, fearless, who studied the life of the time and strove to arouse interest in its needs. For a scientist such an outlook is of high value. Turner owes much of his greatness to the combination in him of scholarly with practical interests. The observation, study and examination of detailed facts were a natural part of his equipment. He had little opportunity to use them in sociological or political effort: in education, in medicine, in natural history were his outlet and service.

This application of his religion to the defence of the oppressed is the theme of the series of books which begins during his exile. In the two earliest, *The Huntyng of the Romyshe Foxe* of 1543 and *The Rescuyng and Seconde Course* of 1545, he is concerned with the development of the contrast between the superstitions of current Catholicism and the simplicities of the New Testament. Then when he became a member of parliament he outlined a more constructive programme. *The Examinacion of the Messe* and the *Huntyng of the Romyshe Wolfe* both contain in addition to criticism definite proposals for the development of education, the encouragement of scholars and pastors, and the remedy of abuses and spoliations. Then the *Spirituall Physic* of 1555 carries the story much further by its ruthless exposure of actual evils and its indictment both of the princes of the Church and of the newly enriched nobility. Observation of specific evils has taken the place of general denunciation; and this led him to the serious study of his country's past and of 'the father of English historical criticism' adumbrated in the edition of William of

1 Cf. Trevelyan, *English Social History*, p. 94: 'He obtained for his natural son four archdeaconries, a deanery, five prebends and two rectories, and only failed in his endeavour to have him succeed in the fabulously rich see of Durham.'

Newburgh's *Rerum Anglicarum* on which he was at work in his last days. Here is a development closely parallel to that which we can trace in his botanical and zoological researches.

Of his scientific interests at this time the clearest account is in the Preface to the 1568 edition of the *Herbal*. In it he writes: 'Above thyrtye yeares ago...beyng yet felow in Pembroke hall in Cambridge wher as I could learne never one Greke, nether Latin, nor English name, even amongest the Phisiciones of any herbe or tre, suche was the ignorance in simples at that tyme, and as yet there was no Englishe Herbal but one, al full of unlearned cacographees and falselye naminge of herbes, and as then had nether Fuchsius, nether Matthiolus, nether Tragus written of herbes in Latin'—a passage widely quoted and not in fact unfair. For the *Grete Herball* printed in 1526 and 1529 by Peter Treveris, to which presumably he refers, deserves his condemnation.

It is in fact a translation of a French original which is itself largely derived from the German *Ortus Sanitatis*; and this again reproduces in a debased form material from *Das Buch der Natur* by Konrad von Megenberg, from Bartholomaeus and from Albertus Magnus. Its woodcuts are largely coarse and abbreviated copies of those in the *Ortus*—and these are so crude and conventional as to be hardly ever recognisable. Thus *Ortus* De Herbis, chs. 276–7, gives two pictures of Mandrakes, naked man and woman with five leaves and flowers growing out of their heads; the *Herball* has caricatured versions of these pictures, although in the text it is stated that 'nature never gave forme of mankynde to an herbe' (ch. 279). It contains 503 chapters or separate items, but these include—besides plants—minerals and precious stones, ivory, amber, butter and soot, many insects, coral, pearls and mummy: in fact all the substances included in the contemporary pharmacy. Of the plants there are no localities and no adequate descriptions: it would be difficult to identify more than a very few species, or even in most cases to guess what the chapters might mean. Many of the illustrations, for example of the Mugwort, ch. 29, Cypress, ch. 97, or Crocus, ch. 104, are obviously misplaced, or at least bear no resemblance to the plants named; and the letterpress gives no assistance. As an introduction to the study of botany it is worse than useless.

Nevertheless, with or without Treveris's help, Turner managed to acquire a considerable knowledge of Cambridge plants. There is a reference to his College and to the orchard immortalised in Ridley's last farewell under Myrrhis (*Herbal*, II, p. 60): 'The one is called in Englishe casshes [*Anthriscus sylvestris*, cf. *Names*, p. E. v*b*]....I never saw greter plenty of it then I have sene in the hortyard of Pembroke hall in Cambrydge where as I was som tyme a pore felow.' To the town he refers under 'Carote' (*Herbal*, II, p. 80*b*): 'Thys wild persnepe [presumably *Pastinaca sativa*] groweth plentuously besyde Cambrydge in a lane not far from Newnam Milles'; under 'Platanus': 'I never saw any plaine tree in Englande saving one in Northumberlande besyde Morpeth, and an other at Barnwel Abbey besyde Cambryge' (*Names*, p. F. iv); under 'Mercurialis' (*M. perennis*) (*Libellus*, p. B. iii): 'I have seen it at Cambridge in the garden of King's Hall, and transplanted a root from there to our own garden.' There are many references to the local names given to plants in Cambridgeshire: *Iris pseudacorus*, 'the Northumbrians call it a seg, the men of Ely and of the Fens a lug' (*Libellus*, p. A. ii); *Ononis arvensis* 'in Cambrydge Shyre a whine' (*Names*, p. A. viii*b*, 'it groweth in many places about Cambrydge', *Herbal*, I, p. 45); 'Germander' (*Teucrium chamaedrys*) is 'Englyshe Triacle' (*Herbal*, p. J. iiii); *Sambucus ebulus* is 'Walwurte' or 'Danewurte' (*Herbal*, p. O. vi*b*, 'it groweth mych about Cambrydg'); *Hordeum murinum* is 'Way bent' (*Names*, p. D. v*b*, cf. *Herbal*, II, p. 17, 'it groweth plentuously in Cambridgeshire about high wayes'); *Petasites vulgaris* 'in Northumbreland an Eldin, in Cambridgeshyre a Butterbur' (*Herbal*, II, p. 83); and the 'Fen Shrub or bushe' (*Myrica gale*) is 'in Cambridgeshyre Gall' (*Herbal*, III, p. 47).

The difficulties that he had to face may be best illustrated by the one passage in his first book in which he gives any detail of his search. In the *Libellus*, p. B. iii, under Narcissus he writes: 'For a long time among us Narcissus has concealed itself under foreign names. There was no herb that gave me more trouble. For after I had seen its picture and outline and studied them as carefully as I could [presumably in Brunfels's *Herbarum Vivae Icones*, I, p. 129, to which he refers in his preface and which has under Narcissus pictures of *N. pseudonarcissus* and *Leucoium vernum*] no one could be found to show me the plant, much less tell me its English name. At last when I was taking a holiday in Norfolk a little girl hardly seven years old met me as I was walking along the road; she was carrying in her right hand a bunch of white flowers; as soon as I saw them I thought to myself Those are Narcissi—for the description of them was still fresh in my mind; and I begged some of them from her. But when I enquired the name no reply was forthcoming. So I asked the folk who lived in the neighbouring cottages and villages, what was the name of the plant. They all answered that it was called "laus tibi": I could get no other name from them. But when I got home I learnt that asphodel was called by many people "laus tibi". Then a little old man whose name is guarinus Asshe [Warren Hash is the name given in the list of "Rewardes" to the Prior and Canons at the Dissolution in November 1538] a canon of Barnwell Priory and well-skilled in herbalism, told me this plant was called French Gillyflower (gelofer). We must use that name until a better is found.'

Turner had in fact a double task in this primary business of identification. The herbalist if he was a man of education had inherited from the

Graeco-Roman world, from Aristotle or rather Nicolaus of Damascus, to whom the *De Plantis* is now ascribed, and from Aristotle's pupil Theophrastus, from Pliny the Elder, and from Dioscorides the author of the *De Materia Medica*, a considerable list of plants with brief descriptions, occasional and traditionalised pictures, and a large and various lore of qualities and uses. From these sources traditional medicine derived its sanctions—and in an age wholly reliant upon authority and almost wholly indifferent to observation such sanctions were of immense influence. The plants of Dioscorides acquired an almost magical significance and his remedies were accepted with blind belief in their value. To identify his species correctly became a matter of the highest importance.

This to an age which had no notion either of the numbers or of the distributions of the vegetable kingdom presented an almost insuperable difficulty. Dioscorides knew the plants of the Mediterranean: some few of these had since been grown and distributed in the gardens of monasteries: some few were to be found wild in Britain. But to fit our native flora into the lists derived from antiquity and the south demanded ingenuity, or unscrupulousness, of the highest order. We know from constant complaints of fraud that many herbalists were content to substitute available alternatives for plants of traditional importance, even if poisoning was the result. We know from the long and keen discussions in Turner's *Herbal* how eagerly he sought the precise identification of Cepea[1] or Pontic Wormwood or the Faba antiquorum.[2] It was perhaps one of his real contributions to science that he came to recognise the value of those plants in the newly discovered lands and in England which were not to be found in the Classics and which Brunfels had dismissed as 'herbae nudae'.[3] But in the main his business was to secure an identification with those already known.

This done there remained the question of an English name. The familiar flowers had of course long been given titles in the vernacular, lovely syllables like Avens and Celandine, Cuckopintel or Lucken Gollande;[4] but these often varied from shire to shire; and the same name was not infrequently attached to several plants that had no other connection. Mouse-ear, Gillyflower, Betony, Lovage, Dittany, Bear-foot—which of the species carrying these names were truly entitled to them? And so

1 Cf. *Herbal*, p. J. iii *b* and I, p. 123, whether it was *Veronica beccabunga* or *Atriplex portulacoides*!

2 Cf. *Herbal*, I, p. 221: 'I wold som body wold shewe us what pulse is there ether in Italy or Spaine, England or Almany, or Franch, which is the olde writers Faba.'

3 Cf. Arber, *Herbals*, p. 55.

4 Turner, *Names*, p. C. ii, identifies this not as is usual with *Trollius europaeus*, but with *Caltha palustris*. If the name means 'locked-up gold-flower' as Britten and Holland, *Dictionary of Plant-names*, pp. 217–18, state, it is obviously appropriate to *Trollius*.

many, known to the herbalists, had no English name at all. 'I could never learne any Englishe name of it. It may be well called after the etimologi of the worde and also of the vertue that it hath, Lous Strife'—and as Loosestrife *Lysimachia* has been named ever since. Indeed, very many of our English names when they are translations of the Latin are directly due to Turner. Goats-beard, Stone Parsley, Hawkweed, Ground Pine are some of his successes. Chokeweed for Orobanche, Pond-plantain for Potamogeton, Thyme-stonebreak for Saxifrage fared less well. The number of his inventions is a remarkable tribute to the novelty and the range of his work. If in many cases his identifications of the traditional plants are hazardous, yet he achieved a real success in bringing the ancient catalogues into a close and often accurate relationship with the actual flora of England, and in indicating the extent beyond which such identification was impossible. How keenly he searched the classics may be illustrated by the section 'Of Segge or Shergres' [Sheargrass] in *Herbal*, p. H. v or I, p. 112. He writes: 'Carex is the Latin name of an herbe, whiche we call in Englishe Segge or Shergres, whereof I finde no mention, nether amonge the Grecianes, nether amonge the Latines, savinge that I have rede of it in Vergil, and in Calphurnius. Calphurnius writeth thus of this herbe: Ipse procul stabo, vel acute [*sic*] carice tectus.[1] I wil stande far away covered wyth the sharpe segge. Vergil also in his *Georgikes* maketh this Shergres to be sharpe, and in his *Ecloges* [iii, 20] he maketh it to growe thycke together in bushes in these wordes: Tu post carectą latebas. Thou lurkedst behinde the segge bushes. Thys herbe that I do take to be carex, groweth in fennes and in water sydes, and hath a shorte roote, red withoute, and manye litle stringes at it. The leaves as they come oute firste are thre square, afterwardes they do go abroade, and represente a long smal knyfe, but not without certayne squares. And the edges of thys herbe are so sharpe that they will cut a mannis hande, and have a certayne roughnes, whiche maketh them to cut the soner: of the whiche propertye the Northen men call it Sheregres. It hath a longe stalke, and thre square, and in the top of that is a sort of little knoppes, in stede of sedes, and floures much lyke unto oure gardine gallingal. I have not red anye use of thys in Physicke. The people of the Fenne countreys use it in for fother, and do heate ovens with it.'

His book on birds though not published till later must have been based on study during his years at Cambridge, and illustrates the problem that he had to face even more plainly than his *Herbal*; for on birds little or no work had yet been done, and identification was a difficult and complicated process.[2] Here he sets out each name in Greek, Latin, English and Ger-

1 T. Calpurnius Siculus, *Ecl.* iii, 74, cf. Vergil, *Georg.* III, 231.

2 How complicated can be seen from the letter of Aloisius (Luigi) Mundella (*Epistolae*, Basel, 1543, pp. 60–77) discussing the identity of Coturnix.

man; then the description from Aristotle in the Latin version by Theodore of Gaza and then from Pliny, adding on occasion similar notes taken from Paulus Aegineta, for example, on the Troglodytes by which he apparently means one of the small Warblers; Aetius, also on the Troglodytes which to him is the Wren; Columella on the domestic fowl; Actuarius,[1] whom Evans (*Turner on Birds*, p. 135), reading Auctuarius which is surely a misprint, translates as 'The Supplement', on the Common Bunting; and finally St Jerome whose letter to Praesidius, now regarded as spurious,[2] supplies the legend of the mourning Pelican (told under Platea); and (to turn from ancients to moderns) Pierre Gilles[3] (Petrus Gyllius) (p. 36), to whom Gesner freely alludes, on the Attagen; and Jean Ruel the botanist to whom Turner refers (p. 130) in regard to the Linnet. He then gives his own comments and identifications, sometimes referring to the Greek of Aristotle and quoting it as against the translation, telling where he has seen the bird in question and often adding some note of real value. Over several—Fulica (pp. 61–4) which, as he argues, is not the Lapwing (*Vanellus vanellus*) 'which our people keep in their gardens to destroy the worms', nor the 'black water-fowl with the white spot on its forehead' (*Fulica atra*), but the small white sea-bird with a black hood and red beak and feet (*Larus ridibundus*) which when a ship is at anchor seizes upon the offal thrown overboard 'crying Keph just as the big gulls cry Cob'; Graculus (pp. 74–6) which is first the Cornish Chough (*Pyrrhocorax pyrrhocorax*), and then 'a Caddo, a Chogh or a Ka' (Jackdaw, *Corvus monedula*), and then apparently an Ibis (*I. comata*, see below, p. 87), and finally, as he saw in the Alps, a Nutcracker (*Nucifraga caryocatactes*); Trynga which he identifies unhesitatingly with the Water-hen, 'common on the lakes which surround the houses of the nobility, and on fish-ponds' (*Gallinula chloropus*), p. 139; and the Pardalus which more tentatively he names Plover and describes as 'running swiftly, whistling as shepherds and vanboys do with pursed lips, and in colour grey with single spots of yellow on each feather' (*Pluvialis apricaria*), p. 108—he produces clear but surprising results. With others—what is Galgulus? and is Onocrotalus really the Bittern,[4] which he has already described under Ardea? is there such a bird as the Phœnix?—he is less successful. Much of his information

1 The name seems to be a title borne by Byzantine doctors: Johannes, John of Zachary, the best-known of them wrote *De Medicamentorum Compositione*, translated by Ruel, Paris, 1539.

2 First printed among the 'letters falsely ascribed' in Plantin's edition of the *Opera S. Hieronymi* (Antwerp, 1579), IX, p. 87. It runs: 'Pelicans when they find their children dead, killed by a snake, mourn; and strike themselves and their sides; and with blood splashed on the bodies of the dead thus revive them.' This is not the traditional story as found e.g. in Isidore, *Origines*, XII, p. 172.

3 Author of *De Vi et Natura Animalium*, Lyons, 1533.

4 Cf. *Avium Praec.* p. 102: he learnt better later; cf. Gesner, *H.A.* III, p. 570.

seems to have been drawn from bird-catchers and fowlers; he is familiar with their use of bird-lime for the small species and of nets for woodcock; and had evidently done some liming himself: for in his account of the Holly in *Herbal*, III, p. 81, he not only notes the value of its berries as 'baytes to entyse Feldefares to come to lymeroddes', but adds, 'if any be desyrous to make byrdlime of the barkes, they may lerne it of me which have made it ofttymes after this manner', and then describes the process. It is significant that whereas in botany he can speak of several Englishmen as competent he names only one of his fellow-countrymen in connection with birds—when he writes of the Shrike, 'I never found any one of our people who knew its name except Sir Francis Lovell', this being presumably the son of Gregory Lovell, Henry VII's standard-bearer at Bosworth and the nephew of Sir Thomas, Speaker of the House of Commons. Francis died in 1551, and there seems no other evidence of Turner's acquaintance with him or of his interest in birds.

But though nomenclature is his chief concern he is far more than a mere identifier of classical species; and some of his most notable observations are East Anglian. The most famous of these is his statement about the Crane (*Grus grus*) on p. 78: 'Cranes nest among the English in marshy places and I have very often seen their "pipers" [*pipiones* or half-fledged young]: this some people born outside England declare to be untrue.' In this Turner is undoubtedly reliable; for the evidence of the *Household Books* of the Lestrange family at Hunstanton, printed in *Archaeologia*, XXV, pp. 411–569, records payment for a Crane in 1519 (p. 426), 'a Crane kylled with the Crossbowe' a few days before Christmas in 1526, a Crane brought in and another 'kylled with the gun' in 1533 (pp. 529–30); and in the accounts of the City Chamberlain of Norwich for 6 June 1543 is a charge for a 'yong pyper Crane' from Hickling.[1]

So, too, he mentions 'White Herons' (Spoonbills, *Platalea leucorodia*) nesting in a heronry, presumably at Reedham, along with the Common Heron (p. 34). Almost equally interesting is his record (p. 64) of the nesting of the Black Tern (*Chlidonias niger*), 'parva avis nostrati lingua sterna appellata,...larus marinis minor et nigrior'. This he describes as so noisy during its breeding-season as almost to deafen those who live in lakes and fens. 'It is', he says, 'constantly on the wing searching for prey and nests in dense reed-beds.' This species ceased to nest in Britain about a century ago. Similarly, his records of the 'Balbushard' (*Circus aeruginosus*) preying upon Coots, and the Bittern (*Botaurus stellaris*) aiming at the human eye are authentic and probably East Anglian.

1 This last is from T. Southwell's edition of Sir T. Browne's *Natural History of Norfolk*, p. 6: other less relevant evidence is collected in Yarrell, *British Birds*, II, p. 502: to them may be added Drayton's inclusion of 'the stately Crane' among the birds of the fens, *Poly-olbion*, XXV, l. 93.

During his Cambridge days Turner produced one other book on the religious problem. This is described by Cooper, *Athenae Cantabrigienses*, I, p. 257, as follows: 'The abridgement of Unio Dissidentium, containing the agreement of the doctors with Scripture: and also of the doctors with themselves. Dedicated to Lord Wentworth Lond. 1538.' It is a version of the very well known Reformation handbook which Turner had obtained for his friend Rowland Taylor.

He also produced his first book on plants, the *Libellus de re Herbaria novus in quo herbarum aliquot nomina greca, latina, & Anglica habes, una cum nominibus officinarum, in gratiam studiosae iuventutis nunc primum in lucem editus*, published by John Biddle in London in 1538. This was reprinted in facsimile and with an excellent account of its author by B. D. Jackson in 1877. It is little more than a glossary and, though written in Latin, can hardly be the Latin Herbal of which he speaks in 1568; it contains 144 plants with their synonyms in Greek and English; sometimes a note of different kinds united under the same heading; sometimes a note of uses or properties. Occasionally there is more. Under Alsine (p. A. ii*b*), after saying 'This is the herb which our women call Chykwede' he adds, 'Those who keep small birds shut up in cages refresh them with this when they are off their feed'. Under Palma he has 'On the Day of Palm-branches as they call it I have often heard priests saying "Bless also these palm-branches" when I could see nothing but sallow boughs. What others saw, I know not. If they were not supplying us with palm boughs, they ought to change their petition and say "Bless these sallow branches". It is a lie to call a sallow a palm.'[1]

The prefatory note 'William Turner to the candid reader' does not seem to have been noticed and is of interest. It runs, 'You will wonder, perhaps to the verge of astonishment, what has driven me, still a beardless youth, and but slightly infected with knowledge of medicine, to publish a book on herbary, when I know that there are six hundred of us Englishmen who in this kind of learning would precede me (as the saying goes) on white horses. I confess frankly that I am unworthy to act as bottle-washer to the most learned Doctor Clement. And I readily give place to many other most learned men. But if it had ever been the wish of Doctor Clement or any other experts in medicine to bestow such help on their fellow Englishmen as Ruellius has done upon the French and Otho [Brunfels] upon the Germans, I should not be going to speak a single word for the present about the names of herbs. But when I saw

1 J. Britten, *Dictionary of Plant-names*, p. 367, has a long note attacking Turner for this 'characteristic protest'. He proves that the custom of blessing sallows and calling them palms is ancient—which is no answer to Turner. For Britten's methods in controversy, cf. Druce, *Comital Flora*; for the inaccuracy of his identifications our study of Turner provides evidence.

that those learned men are making no effort of this kind, I thought it best that I should try something difficult of this sort rather than let young students who hardly know the names of three plants correctly go on in their blindness. So my dear reader take this labour of mine with a smile; and if you make any progress in herbarism by me, nothing will give me more pleasure. If I am caught blundering (and this is very easy) I will gladly be corrected by men of learning. For I am not too proud and pleased with myself to accept gladly the verdicts of the learned. Fare you well.'

It is a pleasant prelude and, in itself, answer enough to Anthony à Wood's charge of conceit and pushfulness.[1] But it is much more. It introduces us to the sources of his knowledge when he had got beyond the *Grete Herball.* Jean Ruel (1474–1537), a physician and professor at Paris, had produced the first adequate Latin version of Dioscorides in 1516 and then wrote a general botanical treatise in 1536. Otto von Brunfels (d. 1534), a Lutheran teacher and doctor, had collaborated with the artist Hans Weiditz to issue in 1530 the *Herbarum Vivae Eicones*, the first and in its illustrations one of the best of Herbals.[2] That Turner had some acquaintance with these is not surprising: indeed his obligation to them was lifelong. What is unexpected is the high importance that he attaches to Dr Clement. For, in spite of the fact that in the dedication to the first *Herbal* in 1551 he wrote, 'There have bene in England, and there are now also certain learned men whych have as muche knowledge in herbes, yea and more than diverse Italianes and Germanes whyche have set furth in prynte Herballes and bokes of simples. I mean of Doctor Clement, Doctor Wendy[3] and Doctor Owen,[4] Doctor Wotton and Maister Falconer.[5] Yet hath none of al these set furth any thyng', Clement's name has never been honoured in connection with botanical or indeed any but medical studies. How Turner came into touch with him remains unknown. He had been taken into the household of Sir Thomas More, apparently soon after he left St Paul's School; had been tutor to the children; and was commended by More as 'a man of charm, well-beloved by all, dear even to those who hate good letters, and especially so highly praised by Linacre as to make me, who have a unique affection for him, quite jealous'.[6] In 1519 he had been professor of Greek at Oxford 'with

1 *Athenae Oxon.* (ed. 1813), I, c. 361, 'very conceited of his own worth, hot headed, a busy body and much addicted to the opinions of Luther'.

2 The original coloured drawings from which the plates were cut have been found and reproduced in facsimile.

3 Thomas Wendy of Gonville Hall, B.A. 1518–19, fellow 1519–24, physician to Henry VIII and his successors. Granted manor of Haslingfield. Died 1560.

4 George Owen, Merton College, Oxford, M.D. 1527. Physician to Henry VIII and his successors. Died 1558. Wrote *A Meet Diet for the New Ague.*

5 See below, pp. 77–8: from this it appears that he had now returned from Ferrara.

6 In *Vita Thomae More*, Cologne, 1612, quoted by Munk, *Roll of R.C.P.* pp. 25–6.

classes larger than anyone had ever had'.[1] In 1522 he had taken up the study of medicine and in 1528 been admitted a Fellow of the College of Physicians: he was elected its President in 1544. In 1526 he had married the famous learned lady and companion of More's daughters, Margaret Giggs,[2] who died two years before him in 1570. They were both strong Catholics and lived abroad during the reigns of Edward VI and Elizabeth.

This tells us nothing about his skill with plants; and the only records of his writings are of translations into Latin from Greek Fathers of the Church. But at least Turner's admiration must mean that he and his associates were deeply concerned with botany, and points a warning against claiming that Turner was altogether a lonely pioneer. Further evidence to the same effect is to be found in the case of the second name on his *Herbal* list, Dr Wendy. For in the long discussion of the identity of the various types of Artemisia in *Herbal*, p. E. i*b*, after describing a Mugwort found on an island near Venice, Turner adds, 'Master doctor Wendy the Kyngs phisycyan dyd examyne the herbe wyth me', which obviously does not mean that Wendy who had been in Italy in 1525 was out there again in 1540, but that Turner had brought home in his *hortus siccus* a specimen of the plant and consulted the royal physician about it. But this makes it probable that there had been some previous acquaintance between them, probably before Turner's exile. The whole circumstances suggest that among leading doctors[3] there had been for some time a desire for more accurate herbarism; and that Turner had been aware of and perhaps encouraged by this desire.

A situation had indeed arisen similar to that in the sphere of religion. New books and fresh study had revealed that in the old world of the Graeco-Roman culture there was a storehouse of knowledge; that this had been misused and debased if not entirely ignored during the Dark Ages; that in consequence superstition and error were almost universal; and that it was possible to correct these by a recovery of the ancient lore. The same motives which made Turner a Gospeller, made him a Herbarist. Yet the appeal to ancient authority, which here as in religion erected the infallibility of that authority into a dogma, and for a time impeded the progress of knowledge by involving the pioneers in punishments for heresy,[4] never prevented Turner as it did John Caius from questioning

1 In *Vita Thomae More*, Cologne, 1612, quoted by Munk, *Roll of R.C.P.* pp. 25–6.

2 Wood, *Ath. Oxon.* I, c. 402, quotes the Epithalamium composed for their wedding by John Leland.

3 It is notable that in his *Spirituall Physic*, p. 59, when citing the heads of the profession Turner names these same men: the passage runs, 'wolde these phisicions be content that doctor Wendy or doctor Owen, doctor Wotton or doctor Huic [Robert Huicke President R.C.P. 1551–2] or I should practise wyth them so?'

4 The classic case in England is that of John Geynes, M.D. Oxon. 1535, who was charged on 22 December 1559 by Thomas Wendy before Caius and the College of

the evidence or using his private judgement. Deeply as he admired antiquity he had too much independence, too real a respect for facts, to allow it to blinker or muzzle him.

The problem of Turner's other writings of which Conrad Gesner, John Bale and in recent times C. H. Cooper have given lists is too intricate for thorough consideration here. The most important and least known account is that contained in Gesner's *Bibliotheca* (on p. 69 of the *Epitome* published at Zürich by J. Simler in 1555); and this may be accepted as accurate since Gesner addressed to Turner his catalogue of his own writings and was in constant correspondence with him. The list seems to contain nothing written later than 1548, but adds several titles to those listed by others. These are—*In Catonis Disticha Moralia*, four books,[1] probably a commentary upon them; *Sententiarum Flores*, an anthology; *In Publii Mimi Versiculos*, presumably another commentary. In addition it records the *De Naturis Herbarum*, the lost work said by Cooper to have been published at Cologne in 1544; the bird-book of the same year; the Hunting of the Roman Fox; the New Dialogue on the Mass; the *Unio Dissidentium*; and the translation of Urbanus Regius; and five whose names are also given by Cooper (*Athenae Cantab.* I, p. 259): the *De Arte Memorativa*; *De Hierosolymorum Excidio*; various Epigrams; a satirical ballad *Pro Standicio* (or as Cooper has it Joanne Standicio) *ad Papam*—these four no doubt writings of his Cambridge period; and an epitaph on Gisbertus Longolius whose Dialogue on Birds he had edited. The work on Cato's Distichs, those popular couplets ascribed to 'Dionysius Cato', or to 'Cato of Cordova', or to 'Cato the Philosopher', but really of uncertain date and authorship, is no doubt a product of his years in Pembroke: for they were an immensely popular introduction to Latin literature and played a large part in the schooling of the period. If ever published, it seems to have left no trace on subsequent editions. The defence of John Standish, who attacked Robert Barnes in 1540 and whose uncle Henry, Bishop of St Asaph, had renounced the papal jurisdiction in 1535, must have given Turner large scope for his jesting; for the Standishes were both notorious trimmers in religion. The list, whether or no it is complete, is at least proof that Turner's activity was not confined to religious controversy and natural history, or to the later part of his life. He had not been idle in Cambridge.

imputing errors to Galen. In 1560 he was ordered under pain of imprisonment to explain his conduct. He gave way, stated his points, acknowledged himself in the wrong, and subscribed to a statement that Galen had not erred as he had stated. The matter is reported in the Annals of the College and by Munk, *Roll of R.C.P.* I, p. 62.

1 This may well refer to the four books of Distichs rather than to four volumes of Turner.

CHAPTER V. WILLIAM TURNER IN EXILE

This period of College work seems to have come to an end in the year 1537, though the exact cause of the change is perhaps impossible to determine. Probably the primary reason was his marriage to Jane the daughter of George Auder[1] an alderman of Cambridge which took place about this time. We know nothing of the lady or her family except that in one of his letters to Sir William Cecil (on 11 June 1549) Turner asks that one of his houses in London may be leased to 'Mistres Auder my mother in law whom I thynk ye know' (cf. B. D. Jackson, l.c. p. iii)—which presumably means that the alderman was dead by that time and that the Auders had been friends of the Chekes and so of Cecil through his first wife.[2] Jane certainly accompanied her husband in his exile; for the son Peter was born in 1542 (so Venn, *Al. Cantab.* IV, p. 276) and there were also two daughters, Winifred and Elizabeth, by the time of his return. She outlived her husband, married again, and founded a scholarship at Pembroke in his memory.

At this time too he was ordained deacon—at Lincoln, then the diocesan Cathedral for Cambridge, on Easter Eve 1536—and received a license to preach from his College next year, probably just before his departure. In his *Huntyng of the Romyshe Wolfe*, p. B. ii, he states, 'I have had commission of God and King Henry the eight and of King Edwarde his sonne and of bothe their counselles auctorite, to read and to interprete the Scripture': and this must date back to his first ordination. On leaving Cambridge he then, according to Wood (*Ath. Oxon.* I, cc. 361–4), toured the country as a Gospeller, spending some time in Oxfordshire 'among several of his countrymen that he found there for the conversation of men and books'—a fact which may account for two obscure allusions in *Names*, one to 'Eliote' under Ligustrum, p. E. 1 *b*,[3] the other to 'Coome parcke' under Astragalus (*Lathyrus montanus*), p. B iiii[4] and Genistella

1 Dr Helen Cam who has very kindly searched the available material reports that there seems to be no record of Auder. He was never Mayor. But the name is known in Cambridge.

2 Mary Cheke who died in 1554, cf. Strype, *Life of John Cheke*, p. 118. Strype, *Cranmer*, I, p. 394, records that Turner was befriended by Sir John Cheke and Sir William Cecil.

3 The allusion is to Sir T. Elyot, *Dictionary*, London, 1545: of Ligustrum, he writes, 'they whiche doo take it for the bushe called Privet be moche deceyved'. Turner replies, 'in Englishe Prim print or Privet, though Eliote more boldely then lernedly defended the contrary as I shal prove in my latin herbal when it shal be set fourth'.

4 Elyot, *Dictionary*, describes Astragalus similarly with Turner: evidently they agree as to its identity, but, as the picture in *Herbal*, p. E. v, makes clear, confuse it with *L. tuberosus*: see below, p. 106. Johnson, *Gerard*, p. 1236, pointed out Turner's mistake and printed pictures of the two species side by side.

(*Genista anglica*), p. H. iii; for Sir Thomas Elyot of Combe near Woodstock was, as we have already seen (cf. pp. 42–4 above), a man who among his other and wide interests had some knowledge of botany and may well have been visited by Turner. At this time, 1536–8, he was living at Combe and hard at work on his *Dictionary*.

That the two men knew one another not only by their books but face to face is made highly probable by a careful comparison of the *Libellus*, the *Dictionary* and *Names*. Elyot at least twice gives certain proof of borrowing from the *Libellus*: of the Hellebores Turner had written in 1638, 'Elleborum album....Radicem vulgus Nesynge powder, herbam autem Lyngwort. Elleborum nigrum....Ego censeo illam herbam quam vulgus Cantabrigiense vocat Bearefote esse verum elleborum nigrum', which appeared in 1545 in the *Dictionary* as 'Elleborus albus, an herbe called lyngwort, the roote whereof is called nesinge powder. Elleborus niger, an herbe called beares foot'; and more strikingly still of 'Bryon thalassion' where *Libellus* has 'a northumbriensibus vocatur slauke', and *Dictionary*, 'is called of northern men slawke'. There are several other clear indications of Elyot's use of the *Libellus*, for example, under Cyperus, Cassutha and Oxytriphyllon; but these are relatively few compared with the very large resemblances between the *Dictionary* and Turner's *Names*. Buglossum, as Borage; Chamaedrys, Germander; Colutea, Sene [Senna]; Gith, Nigella romana; Hypericum, Saynte Johns wurte; Intubus, Endive and Sicory; Lens, Duckes meate; Libanotis, Rosemary; Lolium, Darnel; Lithospermum, Grummel; Ocymum, Basil; Phu, Valerian; Phyllitis, Hartes tongue; Polygonum, Sanguinaria; Ricinus, a Tick and Palma Christi—here are points on which they are entirely agreed, examples sufficient to establish connection. Perhaps the clearest is under Sisymbrium where *Dictionary* has 'whereof be two kyndes...in englysshe water mynte...the other water cresses', and *Names*, p. G. iii*b*, 'Sisymbrium is called in Englishe Water mynte. Sisymbrium alterum in english watercresses'. This last example like the case in which 'Eliote' is mentioned and the case of Heliotropium which Elyot calls 'Cikory or Marigoldes', while Turner says 'they be deceived that holde that our Marigolde is Heliotropium', proves that Turner had certainly seen the *Dictionary* when he wrote *Names*. But the case is less simple than a mere literary contact; Elyot using *Libellus* and Turner then using his *Dictionary*. For there are a number of cases in which Elyot seems certainly to have known Turner's identification not from *Libellus*, but before *Names* was written. Thus, in addition to the list already given, there are the following: of Osyris, which is not in *Libellus*, Elyot says, 'some doo suppose it to be wilde lynne or wylde flax', and Turner that it has no name but 'may be called Lynary or Todes flax'; of Pityusa both agree that it is Spourge and Esula, Elyot adding a note from Actuarius that 'barbarous people call it Turbit',[1] and of Rhus both say Sumach, but Elyot quotes 'Manardus lib I epist iiii sayth that it is also callyd Ros syriacus'; of Halimon both give a similar but independent description which shows that they agree in identifying it as *Hippophäe rhamnoides*; of Limonium each suggests and Turner states that it is a *Pyrola*; of Eryngium while Elyot calls it a kind of Thistle, Turner names it Sea Holly; and of Cyrsium though they agree that it is Buglosse or 'Langdebefe' each takes a different alternative. The conclusion is almost inescapable that the two men had met, talked over their common interest, exchanged information, and then developed its expression independently. Such contact can only have been made during the years immediately preceding Turner's first exile, and at least supports the conclusion that Turner stayed at Combe while Elyot was writing or revising his book.

That Turner's preaching got him into trouble, and that he was arrested and for some time imprisoned is probable—although Wood's account is neither detailed nor unprejudiced. But he probably had on one occasion to appear before Stephen Gardiner; for in *The Rescuyng of the Romishe Fox*, p. N. 7, he desires him 'not to call me heretike without a cause as ye did once in your hall ryght proudly because I weare a cloke and a hat of the new facion'. The suggestion that his exile was a result of the Six Articles in 1539, and the still more baseless supposition that he made some sort of recantation as the price of his release, are pure speculations; and the latter, as Jackson justly insists, is wholly out of keeping with his character and subsequent history. That he was in real peril is clear from his words in the *New Dialogue*, 'Wyllyam Turner unto the Reader': 'Some...wyll desyre to have my bloude as dyvers in Englande at thys daye yet lyvyng have done, God forgive them.' With his friends Bilney, Bayfield and others having been done to death, he could hardly have been secure. But as yet there was little blood-lust and he had not become specially prominent.

Nevertheless in the early days of 1540 and during the progress of the struggle between his enemy Stephen Gardiner and the reputed champion of the reformers Thomas Cromwell his position must have become precarious. When Robert Barnes and other Gospellers were seized, it must have seemed that his preaching work was becoming impossible; and Cromwell's fall, not unexpected though it was, destroyed the hopes of all those who looked for a radical change of doctrine in the Church.

Moreover, Henry VIII's religious policy and perhaps finally his treatment of Anne of Cleves, of which Turner strongly disapproved (cf. *Huntyng of the Romyshe Wolfe*, p. D. vi), may well have increased his decision to develop his interest in medicine and with that object to go

1 Cf. e.g. *Ortus Sanitatis*, De Herbis, ch. 482, 'Turbit arabice...nascitur in littore maris'.

abroad. The precise date of his departure can be fixed within a few days by one of the clearest biographical references in any of his writings. Dealing with the ignorance of the nobility in his *Newe Booke of Spirituall Physic* published in the spring of 1555 but written earlier, he tells how 'about xiii yeares ago it chaunced that I was in Callice [Calais] and whilst I was there the prince of Salerne came thether out of Italy....At that tyme two Englyshe commissioners were sente thether to scoure the towne of traytours, and no depute [Deputy] as yet appointed nor beyng there these two muste welcome the prince...his gentlemen spoke Italian Latin and Frenche too...our gentlemen could not speake one worde againe... one was an Erle and the other a Knyght.' This is an occasion easily identified. In March 1540 Robert Earl of Sussex and Sir John Gage had been given a commission to go to Calais; to send back Viscount Lisle, the deputy, who was suspected of favouring Rome; and to enquire into the doings of Adam Damplip, alias George Bowker, and William Stevens. In April Sir Thomas Wyatt then at Ghent had written to Cromwell warning him that Ferdinand de San Severino, Prince of Salerno, intended to visit England, and suggesting that an 'honest ship' should be provided for him at Calais (cf. *Chronicles of Calais*, Camden Soc. xxxv, pp. 188–9). In fact the prince was delayed and did not get to Calais until early July when he crossed to England with Don Luis d'Avila and spent eight days with the King.[1] The Earl and Sir John Gage were still at Calais, no deputy having yet replaced Lisle. No doubt Turner and his wife had left England at the end of June.

Whether he was ordered to leave the country, or on his own initiative decided to take his wife abroad is unknown. He had crossed the Channel from Dover where he saw Wild Cabbage (*Brassica oleracea*); a 'wonderful great Cole...it hath whyte floures and round berryes' (*Crambe maritima*) (*Herbal*, p. G. ii and G. ii*b*);[2] 'Crithmus or Sampere' (*Crithmum maritimum* (*Herbal*, p. M. iii*b*); and Horned Poppy (*Glaucium luteum*) (*Herbal*, II, p. 77; *Names*, p. C. v*b* and F. i). On leaving Calais he passed along the coast and recorded Sea Trifoly (*Astragalus danicus*),[3] 'I saw it ones in Flaunders by the sea syde about thre myles beyonde Dunkyrke' (*Names*, p. D. iiii, *Herbal*, II, p. 116). Thence he probably struck across to the Rhine, and planted his wife at Cologne or Bonn or possibly Basel where he may well have stayed for some time. But the sequence and dating of events at this period of his life are very obscure.

1 A full record of these events can be pieced together from *State Papers, Henry VIII*, 1540: cf. especially pp. 190, 447.

2 De l'Obel, *Adversaria*, p. 92, who describes and figures this plant as found on the coast of England and says that he received seed from Turner, gives the Isle of Portland not Dover as the locality for it: see below, p. 104.

3 Britten, l.c. identifies this as *A. glaux*: but this is a southern species, unknown in Flanders.

There are very few statements of date which can be taken as definite. Perhaps the clearest is the sentence in the long discussion of Wormwood (*Herbal*, I, p. 6): 'Gerardus de Wijck,[1] twelve yeares ago, when as he was in Colon at that tyme the Emperours [Charles V's] Secretary, taught me fyrste the righte Pontike Wormwode.' This, though its exact date cannot be fixed, may well refer to Turner's earliest visit to Cologne almost directly after his departure from England: for we know that Veltwick was at Worms in conference with Gropper and Bucer in 1540.[2] Turner's own references to his stay at Bologna, 'Lucas Gynus [Luca Ghini] shewed me about a xiiii yeares ago' (*Herbal*, II, p. 18) and 'my master Lucas above xvi yeares' (*Herbal*, II, p. 81*b*), do not help much; for we do not know when the words were written and so cannot fix the year from which the 14 and 16 years are counted: if we count from the date of publication 1562, we get 1548 and 1546 which are both plainly too late. Similarly Gesner's allusion to Turner's visit to him on his return from Italy 'xv years ago or thereabouts' does not profess to be exact. It would seem to place the end of his Italian work about 1540–1. The 'benevolence' which according to Cooper (*Athen. Cantab.* I, p. 256) he received from his College in 1542 might perhaps indicate that at this time he was engaged in academic work and therefore still in Italy. But the birth of his son in this same year shows that he had rejoined his wife in 1541; and as he had certainly not carried her across the Alps we are probably right in dating his sojourn in Italy 1540–1, and on the Rhine at Basel, Bonn and Cologne (cf. *Names*, p. G. v*b*, 'my gardines in Germany') 1541–4.

In any case from Germany he seems to have gone first to Venice where he may have spent some little time. In *Herbal*, II, p. 63, he contrasts the German towns, where 'there is not suche choyse of simples in every place', with 'Venis': in *Herbal*, I, pp. 62–3, he speaks of his visit to an 'islande besyde Venice' 'called Chertosa [La Certosa, cf. Fynes Moryson's description, *Itinerary*, I, p. 161] where as is a Chartarhouse' and of finding the true Pontic Wormwood there; and in *Herbal*, II, p. 59, of seeing

1 Turner calls him Gerardus Delwike in *Herbal*, II, p. 22, and Geraldus Delwicus in the list of Authorities in the *Herbal* of 1568. Gerard Veltwick von Rabenstein, the Hebrew Scholar, was at this time Imperial Secretary and became President of the Council of the Netherlands in 1554–5. In this same passage, *Herbal*, I, p. 6, Turner records that Veltwick had nearly lost his life exploring the Apennines for simples: he adds, 'This noble clerke afterwards was sente by Charles the fyft, Embassator to the greate Turke, and in his jorneye he came thorowe Pontus, and broughte home wyth hym trewe Rapontike'—Veltwick's embassy to Constantinople was in 1545. Amatus Lusit., *Enarr.* p. 298, has the same record, 'Gerard orator of the emperor Charles V to Solyman Emperor of the Turks had with his own hands rooted up Rhaponticum in Pontus'.

2 As this reference to Veltwick does not occur in the unrevised *Herbal* the twelve years probably date from later than 1551 and the meeting may have been on Turner's second visit after his return from Italy, in 1543–4.

Tamarisk 'in an yland betwene Francolino and Venish'. There, too, he got his only sight of an Eagle Owl (*Bubo bubo*) (*Avium*, p. 40).

From Venice he went to Ferrara[1] where he stayed and worked. He speaks of finding both kinds of Conyza 'betwene Cremona and Ferraria by the Padus banke' (*Names*, p. C. iiii*b*), of walking along the Po with his companions when he saw 'the bird we call jaia' (*Avium*, p. 118); of seeing several trees 'by the Floude Padus' (*Names*, p. D. vii*b*); of having only once seen 'Atricapilla, Anglorum lingetta' and then when it was shown to him by Don Francisco of the Holy Council in the house of the Duke of Ferrara[2] (*Avium*, p. 39); of seeing 'Vitex' (*V. agnus-castus*) growing 'at the black freres [the Dominican house of which J. Bauhin speaks[3]] in Ferraria' (*Herbal*, II, p. 165*b*) and Lotus (*Oxalis corniculata*) in the same place (*Names*, p. E. ii); and above all of having worked there with Antonio Musa Brasavola 'som tyme my master in Ferraria'[4] (*Herbal*, II, p. 78). Whether he took his degree of M.D. there[5] or at Bologna is uncertain; for he certainly divided his time between the two places, and on the whole speaks more frequently and warmly of the second. 'Lucas Gynus the reader[6] of Dioscorides [lecturer on Dioscorides] in Bonony my maister' figures pretty freely in the pages of the *Herbal* (e.g. I, p. 17; II, pp. 18, 81*b*, 91*b*, 167*b*): he was a considerable botanist, had an excellent private garden,[7] and seems to have been the first to make a 'hortus siccus' by pressing plants and stitching them on to sheets of paper.[8] It seems evident from one or two references to specimens brought from Italy that Turner himself adopted this method at this time from him: thus for example of Bunium (*Herbal*, I, p. 98) he writes: 'Lucas Ginus my maister at Bonony gave me a pece of it which I have yet to shew' and of the leaves of 'Fistick Nut whereof I have certayn at thys day to shewe well kept in a booke at the lest these seventeen years' (*Herbal*, II, p. 91*b*). It is certain that 'my frende Falconer' (*Herbal*, p. C. vi), the John Falconer to whom reference has already

1 For the journey, Venice—Ferrara—Bologna, in 1594, cf. Fynes Moryson, *Itinerary* (ed. 1907), I, pp. 196–205.

2 Ercole II, duke 1534–59, cf. Amatus, l.c. pp. 141, 154: the great palace-castle of the House of Este was then in its magnificent maturity.

3 Cf. L. Legré, *Les deux Bauhins*, p. 9.

4 Of whom Amatus Lusitanus, who if not at Ferrara with Turner must have come shortly after (he was there for seven years), speaks with affection and respect, cf. *Enarrat. in Diosc.* pp. 7, 15, etc.

5 As Bale, *Script. Illust. Brit.* p. 697, and Fuller, *Worthies*, II, p. 192 following him, explicitly state.

6 In 1539–44 he was lecturing, *De Simplicibus Medicinalibus* (G. B. de Toni, *I Placiti di L.G.* p. 4).

7 The University garden was started by the initiative of Ulisse Aldrovandi in 1567.

8 The first full account of the making of such a 'Hortus Hyemalis' seems to be in Spieghel, *Isagoge*, I, pp. 79–81 (Padua, 1606).

been made, did so: for in *Herbal*, II, p. 11*b*, under Sea Trifoly we read, 'I never sawe it in Englande savinge in Maister Falkonner's boke and that had he brought out of Italy', and in *Herbal*, II, p. 147, under Stachys, 'the other kynd dyd my frende mayster Fauconer shewe me after that he came oute of Italy'.[1] There are also constant references to 'the Mount Appenine besyde Bonony', to the trees and shrubs seen on it, and once to the help of the servant of Luca Ghini who accompanied him there.

From Ghini it is probable that Turner's interest in problems of exact identification and nomenclature, already strong, and afterwards a particular feature of his work, received encouragement and direction. We know little of Ghini; for although there is plenty of testimony to his knowledge and powers as a teacher nothing of his work except one medical treatise was published until in 1907 G. B. de Toni printed *I Placiti di Luca Ghini*, twenty-five pages of notes on many species of plants sent by him from Pisa to Mattioli in 1551. These are not very revealing. Otherwise there are only casual references. One such is illuminating. Joachim Camerarius the younger, who studied at Bologna in 1562 shortly after Ghini's death in 1556, says of him in his *Hortus Medicus*, p. 159, 'he was an excellent judge of simples, and declared that he knew some twenty plants that answered to the description of Seseli Massiliense'. By such a man Turner's sense of the importance of accurate naming would be emphasised, and from him he must have gained the mastery of the relevant material which enabled him to discuss questions like the precise discrimination of Pontic Wormwood with scholarship and in detail. That Ghini was very much more of a botanist than Brasavola[2] is perhaps one reason why Bologna was preferred to Ferrara. Of his character Mattioli gives an attractive picture in a letter dated 12 December 1558 to another of his pupils, Georgius Marius of Wurtzburg (*Epistolae*, p. 342): 'Ghini's death was a heavy blow to me: his intellectual endowment was massive and brilliant: integrity, sincerity, loyalty to his friends were conspicuous; there was never a trace of jealousy. The clearest proof of this is, as you rightly say, that although he had planned certain books on plants and

1 Amatus Lusit. l.c. alludes freely to 'John Falconer the Englishman' whom he knew at Ferrara, cf. pp. 240, 267, 290, 357, 394. He speaks of him as 'a man fit to be compared with the most learned herbarists, a man who had travelled many lands for the study of plants and carried with him very many specimens ingeniously arranged and glued in a book' (p. 337). Falconer when at Ferrara sent to Gesner a note on the Guinea-fowls kept in the duke's garden (*H.A.* III, p. 425). He is often mentioned by Aldrovandi who must certainly have met him.

2 Of Brasavola much may be learnt from Amatus Lusit. *Enarrationes*; for Amatus was at Ferrara in 1541 (p. 187) and in 1545 (p. 482), working there six years (p. 141). Of Brasavola he speaks with respect and affection, but for his medical skill and large practice, 500 to 1000 cases a year (p. 307). Brasavola's books deal almost entirely with drugs and diseases; and his early dialogue, *Examen Simplicium* (Lyons, 1537), is heavily criticised by Mundella, *Epistolae*, pp. 179–263.

pictures to illustrate them, when he read my Commentaries he not only wrote to congratulate me on anticipating him and lessening his task but sent me a number of plants and pictures which I have used in my Dioscorides.'[1] Ghini moved to Pisa to be prefect of the new garden there in 1547. He died in 1556.

How long Turner stayed at these two Universities can only be a matter of conjecture. If the dating of his exile as suggested above is correct, the time would be only a year at most; and this may seem absurdly short considering the large influence which it had upon his work and the space which Italian records occupy in his writings. But there is little reason to suppose that it was longer. A few weeks in either place could in fact have given him all that he has recorded. And we know from the contemporary case of Stephanus, the companion of Valerius Cordus of whom details are given by H. Schreiber of Nuremberg in a letter of December 1544,[2] that a visit for the purpose of attaining a degree need not be long. Stephanus visited Ferrara, saw Brasavola and stayed a few days: he then went to Bologna, met Ghini and others, took his doctorate and left in less than a month for his return to Germany. Turner doubtless stayed longer than this, particularly at Bologna: but there is little to suggest a lengthy visit. That he left Ferrara before 1541 is probable from the fact that Amatus Lusitanicus began his work there with Brasavola in that year (*Enarr.* p. 187) and that Turner almost certainly never met him.

An allusion under Rhus in *Names*, p. F. vi*b*, gives the clue to the route by which he left the Italian universities to explore Switzerland: 'Thys kinde did I firste see in Bonony, afterwarde besyde Cremona, laste in the rockes besyde Lake de Come.' On this occasion and on his way to Como, he went up the Po as far as Milan and 'sawe Ryse growing in plenty' there (*Herbal*, II, p. 73), or as *Names*, p. E. viii, records it, 'in watery myddowes betwene Myllane and Pavia'.[3] He tells us also that at Pavia 'faselles [*Phaseolus vulgaris*] grow in great plentie' (*Names*, p. F. ii*b*), and that 'I never saw better hoppes [*Humulus lupulus*] then I saw growyng wylde a litle from the wall that goeth from Chertosa [Certosa] to Pavia by a litle rivers syde' (*Herbal*, II, p. 43). It may well have been here that he saw the 'Albardeola' (*Avium*, p. 34) which he distinguishes from the 'Shovelard' and which was probably the Little Egret (*Egretta garzetta*). Two other allusions refer to this visit to Como; he writes of Phalaris (probably *P. canariensis*) in *Herbal*, II, p. 85: 'The first tyme

1 The story is also told in the Life of Aldrovandi by Fantuzzi and by Calvi in his *Commentarium Historicum Pisani Vireti* (Pisa, 1777). It is the more remarkable because on Mattioli's own evidence, *Comment.* Ep. Nunc., the two men had never met.

2 Printed in C. C. Schmiedel, *Stirpium Descript.* (Nuremberg, 1753).

3 This seems to have been its chief centre of cultivation in Europe, cf. G. Bauhin, *Theat. Botan.* (Basel, 1658), c. 482, 'around Milan whence it is principally brought to us'.

that ever I saw thys herbe was in the citie of Come where as the chefe Physician of the citi no lesse gentle than well learned shewed unto me and my felow master Johan Walker'—possibly of King's College, Cambridge, and later physician to the Duke of Somerset:[1] he also saw *Aristolochia longa* in fruit 'besyde Lake de Come' (*Names*, p. B. ii). From the lake he seems to have gone on to 'Clavena' (Chiavenna), 'a little cyty as we enter in at the foote of the mountaynes besyde an olde castell' where he saw 'Lotus urbana' (*Oxalis corniculata*) (*Names*, p. E. ii and *Herbal*, II, p. 42)[2] and explored the mountains; for he writes of 'Monkes hode' (*Aconitum napellus*), 'I have sene it in great plentie upon the alpes betwene Clavena and Spelunca [Splügen]' (*Names*, p. A. v*b*), or as he expresses it in *Herbal*, I, p. 20: 'Thys kinde groweth verye plenteouslye in the very top of the Alpes betwene Splengen [Spleunge in *Herbal*, p. B. 1*b*) and Clavenna.'[3]

From Chiavenna he made his departure from Italy, and crossed Switzerland. There is a reference under Veratrum (*V. album*) in *Names*, p. G. vii*b*, which clearly indicates such a journey; he describes it as growing 'in the Alpes betwene Cureland and Lumberdy'; and this is borne out by his note on *Gentiana lutea* (*Herbal*, II, p. 76), 'I have sene it in the Alpes growinge betwene Italy and Germany'. We know that he spent some time in the mountains and eventually got safely to Chur and from there to Zürich. The records contain several passages of interest. One is in the *Avium Praecipuarum*, p. 42, and describes how he met the shepherd whose remarks about the Goatsucker he had reported: 'When I was in Switzerland in the mountains seeking herbs I saw an old man feeding goats. I asked him if he had ever seen a bird the size of a Blackbird, blind by day but able to see by night, which sucked goats' udders and infected them with blindness. He said that fourteen years ago[4] he had seen plenty of them and had had as many as six of his goats blinded: but now all these birds had gone from Switzerland to Lower Germany, where they not only stole milk and blinded goats but killed sheep. I asked him the name of these birds; and he answered "priest" ["paphum"

1 Cf. Venn, *Al. Cant.* IV, p. 317.

2 This reference is to a picture and description which make the plant's identification certain: Britten, *Turner's Names*, p. 102, presumably following G. Bauhin in Mattioli, *Opera*, p. 810, identifies it as *Melilotus caerulea*. Unfortunately, Turner has himself caused some confusion (e.g. in J. Bauhin, *Hist. Plant. Univ.* I, pp. 234–5) between this 'Lotus urbana' and the 'Lotus arbor' or Celtis. In *Names*, p. B. viii, he says that he saw the tree in Italy: but in *Herbal*, p. H. v*b* and I, p. 116, he adds a note that he saw it 'in Clavenna'.

3 It is interesting to note that Turner's route was identical with that followed in reverse by Gilbert Burnet in 1685—Zürich, Wallenstadt, Chur, Thusis, Splügen, Chiavenna: cf. G. R. de Beer, *Early Travellers in the Alps*, pp. 68–70.

4 The cantons which had renounced Catholicism first did so in 1526–7. Hence 1541 is the probable date of this incident. It could be fixed more conclusively if we knew in what part of the country he met the old man.

is no doubt pfaffe]. Perhaps he was making fun of me!'—he certainly was! A more important reference is in his record of the Larch (*Herbal*, II, p. 28). He saw it very abundantly growing at Chur in the mountains, and so was able to testify that the old authors who said that it would not burn were mistaken: but as he only saw it in the summer he could not himself state of his own knowledge that it was deciduous. But he adds: 'Antonius Traversus a ryght Gentleman of the countre of Rhetia, when as I lay in hys house, restyng me after my great labours that I had taken in seking of herbs in the alpes, told me....' So, too, he records 'the thirde kinde of Cinkfolye [? *Potentilla grandiflora*] growing upon the walles of a citye called Cour in the land of Rhetia, a litle from the mayn alpes' and describes its hairy leaves and stalks, yellow flowers and strawberry-like seeds (*Herbal*, II, p. 110*b*). It was in the Rhaetian Alps and presumably on this visit that he saw a dark bird with white spots like a Starling and gave it the name of Nutcracker (*Nucifraga caryocatactes*) (*Avium*, p. 76). He found a few plants but, it must be admitted, did not make much use of his opportunity: the typical Alpine flora was of course unknown to his authorities. But he found *Polium teucrium* 'in the Alpes of Rhetia besyde Cure' (*Names*, p. F. iiii) and *Ajuga genevensis*, 'Geagged Bugle', in 'Swycherland' (*Names*, p. G. vii); and records under 'Chamepeuce' (*Lycopodium selago*), 'This whose figure ye se grewe in the top of the alpes wher as I gathered it my selfe' (*Herbal*, I, p. 130). On the way to Zürich he must have noted the Cyclamen growing 'beside Wallense' (*Names*, p. C. vi*b*) and perhaps here saw the 'Wood Crow' or Waltrap, the most interesting of all the birds which he reported. At Zürich he met Gesner and may be said to have brought his apprenticeship in science to an end.

His sojourn in Italy was of the highest importance to Turner as a botanist. It brought him into touch not only with the most vigorous scientific centre of the time—for Bologna was probably ahead of Montpellier or of any of the German Universities—but in consequence with the leading writers and workers in the field. He became a figure in the learned world, and could thenceforth comment with knowledge upon the whole literature of herbarism. 'Following the autorite of my masters, of whome I learned first the knowledge of herbes, who were Antonius Musa, Fuchsius and Ruellius'—so he writes in *Herbal*, II, p. 67*b*. In the original dedication of the *Herbal* of 1551 he is still more explicit: 'Hermolaus Barbarus therefore, Nicolaus Leonicenus [Niccolo da Lonigo], Joannes Manardus and Antonius Musa Italianes, Otho Brunfelsius, Leonardus Fuchsius, Conradus Gismerus [Gesnerus] and Hieronymus Bochius [Jerome Bock] Germanes, and Johannes Ruellius the Frenche man have greatly promoted the knowledge of herbes by their studies.' If in this list he leaves out Mattioli and Rembert and Amatus Lusitanicus, the reason is that in 1551 they had not yet become known. In his later publications he speaks freely of them. Plainly the Englishman felt himself on a level with these other students; and his independence of outlook is fully expressed. He can examine and dismiss 'one Riffius'[1] who differs from Fuchs about the Hellebores by an appeal to the meaning of Dioscorides (*Herbal*, pp. L. v and v*b*). Even from those whom he most revered he was not afraid to differ: 'Antonius Musa som tyme my master in Ferraria, and Leonardus Fuchsius my good frende in Germany hold that Feverfew is not Parthenium....But though they ar both my frendes, yet I will hold with the truthe rather than with them, when as I judge they hold not with it': so he wrote in *Herbal*, II, p. 78*b*.

There seems no evidence that he had as yet met any of these except Luca Ghini at Bologna and Antonio Musa Brasavola at Ferrara, and his own fellow-students Falconer and Walker. With the greatest of contemporary naturalists, the very learned and very religious Conrad Gesner, he made contact at Zürich[2] after leaving the south. 'Fifteen years ago or thereabouts', wrote Gesner in his *De Herbis Lunariis* (Zürich, 1555), p. 34, 'an Englishman returning from Italy greeted me (whether it was Turner a man of excellent learning both in medicine and in many other fields of study, or someone else I hardly remember): among other pictures of the rarer plants which he had taken care to get painted[3] he showed me a painting of Elleborine.[4]...He said that...in England it was commonly used by country folk'—whether or not this alludes to Turner, certainly a visit was paid. For Turner refers to it in his *Avium Praecipuarum*, p. 125, under 'Platea or Shovelard', 'Conrad Gestner, a man as learned as honest when I was at Zürich first gave me knowledge of this bird, and that it was called in German "lefler" [Löffelgans or in Dutch lepelaar] because it had a beak like a spoon'. And there are many allusions in the *Herbal*. 'The fyrst that ever I sawe of thys kinde [*Trifolium medium*] grewe in Doctor Gesnerus gardin in Zürich' (*Herbal*, II, p. 158).

The effects of the visit were lasting; and thenceforward the two men regarded one another with the highest respect. Turner evidently kept up a correspondence: for we learn from *Herbal*, II, p. 122*b*, that Gesner had sent him seed of Rue from Zürich; and the short treatise on Fishes, the only work on the subject that he has left, was sent as a letter to the Swiss

1 Walther Hermann Ryff who edited Dioscorides, *De Re Medica*, at Frankfort in 1543 and criticised Fuchs's work of the previous year. Fuchs replied with his *Apologia*, Basel, 1544.

2 This must have been very soon after Gesner's return from Montpellier, and the taking of his doctor's degree at Basel, 1540–41.

3 There seems no other evidence of this: perhaps Gesner is confusing pictures with dried and mounted specimens.

4 The identity of this plant is discussed by Gesner in his letter to Caius of August 1561, *Ep. Medic.* p. 136*b*; cf. below, p. 146.

savant. Gesner on his part in his letter to Caius in 1561 speaks of onion bulbs and other gifts received from Turner[1] and lists him as one of his authorities both as a writer and as a helper. He always quotes his opinion prominently and attaches great weight to it. Indeed, most of the *Avium Praecipuarum* is incorporated in the third volume of the *Historia Animalium*; and there are further details about birds, as for example the tame Pelican at Malines (*H.A.* III, p. 570), reported by letter and at a later period.[2] Moreover, in 1562 when his life was drawing to its end, it was to Turner 'a man of eminent learning and piety, very expert in Theology and Medicine, Dean of Wells' that Gesner addressed the letter published at Zürich in which he gives a list and brief account of his own writings, sixty-seven already published and eighteen in preparation, the letter which tells how he hopes to use Caius's treatise, *De Canibus*, and has collected pictures and other material for books on Serpents and Insects. So, too, in regard to plants, though Turner's *Herbal* was published in English, Gesner was clearly acquainted with some at least of its findings—for example, with the plant called Paulis Betony (*Herbal*, p. F. v) and with his identification of Cantabrica (*Herbal*, p. H. ii and Gesner, *Hort. Germ.* p. 285*b*). It may, of course, be from letters that these points are derived, but it is clear from Gesner's note in the list of authorities prefixed to Kyber's translation of Bock's *De Stirpium* that Turner had sent his friend the 1551 edition of his *Herbal*.

Thence Turner must have moved westward to Basel, perhaps finding *Anthemis tinctoria* on his way 'in a corne fielde betwene Basyle and Surike' (*Names*, p. B. i). Here he may well have met the great German herbalist and doctor Leonhart Fuchs, who though living at Tübingen was publishing his *De Historia Stirpium* with the house of Isingrin there in 1542, and whom Turner certainly came to know personally about this time. It is probable that he also met the brilliant young botanist Valerius Cordus; for in *Herbal*, II, p. 86*b*, after stating his first opinion of the Hartstongue (*Scolopendrium vulgare*), he adds 'but admonished by Cordus many years before Matthiolus had ether writen in Italian or Latin, I left my former opinion'—which, as Cordus published nothing before his death in 1544 and as Mattioli's book appeared in that year, strongly suggests a personal contact. Cordus was then on his way to Italy.[3]

At Basel Turner's book, *The Huntyng and fyndyng out of the Romyshe*

1 Cf. *Epist. Medic.* pp. 133*b*–6*b* and below, p. 146.

2 Cf. Gesner, *H.A.* III, p. 227 (on Greenfinch); p. 244 (Fieldfare); p. 557 (Wagtail); p. 567 (Goatsucker); p. 607 (Partridge); p. 683 (Fieldfare); these in addition to others given by Evans, *Turner on Birds*, pp. xi–xiv.

3 Turner's other reference to him, *Herbal*, III, p. 29, looks like a reference to his posthumous book, *Historia Stirpium*, published by the influence of Gesner in 1561. In this book, p. 113, he names Hartstongue 'Phyllitis' but says that it is also called Hemionium.

foxe which more than seven yeares hath bene hyd among the bysshoppes of Englonde, was printed under the pseudonym of William Wraghton. It was dedicated to Henry VIII on the date 1 May 1543; and is a vigorous argument in twenty-nine sections against Roman doctrine and practice. It rehearses the chief counts of the Protestant indictment—images, communion in one kind, canon law, the celibacy of the priesthood, and the rest—and states the Scriptural case against them. The language is outspoken, but not intemperate, and by comparison with that of many other controversialists on both sides fair in presentation. That it was carried into England and attracted some attention is clear from the fact that Stephen Gardiner, then Bishop of Winchester, published a reply to it. This is mentioned in his *Declaration* against George Joye and is quoted very fully in Turner's answer.[1] From this it is printed by J. A. Muller, *Letters of S. Gardiner*, pp. 480–92, but no original copies of it survive. It seems to have been written in 1544; for as we shall see Turner published his answer in March 1545. Turner quotes it by the title *The Examination of the Hunter*.

From Basel he went up the Rhine to Cologne, perhaps travelling slowly and with halts at Bonn or elsewhere. How much of the botanising in Germany that is recorded in the *Herbal* belongs to this period is not easy to decide. Some references in the *Names* must plainly do so: 'Alcea' (*Malva alcea*) 'at Bon [Bonn] by the Rhene side'; 'Apios' (*Conopodium denudatum*) 'betwene Redkyrke and Colon'; 'Asplenum' (*Ceterach officinarum*) 'besyde Embis [Ems] bath and besyde S. Goweris [St Goar]'; 'Germander' (*Teucrium chamaedrys*) 'over agaynste Byng [Bingen] besyde Erenfielde [Ehrenfels]'; 'Ervum' (*Vicia ervilia*) 'aboute Mense [Mainz]'; 'Peucedanum' (*Trinia glauca*) 'beside Erensfield over agaynste Byng and also in the middowes beside Mence, called other wyse Maguncia'; 'Napus' (*Brassica napus*) 'plenteously at Andernake [Andernach]'. Bonn is much the most frequently mentioned: 'plentuously aboute Bon' occurs many times: *Meum athamanticum* 'eighte myle above Bon in a fielde besyde Slyde' and *Scandix pecten* 'betwene Bon and Popelsdorp in a corne fielde' are cases in which details are added. The *Herbal* of 1551 adds of 'Colocasia' (?*Nelumbium speciosum*) 'three Englyshe myles from Bone besyde Seberge [Siegberg]' (p. L. iiii); and of 'Hethe Cypres' (*Lycopodium alpinum*) 'in a heathe beyonde Bon in the syde of a mountayne' (p. J. iv). He further reported in a letter to Gesner (*H.A.* III, pp. 214, 567) that he had 'seen the Goatsucker [*Caprimulgus europaeus*] at Bonn where it was called "naghtrauen"'. Surely, Turner must have stayed some time in the neighbourhood. And at Cologne he certainly practised medicine; for under 'Myrica' (apparently *Tamarix gallica*) he writes: 'The Poticaries of Colon before I gave them warning used for thys the bowes of ughe.' It is probable from references under 'Cytisus' (*Names*,

1 *The Rescuynge and Second Course*: see below, p. 91.

C. vii and *Herbal*, I, p. 197), Heliotropium (*Names*, p. D. iv*b* and *Herbal*, II, p. 13*b*) and 'Siliqua' (*Names*, p. G. iii and *Herbal*, II, p. 136*b*) that he had a garden at Cologne on this visit as he certainly had in his second exile.

Of his stay at or near Cologne the clearest evidence is in the account of Epimedium included in the *Herbal* of 1551 but not in the later edition. The record is worth quoting at some length. 'This herbe is strange and yll to fynde: how be it I found a certaine herbe in Germany besyde the Byshop of Colens place: called Popelsdop, by a brokes syde very well agreeyng in all poyntes savyng in one with the description that Dioscorides maketh of epimedum. In the moneth of July I sawe thys herbe havyng ix or x leves comyng out of an roote...and at that tyme I coulde fynde in it nother floure nor fruyte. The nexte yere folowyng in the myddes of Marche in the same place, I founde the same herbe, wyth leves, stalke and floure, lyke unto wylde valerian, and twoo handbredes from that place I found two or thre leves lyke unto violettes commyng out of the same roote, so that out of the one end of the roote came leves lyke violets, out of the other ende leves, stalke, and floures lyke Valerian. But Dioscorides describeth...' (*Herbal*, p. P. v*b*). It must be admitted that the wood-cut which accompanies this description does not bear a very clear resemblance to *Epimedium alpinum*: but the record proves that he was in the neighbourhood for at least the greater part of a year.

As regards the date of his stay the only direct evidence seems to be under 'Amomum' (*Anastatica hierochuntis*) in *Herbal*, p. C. iii*b*, 'I sawe about VI yeres agoo at colon a litle schrube'. This, if written in 1550, the latest possible date, would mean that he was at Cologne until 1544; and this fits into the time-table that we have suggested.

It was while he was in the city that his bird-book was written. This is made clear in a very interesting passage under the 'Onocrotalus' (*Avium*, pp. 101–2) where he is describing the anatomy of a Bittern which he dissected and identified as this bird. He says that he was engaged upon this when the first pages of his book were being printed, and was being helped by Johann Echthius (Echt),[1] a man of great learning, a zealous investigator of the secrets of nature, and a physician of great renown at Cologne; by Cornelius Sittardus, a doctor of medicine;[2] by M. Lubertus Estius, a professor of arts; and by Conrad Embecanus, a man of sound learning and a proof-reader in the press of Gymnicus.

1 Aldrovandi, *Ornithologia*, I, p. 108 (ed. 1610), speaks of him as a famous doctor and records that his son, Johann Bachovius Echt, had told him a story of fishermen using Heron's fat, as English fishers used that of the Osprey, to smear on their bait. He is probably the Johann Echt who matriculated at Cologne in 1500: cf. H. Keussen, *Kölner Univ. Matrikel*, II, p. 496.

2 'Cornelius Sittardus of Cologne', then described as a student of medicine, was at Bologna when Valerius Cordus was there, as Schreiber states in the letter printed by Schmiedel (Nuremberg, 1753): this must have been in 1542 when, according to Johann Crato, Cordus arrived in Italy, shortly before the present occasion. Gesner (*Ep. Med. passim*) usually spells the name Sighartus (Sieghardt).

It was by Johann Gymnich that this book, his *Avium praecipuarum quarum apud Plinium et Aristotelem mentio est, brevis et succincta historia*, was published.[1] Dedicated to Prince Edward, the long and fulsome introductory letter contains a point of historical interest. In complimenting Henry VIII on his care for the education of his children he states that Henry appointed George Folbery as instructor of the late Duke of Richmond, Henry Fitzroy who had died in 1536. This is mentioned as a report by Cooper, *Athenae Cantabrigienses*, I, p. 76, but dismissed as unsubstantiated by the author of the notice of Folbery in the *D.N.B.* Turner, whose intimate acquaintance with Folbery has already been described, can hardly have been mistaken on such a fact, especially as in the same sentence he alludes to Edward's own tutor, 'a man of preeminent learning', who was in fact his friend Richard Cox.[2]

Scarcely less interesting though for a different reason is the account in the Peroration of the book itself. Turner begins by acknowledging that the reader may well criticise him for indulging in conjectures rather than affirmations; and for saying nothing of the habits or medicinal uses of birds. He defends himself by pointing to the unfamiliarity of the subject—on which indeed almost nothing had been published since Pliny—and to the circumstances of his life. He was in ill-health: he had no money for travel or for investigation: he had no time—the whole book had been written in two months. So he pleads that men of wealth and influence may give encouragement to the study and that meanwhile any who have knowledge or specimens will inform him. In fact he need not have been so apologetic: the book is an excellent piece of pioneer-work: most of his identifications are sound and all are careful; his descriptions especially of habits are very good; if he includes the Phœnix, it is a bare quotation from Pliny; if he accepts the Barnacle Goose, it is on evidence that he has tried to check; and unlike his friend Gesner he refuses to include the Bat.

At beginning and end of his book he indicates his intention of further publication on the subject. Birds evidently interested him; and although his intention was never fulfilled there are plenty of allusions in his later works which show the maintenance of his observations. He must surely at some time have kept small birds in cages; for not only does he describe in the *Avium*, p. 86, how Goldfinches[3] draw up the little pots containing

1 Reprinted by G. Thackeray in 1823; edited with translation and notes by A. H. Evans, *Turner on Birds*, Cambridge, 1903.

2 John Cheke succeeded him in July 1544. The preface of *Avium Praec.* is dated 9 February in that year.

3 This detail is given under the section on the Greenfinch, but the Goldfinch is mentioned and probably meant; so Ray, *Ornithology*, p. 257, referring to Turner. Cages with food and water vessels drawn up on a cord are still not uncommonly used for goldfinches, redpolls and siskins.

their food or water, but in the *Herbal* there are many allusions to their welfare. 'If any man were desyrous to fat or fede in cages any small byrdes, it were good to sow good plenty of panic and millet' (II, p. 76*b*); and under 'Sesame', 'Men fede byrdes wyth the sede of it, these namelye, syskennes and linnettes and golde finches and byrdes of Canaria' (II, p. 134*b*); and, delightfully, of 'Chikwede', 'They that kepe lyttle byrdes in cages, when they are sycke, gyve the byrdes of thys herbe to restore them to their health agayne' (*Herbal*, p. B. vii). So, too, there are more general notes—of Teasel, 'In the begynnyng of wynter the golde finches use mych to haunt this herbe, for the sedes sake whereof they are very desyrus' (*Herbal*, p. O. v); of Yew, 'The byrdes that eat the berries of the Italian Ughe are made black' (II, p. 150*b*), and of Juniper, 'a lytle from Bon where as at the tyme of yeare the feldefares fede only of Junipers berries [sweet and strongly aromatic in flavour], the people eate the feldefares undrawen wyth guttes and all, because they are full of the berries of Juniper' (II, p. 25).

To the book's records of English birds allusion has already been made; and the best example of his powers as an observer—his account of the Great Grey Shrike (*Lanius excubitor*) based on his knowledge of it in Germany—will be cited when we come to consider his ability as a naturalist. Here we need only comment upon his account (pp. 75–6) of the Waltrap of Switzerland which he had seen and handled, presumably when he was crossing the Alps—'long bodied, a little smaller than the stork, with short thick legs, a red bill slightly curved and six inches long, a white spot on the head, and this spot so far as I remember naked of feathers. If it be web-footed and sometimes swims I should unhesitatingly declare it to be the third kind of Graculus [i.e. as described by Aristotle, *H.A.* VIII, 48 and mentioned by Turner]; but though I had the bird in my hands I do not remember whether it was web-footed and bald: so until I have fuller knowledge, I make no decision.' This passage anticipates the account of the 'Wood Raven' given by Gesner; and plainly refers to the same bird. That this bird is the *Ibis comata* (*Comatibis eremita*) has been demonstrated in a full survey of the evidence in *Novitates Zoologicae*, IV, pp. 371–7, by W. Rothschild, E. Hartert and O. Kleinschmidt. It is a bird which has long ceased to nest in Europe, but is still found from the Euphrates to the Sahara. It nests, as Gesner describes, in old ruins and cliffs, has a long red beak, an iridescent black plumage, and a crest such as would justify Turner's assumption that it was the ancient Phalacrocorax. It is an insect-feeder, but like other Ibises probes the mud of marshes for worms. In immature plumage—and presumably Turner handled it when taken from the nest in the way told by Gesner—it has white spots on the feathers of head and neck, and like Gesner's picture is not bald-cheeked. From his account it appears that its nesting-colonies were known from the Danube and Italy as well as from Lake Maggiore and Switzerland. There seems to be no evidence as to when or why it disappeared from its sixteenth-century haunts.

Another European bird which he records is the second of the Kites. The Red Kite (*Milvus milvus*), p. 95, was then of course common in England: he says here, 'it often snatches food from boys' hands in our cities and towns'; and in his *Spirituall Physic*, p. 54, gives an excellent account of its nesting-habits and its food. But 'the other kind, smaller, blacker and more seldom seen in cities [*Milvus migrans*] I do not remember ever to have seen in England though very frequently in Germany'. So, too, the Golden Oriole (*Oriolus oriolus*), p. 142, whose 'round nest, slung on a branch at the top of a tree' and 'note like that of the big flute which takes the bass part in accompanying' he describes evidently from knowledge, and the Hoopoe (*Upupa epops*), p. 143, whose 'crest running lengthwise from beak to nape it depresses or erects at pleasure as a horse pricks up or sets back its ears', are both birds seen in Germany very often but never in Britain.

In the same year he is said to have prepared for the press the *Dialogus de Avibus et earum Nominibus Graecis*, the last work of the brilliant Dutch scholar Gisbertus Longolius (Gilbert de Longueil), who died at Cologne in 1543 at the age of 36 after his return from Rostock where he had been offered a professorship. Turner's precise connection with the book is not easy to discover: but as it was published by Gymnich and dealt with a subject so close to his own it is probable that he merely put together and corrected a script which even as printed is incomplete.

Nevertheless, his stay in the Rhineland was not on this occasion much prolonged: for he certainly must have gone on to East Friesland early in 1544, since he stayed there 'ad integrum quadrennium', for a full four years.[1] This is a point mentioned in his letter to Gesner on Fishes, 'In East Frisia in which I spent four full years I bought two whole Porpesses and dissected them': it seems to have been unnoticed by Jackson and other students of Turner. The longest account of his stay is in *Herbal*, III, p. 13, where under Bistort he gives the story of an attack of scurvy ('scruby ill') and consequent loosening of teeth among the Frisians. 'When I was the Erle of Emden's Phisitian[2] [Enno II who died in Sept. 1546 and was succeeded by his widow Anna as guardian of his six children] which is the Lord of East Friseland, divers asked me councell for that disease, and by the helpe of God I did heale them, and perceived that it was the verye same disease that Pliny spake of...that the inhabitores of

1 Cf. *Spirituall Physic*, p. 3, 'in low Germany where as I have ben moste'.

2 Ubbo Emmius, the historian, in his *Rerum Frisicarum Historia*, 6th Decad. Bk. LIX, p. 354 (ed. Emden, 1616), says that he lived in the house of the widowed Countess.

Freselande taught the Romaines to heale their disease with a herbe called Britannica.' So he looked for this herb 'with greate payne, and also with some cost by al the sea syde of East Freselande....Then I sayled to an Iland called Iuste, and there I found it not, and after that I sayled to an other Iland called Nordeni [Norderney], and sought there also but I could not find it....I wente to seke it in medowes and woddes, and in a woode I found Bistorta in such great plentye as I never sawe in no gardine of England....I judge that Bistorta is the Britannica[1] that he maketh mention of, and it is surelye proved by experience that Bistorta healeth the disease.' Under 'Cepea', a plant that includes 'Brooklyme' (apparently *Veronica beccabunga*) and 'Sea Porcelline' (*Atriplex portulacoides*), he records a similar success with the former: 'I have proved it my selfe by experience, that brooklyme is very good for a disease that reigneth much in Freseland called the Scourbuch. I sod the herbe in butter milk, the chese and butter taken away, and gave the pacientes it so. I made them eate it diverse ways whereby they were within a shorte tyme healed', and he found the latter 'oft in Freselande by the sea syde within the sea bankes, in such places as the sea commeth to every springe tyde' (*Herbal*, I, p. 123, cf. p. J. iii*b*). His practice was evidently well appreciated: 'I knowe two of my pacientes to whome Cassia was a poison...one was a gentle man of Freseland, the Jonker of Aldersham' (*Herbal*, III, p. 21).

He evidently had an interesting time medically as well as botanically, covering a large area and being thrown onto his own resources. He describes in *Herbal*, II, p. 79*b*, one successful experiment with Feverfew (*Chrysanthemum parthenium*) 'with a sike woman and the present necessite required purgation; and there was no potecaries shop at hand (for there is but one citie in all East Fresland where there are any potecaries in, and that is called Emden)'. This solitude and the consequent need for ingenuity brought him through an adventure typical of the sort of risks run by physicians in his day. Quoting 'Symeon Sethy a later Grecian' to the effect 'that opium is poyson' (*Herbal*, II, p. 78), he adds, 'For allthoughe sum be very bold in occupying it, I taught by experience how jeperdus it is, dare not without grete warnes geve it in to the body. For once in East Friesland, when as I wasshed an achyng tooth with a little opio mixed with water, and a litle of the same unawares went down, within an hour after my handes began to swell about the wrestes, and to itch, and my breth was so stopped, that if I had not taken in a pece of

1 Amatus Lusit., *In Diosc. Enarr.* p. 373 (Venice, 1553), repeats this identification, saying that he has got it 'by enquiry from learned men' (Amatus never quotes or names Turner). For this identification Mattioli, *Apol. adv. Amathum* (ed. G. Bauhin, p. 36), condemns him: it was not generally accepted; for Gesner, De l'Obel, Dodoens and others identify Pliny's 'Britannica' with *Cochlearia officinalis*.

the roote of masterwurt [*Peucedanum ostruthium*], called of som pilletory of Spayn with wyne, I thynck that it wold have kylled me.'

His visit to the islands gave him several plants. Of 'Spartum' (*Ammophila arundinacea*), the bent of which hats were made in Northumberland (see above, p. 50), he says, 'I have also sene it in ii ylandes of Easte Freslande whereof the one is called the iust and the other nordenie, there men use thys rishe only for to make ropes of it and to cover houses wyth it' (*Herbal*, II, p. 144*b*); of 'Smallage' (*Apium graveolens*), 'I have sene it growe oft tymes by brook sydes and in a certayn yland of East Freseland, called Nordenye' (*Herbal*, p. P. i); of 'Limonium' (*Pyrola rotundifolia*) 'in an Iland of east Freseland called nordeney' (*Names*, p. E. i*b*); of 'Prickwylowe' (*Hippophäe rhamnoides*) 'in the Ilandes of east Freselande where as the inhabitants make veriuce of the red berries' (*Names*, p. D. iiii*b*); and of 'Sothernwod' (*Artemisia abrotanum*), one of the Wormwoods whose identification gave him such trouble, 'I truste I shall take that dout awaye, for when I walked by the Sea syde in East Freseland ["at Norden", *Names*, p. A. iiii*b*], in the springe of the yeare, I founde in diverse places of the Sea banke Sea Wormwod, as yet uncummed furth, but only beginning to spruyt out' (*Herbal*, I, p. 14). He records also *Asparagus officinalis* and *Laminaria saccharina*.

On the mainland he records 'Carawayes in greate plentye in Freseland in the medowes ther betwene Marienhoffe and Werden' (*Herbal*, p. H. iiii*b* and I, p. 111). Under 'the oke tre' is a very interesting note, 'I have not sene any galles in England growing upon oke leves. But I have sene them not only in Italy, but also in great plenty in East Fresland, in a wod a litle from Aurike. The galles of Italye come to perfection, and are at leyngth harde, but they of Freslande, beyng ones taken with cold wether and moyste are never harde but soft. Not withstanding I have proved that they serve well to make ynke of....If there be such diversite of okes that som will beare galles, and som will beare none; it were wel done, to fetche som from Freslande and to plant them in som hote sunnye place of England, to se whether the ayr of the countre or kind of tre or no is the cause that galles grow upon som oke leves, and not upon other som' (*Herbal*, II, p. 109)—a suggestion which aptly illustrates both the acuteness and the limits of his knowledge. Near Aurich he also found 'Raspeses [*Rubus idaeus*] plentuously in the woddes' (*Names*, p. F. vii*b*) and 'Salix greca' (*S. vitellina*, *Herbal*, II, p. 125*b*). A further note in the *Herbal*, III, p. 37, on 'the herbe called Kali' (*Salsola kali*) raises a new subject: 'The older that the herbe is, the longer are the leaves; at the lenght growe oute rounde knoppes, wherein are verye smal sedes, whiche the Larkes in East Freseland eate in winter.'

During this time he was also working at his Latin Herbal, the task which he had undertaken in the dedication of his bird-book in 1544 and which occupied him, as he states in the dedication of his *Names of Herbes*, for two full years.

Early in his stay in Friesland he must have finished his dialogue[1] entitled *The Rescuyng of the Romishe Fox otherwyse called the examination of the hunter aevised by Steven Gardiner and The Seconde Course of the Hunter at the romishe fox and hys advocate and sworne patrone Steven Gardiner*; for this is dated in March 1545, and published as from Winchester 'by me Hanse hit prik'. It was probably printed at Emden and smuggled over to England for distribution. Once again its author calls himself Willyam Wraghton: but this time his work is thrown into the form of a discussion, the author speaking his own views under the role of the hunter against Stephen Gardiner the rescuer. He quotes Gardiner's work paragraph by paragraph and answers it, fastening especially upon the weakest point in his opponent's position when he attacks him for maintaining the full Papal doctrine and practice while repudiating the Papal supremacy. But he does not confine himself to Gardiner's words. His indictment ranges over a wide field, and is outspoken and uncompromising. On the back of the title-page is a fox with a crozier and cropped ears, and lines headed, 'The bannished fox of rome speakethe'; and Strype's comment that 'he delivered his reproofs and counsels under witty and pleasant discourse' is very relevant. Though some of his criticism is too direct for modern taste, he is less abusive than Gardiner,[2] and seemly and considerate as compared, for example, with the contemporary and anonymous pamphlet, *Yet a course at the Romyshe Fox compyled by John Harryson*—probably by John Bale; or, indeed, with Erasmus or Luther or even, on occasion, Sir Thomas More.

More important than his book was the friendship that he made at this time with John a Lasco (Jan Laski), the Polish baron then living in Emden under the protection of the Countess Anna. A Lasco, trained at Bologna and Basel, and accepting the principles of the Reformation, had eventually been banished from Poland and had gone to East Friesland in 1542. There he had done a remarkable work and become one of the most eminent of Continental Protestants. Having enjoyed the acquaintance of Erasmus, of Zwingli and of Melanchthon he represented a different and in some respects more radical and liberal doctrine than that of Luther; and it seems that he and Turner found that they had very much in common. Their meeting had far-reaching consequences upon the development of religion in England.

1 A. Koelbing in *Cambridge History of English Literature*, III, p. 81, suggests that Turner borrowed this form from the German *Fastnachtsspiele*.

2 'Whether a man call hym fool, proud, arrogant, glorious, disdaynful, spitful, haytful, unlearned, untaught, busy, partly lyer, wrangler, seditious, malicious, or many other of that sort, he can not speake amiss.' Gardiner's *Examination*, *Letters*, p. 492.

From Friesland he seems to have moved to Gröningen though his stay there may have been short. It is alluded to when he is discussing the properties of 'Scurby wede' (*Cochlearia officinalis*) in the revised note on 'Brassica marina' in *Herbal*, I, pp. 90–1: claiming that he has proved it good not only against scurvy but against dropsy, he adds, 'the noble clerke Reinerus the Rector of the scole of Gruninge [the University was not founded till 1614], and Henrike Herbart can bare witnes'.

B. D. Jackson, whose survey of the place-names in the *Herbal* is a useful piece of work, places the visit to Louvain, Antwerp and Brabant after these years in Friesland. Certainly he went there during his first exile, and after his Italian travels and stay in Cologne; and it is a reasonable guess that when the news of Henry VIII's death came to him and with it the hope of a return to England he may have moved nearer to the Channel, first to Brabant, then to the town of Bergen op Zoom from which his letter to Gesner about the bird called 'Culicilega' (*H.A.* III, p. 557) seems to have been written.

On the way he certainly called at Malines and saw the Pelican which had been kept alive there for some fifty years. Of this he sent a brief but adequate description to Conrad Gesner; and this is quoted along with one of the same bird from Johann Culmann of Goppingen in the *Historia Animalium*, III, p. 570, and thence by Aldrovandi, *Ornithologia*, III, p. 46.

That the visits to Louvain and Antwerp were closely connected is clear from an interesting reference under 'Adianthum' (*A. capillus-veneris*) in *Herbal*, p. B. iii: 'Many erroures have ben about this herbe. I have sene some Pothecaries in Antwerpe use for this herbe Dryopteris, in Louan other use Walle Rue....And our Pothecaries in England use Trichemanes whiche they calle maydens heyre.' Another reference (p. J. v*b*) under 'Chamepeuce' deals also with 'the Pothecaries of Louane'. Probably he only went over for a brief visit: is it possible that he went to see Rembert Dodoens?

In Antwerp we hear chiefly of the famous garden of Pieter Coudenberg,[1] 'a faythfull and lerned apothecari', where he saw many rare and interesting plants: *Artemisia ?arborescens*, 'a bushe or two of this kinde groweth in Antwerpe in Peter Coudenberges garden where as are manye

1 His shop was at the sign of the Old Bell—so Gesner, *Horti Germ.* p. 243: his garden, then recent, was at Burgerhout—so De l'Ecluse, *R.P.H.* p. xvi. In 1568 Coudenberg, dedicating the edition of the *Dispensatorium* of Valerius Cordus which he prepared for Plantin, speaks of himself as having been a gardener for twenty years and as growing 'six hundred kinds of exotic plants'. The notes by him headed with his initials deal mainly with the names and identification of Cordus's plants and only occasionally mention localities or his garden (e.g. p. 60). They show good knowledge of the literature, Fuchs, Mattioli, De l'Obel and others. Before its publication further notes were added after Coudenberg's death by De l'Obel: cf. pp. 134–6.

other straunge and holsom herbes, hard to be found in anye other place of Germanye besyde'[1] (*Herbal*, I, p. 9); 'whyte Henbane' (*Hyoscyamus albus*) (*Herbal*, I, p. 31); 'Oregan' (Origanum) (*Herbal*, II, p. 69); 'Male Peony' (*Peonia corallina*) (*Herbal*, II, p. 84*b*); and 'Vitex' (*V. agnus-castus*) (*Herbal*, II, p. 165*b*). He alludes also to another garden, saying under 'Sison' (*Names*, p. G. iii*b*), 'The best kinde groweth in Anthony the Poticaries gardine in Anwerp'. Of a third Antwerp herbalist he is less complimentary: of 'Mew' (*Meum athamanticum*) he writes, 'I saw it also ones in Antwerp, in Apothecaries gardin, but the pothecari named Petrus de virulis called it peucedanum, not without a greate error' (*Herbal*, II, p. 57). Otherwise he has little to report: 'I have sene Papyrus diverse tymes in Anwerp wherein was sugar and diverse other merchaundise wrapped' (*Names*, p. F. i); 'I have sene herbe Aloe also in Antwerpe in shoppes and there it endureth long alyve as Orpyne doth and houseleke' (*Herbal*, p. B. vi). Probably he actually sailed from Calais; for the last note in the *Names*, p. H. v*b*, under 'Vulvaria' (*Chenopodium vulvaria*), is 'I did se it also in my Lorde Cobham's gardine at Calice', and George Lord Cobham was deputy there from 1544 to 1550.[2] Perhaps it was on his return that he noticed that 'Chestnut trees growe plentuously in Kent abrode in the feldes' (*Herbal*, p. H. vi).

CHAPTER VI. WILLIAM TURNER IN MATURITY

Turner came back to England a marked man. In the last year of Henry's reign when the influence of the Crown was exerted against all champions of the Reformation, on 7 July 1546,[3] his books on religious subjects had been condemned by name; and their author thus publicly branded as a champion of Protestantism. As such, or more probably owing to his connection with the Wentworths, he was offered a position as physician

1 This record is quoted by J. Bauhin, *De Plantis Absinthii*, p. 38; to it Coudenberg referred in a letter to Gesner, quoted in Gesner, *Horti Germaniae* (Strasburg, 1561), p. 244: he wrote: 'I gave a sprig of Roman Wormwood formerly to William Turner who inserted its picture in his English Herbal: but then the plant died for me without seed, nor could I for all my care recover it. The whole structure of it was like Common Wormwood, but the leaves were a little smaller. I had received it from Rome.' This description hardly agrees with Turner's l.c.: for he says that Coudenberg's plant had longer, whiter and bitterer leaves than the Common.

2 Cf. *Chronicle of Calais* (Camden Soc. XXXV), p. xxxviii.

3 *Wriothesley's Chronicle* (Camden Soc. N.S. XI), I, p. 169.

and perhaps chaplain in the household of the Lord Protector, Edward Seymour Duke of Somerset; was incorporated as a Doctor of Medicine at Oxford; and took himself and his family to a house of the Duke's at Kew[1] belonging to his Shene property[2] and opposite Syon House, his magnificent seat on the Thames between 'Thistleworth' (Isleworth) and Brentford.[3] Here, as he describes it in the preface to the first *Herbal*, published in 1551, 'I have more than iii years bene a dayly wayter ["in attendance"—in the sense still preserved in Lady-in-waiting] and wanted the chefe parte of the day most apte to study, the morning, and have bene long and sore vexed with sycknes....For these thre yeares and a halfe I have had no more lyberty but bare iii wekes to bestow upon the sekyng of herbes, and markyng in what places they do grow.' Yet he did a great deal of work, and at first was not unhappy.

There was in fact much to encourage him. Somerset if rapacious was a man with a rare power of calling out affection; genuinely concerned to protect the poor from oppression, to mitigate the savagery and intolerance of the times, and to secure liberty of conscience. His household at this time included William Cecil, his secretary, Thomas Becon—Turner's Cambridge contemporary—his chaplain, and the young Thomas Norton, afterwards Cranmer's son-in-law, his amanuensis.[4] Nicholas Grimald, Ridley's chaplain, and Thomas Some, who edited Latimer's sermons in 1549, were in the neighbourhood. Randall Hurleston, a young fellow of King's College, and Robert Thomlinson were also friends of Turner at this period, who with the others contributed complimentary verses to his *Preservative* in 1550. For a year or two it may well have seemed that his dreams were coming true, and that the advocates of reform in religion and of social justice in political and economic life were on the road to success.

From a passage in his *Spirituall Physic*, p. 44*b*, where there is an account of his having seen 'gentlemen of the fyrst head slepe whylse learned men dispute in the parlyment house when I was a burgesse of late of the lower house', and from the fact that in *The Huntyng of the Romyshe Wolfe*, p. A. 4, the Hunter who plainly represents Turner says that he has been a member of the Lower House for five years, it is clear that after his return Turner sat in the Commons. B. D. Jackson argues that he was not in the Commons but in Convocation on the ground of his clerical status: but although he had been given a licence to preach by his College and been ordained deacon in 1536, he had married, studied and practised

1 He wrote to Cecil from here in June 1549.

2 The Carthusian Priory at Shene had been granted to Somerset on its dissolution.

3 Formerly a convent, and given by Edward VI to Somerset, Syon House had been built and its gardens laid out by him. He was interested in plants. See *Annals of the Seymours*, p. 390.

4 Cooper, *Ath. Cantab.* I, p. 485.

medicine, and lived as a layman, and was not ordained priest until 1552; and there is a reference to Convocation in *The Romyshe Wolfe*, p. A. v*b*, which shows that it was not his place. Foster, *Alumni Oxonienses*, states[1] that he was member for Ludgershall (Wilts.), probably through Somerset's possession of Amesbury. In parliament he was doubtless one of the Commonwealth men led by Hugh Latimer, John Hales and Thomas Lever whose principle that 'it may not be liefull [lawful] for every man to use his own as hym lysteth, but every man must use that he hathe to the most benefyte of his Countreie',[2] he cordially supported; and whose indictment of existing evils, and programme that the riches of the monasteries ought to be devoted to education and the building of schools, as set out by Henry Brinklow in his *Complaynt of Roderick Mors*,[3] or by Robert Crowley in his *Informacion and Petition*,[4] he advocated in the two books above mentioned. When in May 1548 Somerset issued his proclamation on Enclosures and appointed his Commission, it seemed that at last the spoliation of the peasantry was to be noticed if not relieved by the Crown.

It was probably at this time, when Somerset's policy was not yet frustrated by Warwick, that Turner had the honour of a meeting with the Princess Elizabeth and of being addressed by her in Latin. In the dedication to her of his final *Herbal*, dated 5 March 1568, he recalled this event in the words 'as for your knowledge in the Latin tong xviii yeares ago or more, I had in the Duke of Somersettes house (beynge his Physition at that tyme) a good tryal thereof, when as it pleased your grace to speake Latin unto me'.

One of the first acts of Turner after his return was to bring to the notice of the Protector and of the Archbishop the work of his friend John a Lasco in East Friesland, and to urge that he be invited to come over to England.[5] Somerset wrote to the Countess Anna begging her to give him leave to come. Cranmer on 4 July 1548 sent an equally warm invitation to him in person and urged him to persuade Melanchthon to come also.[6] A Lasco accepted the call, travelled secretly to Calais, crossed to England early in September, and stayed with the Archbishop at Lambeth. He visited Windsor, corresponded with Peter Martyr then working

1 No authority is quoted for this.

2 Cf. Lansdowne MS. 238, quoted by A. F. Pollard, *England under Protector Somerset*, p. 216.

3 Cf. Early English Text Soc. edition, pp. 48, 52, and Pollard, l.c. p. 215.

4 Cf. Strype, *Memorials*, II, i, pp. 217–20 and E.E.T.S. *Select Works of R. Crowley*, pp. 151–76.

5 The primary evidence is Ubbo Emmius, *Rerum Frisicarum Historia* (Emden, 1616), Bk. LIX, p. 354. It is set out by H. Dalton, *John a Lasco* (tr. by M. J. Evans) (London, 1889), pp. 361–77; F. de Schickler, *Les Églises de Refuge en Angleterre*, I, pp. 9–12; G. Pascal, *Jean de Lasco* (Paris, 1894), pp. 194–256.

6 Gabbema, *Epistolae* (ed. 1669), pp. 108–9 and *Original Letters*, I, pp. 16–17.

at Oxford, and spent some six months in the country before sailing from Yarmouth to Emden. His expenses were paid by the Privy Council who granted him £50;[1] and he was presented by Turner to Somerset[2] and so to the King. On his departure he sent a long letter of thanks to Cecil with a special message of greeting to Turner.[3]

In May 1550 he came back with his wife and children and a band of German Protestants; and in July received from the Crown by letters patent permission to establish the Church of the Strangers in London and to become its superintendent. In 1552 he issued his *Brevis de Sacramentis Tractatio* printed by the publisher of Turner's *Herbal*, Stephen Myerdman, a member of his own congregation and friend of the Dutch printer Egidius van der Erve, with whom also Turner became acquainted and to whom after his banishment from England he sent his last two controversial writings.[4] A Lasco stayed in England until the expulsion of the Continental Reformers three days after Cranmer's arrest. Of his influence upon the Archbishop's doctrine and work and so upon the shaping of Anglican theology this is not the place to speak: it was sufficiently large to make Turner's action in bringing him to England a matter of high historical importance.

Otherwise Turner's public life seems to have been disappointing. Looking back on it in the passage of *The Romyshe Wolfe* already quoted he could only write: 'In all my tyme (althoughe there were good actes made for the establishing of religion) yet there was always some that either sought theyr owne private lucre as the noble lordly and knightly shepe-masters dyd...or else sought verye earnestly the Kinges profit wherin they intended alwayes to have not all the smallest parte.' The only other allusion to definite legislation is on p. E. vi of the same book, 'There was an act of parliament made for the destruction of rookes which destroied the corne'—an early example of the sort of law in which he must have been interested. But otherwise he seems to have been disappointed in the course of public policy and bitterly hurt by the greed and ignorance of the legislators. As we shall see, his last book on a political or religious subject, his *Spirituall Physic*, is a strong and sustained warning to the nobility and gentry of England as to the diseases from which they are suffering and the remedies which must immediately be taken if the whole community is not to suffer disaster.

He had himself brought from Friesland the Latin Herbal which he intended to publish. But when he discussed the project with physicians,

1 Afterwards he was given £100 a year.

2 Cf. Strype, *Cranmer*, p. 336.

3 Dated 5 April, Emden: printed by A. Kuyper, *J. a Lasco Opera*, II, p. 622.

4 So F. S. Isaac, *The Library*, 4th series, vol. XII, pp. 341, 343, quoted by C. H. Garrett, *Marian Exiles*, p. 315.

presumably Thomas Wendy and the others whose names he recorded, they advised delay 'tyll I had sene those places of Englande, wherein is moste plentie of herbes, that I might in my herbal declare to the great honoure of our countre what numbre of sovereine and strang herbes were in Englande that were not in other nations'. So publication was postponed—indeed was never undertaken; and Turner set himself to wait 'tyl that I have sene the west countrey which I never sawe yet in al my lyfe'. Meanwhile he studied the gardens of Syon House,[1] of the famous Dr Richard Bartlot (cf. *Herbal*, p. A. vi*b*)[2] and 'of the barbican in London' (*Herbal*, p. K. ii); observed the plants of the neighbourhood, 'Ornithigalum or dogleke' (*Ornithogalum umbellatum*) 'besyde Shene herde [Sheen "hard" or quay] by the Temmes syde'; 'Cammomyle' (*Anthemis nobilis*) 'viii myle above London in the wylde felde, in Rychmonde grene, in Brantfurde grene and in mooste plenty of al, in Hunsley [Hounslow] hethe' (*Herbal*, p. D. ii); 'Herbe Ive' (*Plantago coronopus*) 'muche aboute Shene in the hyghe waye' (*Herbal*, p. M. ii); 'Cirsium' (*Helminthia echioides*) 'abroade in the felde in great plentie betwene Sion and Branfurd'; 'Paulis Betony'[3] (*Veronica serpyllifolia*) 'in dyverse woddes not far from Syon' (*Herbal*, p. F. v); and had a garden of his own at Kew in which he grew Ciche (*Cicer arietinum*) (*Herbal*, p. K. ii).

To this time belong the further reference to Lord Cobham's gardens in *Names*, p. G. iv*b*, of *Spartium monospermum*, 'it is founde nowe in many gardines in England, in my Lordes gardine at Shene, and in my Lorde Cobhams gardin a litle from Graves Ende' (with which compare *Herbal*, II, p. 144); and perhaps the delightful testimony to *Phalaris canariensis*, 'I saw it in England taken for mil, for they that brought Canari burdes out of Spayn brought of the sede of Phalaris also to fede them with. Whereof when I sowed a litle, I found that it was the right Phalaris which I had sene in Italy before' (*Herbal*, II, p. 85): is this the earliest reference to the import of caged Canaries?[4]

Pending the Latin Herbal 'certeine scholars, poticaries and also surgeans, required of me...at the least to set furth my judgement of the names of so many herbes as I knew'. There seems to have been nothing in English to help them to connect the traditional simples and herb-women's

1 He has been credited with the laying out and planting of his garden, cf. B. D. Jackson, 'Notable Trees and old Gardens of London' (*South Eastern Union of Science Socs.* 1917), where it is said that there still exist at Syon House Mulberries planted under Turner's superintendence (p. 62).

2 An original colleague of Linacre when the College of Physicians was founded in 1518; President 1527, 1528, 1531 and 1548. He lived in Blackfriars and died in 1557 (Munk, *Roll of R.C.P.* I, p. 23).

3 'Betonica Aeginetae secundum Guil. Turnerum', Gesner, *Hort. Germ.* p. 250.

4 It is nearly a century earlier than the first reference in *New English Dictionary*, II, p. 60.

lore of the countryside with the plants prescribed as medicines by their classical authorities; and even the common herbs regularly used in their drugs were often wrongly identified or replaced by quite different species. No wonder his friend Master Falconer who had studied at Ferrara and the doctors who were beginning to develop a knowledge of pharmacy were eager to make use of one who was familiar not only with the writings of antiquity but with the recent work of French, German and Italian herbalists. So he 'made a litle boke which is no more but a table or registre of suche bokes as I intende by the grace of God to set furth here after.... Thys conteineth the names of the most parte of herbes, that all auncient authours write of both in Greke, Lattin, Englishe, Duche and Frenche, I have set to also the names whiche be commonly used of the poticaries and common herbaries....I have shewed in what places of Englande, Germany and Italy the herbes growe, and may be had for laboure and money.' It was a 'litle boke' packed with information. As the preface was dated 1548,[1] March 15, he cannot have wasted much time over its composition.

The Names of Herbes in Greke, Latin, Englishe, Duch and Frenche wyth the Commune Names that Herbaries and Apotecaries use is in fact a condensed first draft of the *Herbal* which ultimately replaced it. It is like his previous *Libellus* arranged alphabetically in the sequence of the Latin names: but gives many more plants and much fuller information. The 'newe founde Herbes wherof is no mention in any olde auncient wryter' are placed separately at the end. It is dedicated 'to the mooste noble and mighty Prince Edward by the grace of God Duke of Summerset'— and nine lines more of titles; a souvenir of his patron's brief but splendid hour of supremacy. It is of course a classic in the history of English botany, the earliest recognisable record of at least 105 of our plants, over and above those already identified from his *Libellus*. The colophon gives John Day and William Seres of London as the printers. A reprint, edited with a brief introduction, an index and an identification of the plants mentioned, by J. Britten, was published for the English Dialect Society in 1881.

It must have been about this time and very soon after his return to England that he sent to Gesner a letter which discloses his interest in another department of study. In the first book of the *Historia Animalium* on p. 748 under the heading *Mus araneus* are the words: 'About this Mouse I will set out the words written to me in a letter two years ago [i.e. in 1548 or 1549] by William Turner, the very learned English doctor. "The Mus araneus", he writes, "I think that I have seen in England: it is called in our language a Shrow or a Shrew. It is almost black (with

1 This is no doubt the 'Old Style'; and the book's date of publication should be given as 1549.

us in Switzerland dark red-brown) with a very short tail and a somewhat long and sharp snout. I send you its head so that you can judge the better of it (he sent the head not of our 'mizet' which we have depicted here but of another mouse very small with teeth just like those of other mice, whereas the araneus is widely different as I show below). If we lay a curse on anyone we say 'I beshrew thee' that is I lay upon you this mouse or its bite or whatever evil there is in it." So says Turner.' From this and from the description and picture of the teeth of Gesner's Shrew (*Sorex araneus*) which are attached, it looks as if Turner had sent the head of a Harvest-mouse (*Micromys minutus*) instead of a Shrew—though his description is obviously of the latter. That he distinguished clearly between the two is proved by his words in *The Huntyng of the Romyshe Wolfe*, p. A. vii*b*, where he contrasts the 'blacke house mouse' with the 'feld mouse with the long snout much like unto a shrowe'.

In the same year he revised his first book adding a 'Preface of the Translator' and a few minor references to Catholic authorities, omitting the original preface, and publishing it as *The Olde Learnyng and the New compared together* (Robert Stoughton, London). He also wrote a preface for 'my scholar sumtime and servante' Robert Hutten's English version of *The Sum of Divinitie drawen out of the Holy Scripture*.

In 1549 this time of hope, security and activity came to an end with the failure of Somerset's efforts for the relief of the poor and with the consequent risings of the peasantry. From Kew on 11 June Turner wrote to Cecil a letter which strikes a note of disappointment contrasting with the optimism of the preface of his *Names*. He had been invited by Robert Holgate, Archbishop of York, lately married to Barbara Wentworth by the advice (as he afterwards declared) of the Protector, to go to him in Yorkshire. He explained to Cecil that he was very ready to do so or, as Somerset wished, to go to Winchester, 'if that I might have a lyvyng for me and myne there': but that he could not go on the bare chance that at some time a prebend might be available for him. 'The love that I bear unto my wyfe and childer will not suffer it. My chylder have bene fed so long with hope that they ar very leane. I wold fayne have them fatter if it were possible.' The 'good duke's' arrest on 12 October made these desires temporarily hopeless.

Nevertheless, in February 1550, at the time of Somerset's release, he was given the prebend of Botevant[1] by the Archbishop of York—though he had then to wait for license to preach. 'Whiche obteyned' (as he says in the preface of his *Preservative*) 'I began to rede and so to discharge mi conscience...I devised a lecture in Thistelworth'—and in consequence became entangled in a controversy on Pelagianism, apparently with a

1 Le Neve, *Fasti*, III, p. 176, says he held this for just over two years—until he settled down in Wells.

Robert Cowch[1] or Cook—so Cooper, *Ath. Cantab.* I, p. 207. His *Preservative or triacle agaynst the poyson of Pelagius lately renued and styrred up agayn, by the furious Secte of the Annabaptistes*, dedicated to Hugh Latimer, was printed for Andrew Hester in London, 30 January 1551.[2] He apologises at the end of this book for its incompleteness, on the ground that 'I have so much ado wyth the setting out of my Herball': but although the controversial portion, apart from an exposition of the patristic evidence for infant-baptism, is not of the highest value, the work is of interest in showing Turner's considered proposals for the improvement of conditions. These were very similar to those then advocated by Latimer in the remarkable sermons of this period,[3] and were developed in his later religious writings with fuller detail but on the same general lines. First, he wholly rejects 'fyre and faggot' as instruments for the restraint of error. Then he sees the danger of illiteracy and superstition, the lack of sound learning, the failure to replace the monastic houses and schools, the short-sighted greed with which their endowments have been squandered. So he urges the establishment of more schools, the provision of money for the encouragement of learning, and the restoration to the Church of means sufficient for the carrying on of its work.

In the dedication of this book after describing himself as one of Latimer's disciples he says 'though the stryfe agaynst Gods enemies were common to me as to many, yet I had specially to do with a bucke,[4] with a certayn man that had a name of the colour of madder,[5] with the fox and his foster,[6] with a certayn wytche called Maystres Missa'. These more or less cryptic allusions to his controversies plainly suggest that *A New Dialogue wherein is conteyned the examination of the Messe*—a book issued at first without date or place as 'imprynted by me Richarde Wyer' and then 'at London by John Day and William Seres' the printers of *Names*—had already been issued, perhaps by T. Raynalde, London, 1548. It must certainly be later than the deaths of Barnes and the others, and probably (if an allusion to seven years since he answered 'a learned man of England'

1 So the Acrostich in the prefatory matter spells his name.

2 At the beginning and end are a number of commendatory verses.

3 E.g. the Sermon of the Plough, preached at St Paul's, January 1548 (ed. Parker Soc., especially pp. 68–9); and those before Edward VI in March of the same year (l.c., especially pp. 119–20, 178–9, etc.).

4 Probably Henry Hart, author of two treatises in 1548 and 1549, and 'the principal of all those that are called Freewill men [i.e. Pelagians]...he hath drawn up thirteen articles to be observed by his company' (Foxe, *Actes and Monuments*, VIII, p. 384). He is the chief object of Turner's criticism in this book.

5 John Redman (1499–1551), fellow of St John's College, Cambridge, 1530; first Master of Trinity 1546; a supporter of Henry VIII and the Six Articles; no doubt known to and criticised by Turner; and Latimer's chief critic at Cambridge in Turner's time.

6 Cf. his *Huntyng of the Romyshe Fox*, 1543, and *Rescuyng*, 1545.

refers to the *Romyshe Foxe* and Gardiner in 1543) appeared first in 1550. Strype describing the events of 1548 speaks of it as appearing about this time.[1] Latimer referred to it in his Conferences with Ridley in 1556.[2]

The events of these years have been largely misinterpreted by the students of Turner's life. Even Jackson, generally sympathetic, represents him as sponging upon the courtiers for preferment and has gathered all the evidence not only of the letters in the Lansdowne papers but from Cooper and elsewhere to show how eagerly he sought for a lucrative position. The facts do not bear out such a picture, if the political circumstances of his patron are taken into account. He had returned to England as a champion of the Reformation and been called to the service of the great Protector who almost alone among the nobles was its convinced and devoted advocate. He had seen and rejoiced in the first years when his cause triumphed. He had seen the disgraceful intrigues which finally led to Somerset's arrest; and was himself involved in his fall. Unlike many others he stuck loyally to his patron, and until Somerset's release in February 1550 must have been in a position of acute anxiety and perhaps danger. Then for a time the Seymour fortunes revived and he benefited by them. In June the Privy Council proposed him as Provost of Oriel: but the Fellows not waiting for instructions had elected Joseph Smyth on the very day on which the post fell vacant. Eager to get away from the Court to any place where he could live quietly and free from want, he then applied unsuccessfully to Cecil for the Presidentship of Magdalen.[3] In addition, Wood in *Athenae Oxonienses*, I, c. 36, has stated that he was made a Canon of Windsor—a confusion between him and Richard Turner who went there in 1551—and Cooper in *Athenae Cantabrigienses*, I, p. 256, that he asked Cecil for an archdeaconry—a statement for which there seems to be no evidence. He was certainly unhappy. Somerset's influence was waning; the causes for which he cared were being betrayed; between concern for his children's welfare and the breakdown of his own health he may be pardoned if he expressed his outspoken disgust at his misfortunes and at the evils of the time. But he was no place-hunter: escape from what he called 'swete lipped Londoners and wanton courtiers' (*Herbal*, II, p. 76), into a spot where he could serve God, get on with his work, and bring up his family was all that he asked.

At the end of 1550 the poor man wrote again to Cecil, now since September Secretary of State: '...when Christes Churche nedeth me it will lawfully call me without any of my laboryng. In the mean time if that I had my helth I were able to get my lyvyng with my service not only in England but also in Holland, Brabant, and in many places of Germany, although sum thynk it wold not be in England as yet till I were better

1 *Memorials*, II, i, p. 215.
2 *Works of Ridley*, p. 108.
3 *Calendar of State Papers, Domestic*, 1547–80, p. 29, letter of 27 September.

knowen. Wherefore...I pray you heartily seyng that I cannot have my helth here in England and am every day more and more vexed with the stone, help me to abteyne me lycence...that I may go into Germany and cary ii litle horses[1] with me to dwell there for a tyme....I will also finishe my great herball and my bookes of fishes, stones and metalles, if God sende me lyfe and helth.'[2] He had in fact been doing a little work on fishes though only one of his notes refers directly to the Thames, 'Smelts come up the river beyond London to Kew and Brentford'.

In 1551 his prospects improved.[3] On 24 March by royal mandate he was appointed Dean of Wells in place of John Goodman deprived;[4] and though he found the Deanery and its estates let off to Goodman's cousin, and so for some time was 'penned up in a chamber of my lorde of Bathes with all my servantes and children as shepe in a pyndfolde', yet he seems to have got rid of the intruders; was granted a dispensation from residence 'whenever he may be occupied in preaching the Gospel in any part of the Kingdom' on 10 April 1551; was ordained priest by his friend Nicholas Ridley, then Bishop of London, on 21 December 1552;[5] and settled down for at least a few months of comfort. That even so life was not easy is clear from a passage in his *Spirituall Physic*, p. 51*b*: 'when as I had but 74 lib to spende in the yeare, my fyrst frutes yet unpayd, yet they never gave me a cup of ale undeserved in al their lyves. I have yet copies of their begging letters here in Germany to be witnesses.' Nor had he a very high opinion of his bishop the 'Maister Barlowe which wrote a noughtye and false lyeng boke [his *Dialogue* against Luther, 1531] compelled by force to do so' (*Spirituall Physic*, p. 40*b*).

In the three weeks in 1550 which as we have seen were all the holiday that he had had for botanising, he seems to have gone to Purbeck and explored the coast and neighbourhood. This gave him a considerable number of new records. One of these 'not sene in England saving only besyde Porbek' he at first, in *Herbal*, p. G. ii*b*, identified with Soldanell (*Convolvulus soldanella*) thinking this to be the 'Brassica marina of Dioscorides'.[6] In the revised edition (*Herbal*, I, p. 90) he changed his mind, and reporting that he had 'sene it in England at Westchester, at Portlande

1 The export of horses was at this time strictly forbidden.

2 *Cal. State Papers*, l.c. p. 31 of 28 November.

3 He was urged to stand for the presidency of the College of Physicians, cf. letter to him from Walter Bower (*Cal. State Papers*, p. 32) dated 20 January 1551: but was already in touch with the Archbishop over the Deanery of Wells: cf. his letter of 5 January to Cecil in *Camden Soc.* N.S. XXXVII, 1885, pp. 133–4.

4 Goodman and his Bishop William Barlow had been quarrelling for some time: now the Bishop had deposed the Dean! (*Cal. State Papers*, pp. 28, 32).

5 For the official entry in Ridley's Ordination Book, cf. W. H. Frere, *The Marian Reaction*, p. 207: Garrett, *The Marian Exiles*, p. 315, has misread the date as 1550.

6 In this opinion Gesner, Mattioli and Cordus agreed with him; cf. Cordus, *Hist. Stirpium*, p. 205*b*.

and at Porbeke', separated it from Soldanell, gave its name as 'Scurby Wede' and identified it with *Cochlearia officinalis*. He admits that another plant, also a denizen of the coast in England and in Flanders, 'called of the Herbaries Soldana or Soldanella', by description obviously *Convolvulus soldanella*, might be Dioscorides' species.[1] So, too, into his account of Cepea by which as he supposes Dioscorides means 'Brooklyme' (*Veronica beccabungi*) and Pliny 'Sea Porcelline' (*Atriplex portulacoides*) he interjects 'I found the same herbe [*A. portulacoides*] of late besyde the Ile of Porbek' (*Herbal*, p. J. iii*b*). He dates his visit by his note on the Sea Girdle or Alga (*Laminaria saccharina*), 'Thys herbe is plentuously sene in Purbek by the see syde after a great tempest hath ben in the see, whiche commonly lowseth suche see herbes and dryveth them unto the syde.... I saw the See gyrdell this year in July' (*Herbal*, p. K. iiii*b*). He seems to have visited Poole and writes of it 'the Stechas that Dioscorides writeth of [? *Lavandula vera*] is very plenteous in the toun of Poule and in diverse places of the West countrey where as it is called Cassidonia or Spanish lavendar' (*Herbal*, II, p. 148). Of Iris (*I. foetidissima*) he says, 'I have sene a litle flour delice growyng wylde in Dorsetshyre' (*Herbal*, II, p. 23) and, 'This herbe is called in the yle of Purbek Spourgewurt' (*Herbal*, II, p. 171*b*); of 'Lonchitis' (*Blechnum spicant*), 'I have sene the herbe oft both in Germany and in diverse places of Sommerset shyre, and Dorset shyre. It is muche longer then ceterach and the gappes that go betwene the teth, if a man may call them so, are much wyder then the cuttes that are in ceterach. And the teth are much longer and sharper' (*Herbal*, II, p. 41*b*); and of *Vinca minor*, 'it groweth wylde in the west cuntre' (*Herbal*, p. K. vi*b*).

How many other of the records of Western plants belong to this period it is perhaps impossible to say. But the second part of the *Herbal* must mainly have been prepared for publication before his return from his second exile, and it seems probable that those contained in it were made during 1551–2. If so, the note about Peonies, 'the farest that ever I saw was in Newberri in a rych clothier's garden' (*Herbal*, II, p. 84*b*) suggests one of his journeys from London to his deanery; those about 'Scorpion's tayle' (*Heliotropium europaeum*) and 'Hordeum hexastichum' that he had begun to enjoy his own garden (*Herbal*, II, pp. 13*b*, 16); and that about Thlaspi 'in maister Hambridges gardin' that he was making friends with his neighbours (*Herbal*, II, p. 152*b*). He had certainly been to Bath; for the treatise on baths written during his second exile refers to a visit, and his records of 'Rammes' (*Allium ursinum*) and of 'Middow Saffron' (*Colchicum autumnale*) there both occur in the *Herbal* of 1551 (pp. B. v, L. iii*b*). The visits to Glastonbury (*Herbal*, II, p. 137, *Sinapis nigra*, 'in

1 The confusion among herbalists between *C. soldanella* and *C. officinalis* seems to have persisted until Johnson's time: cf. *Gerard*, p. 839.

the corne a little from Glassenberrye'), to Bristol (*Herbal*, II, p. 83*b*, *Trinia glauca*, 'I found a root of it at Saynt Vincente's rock a little from Bristow'), and to Chard (*Herbal*, I, p. 36, *Sherardia arvensis*, 'in Maister Baylies marchant of Chardis Lordship' and *Herbal*, II, p. 128*b*, *Spiranthes autumnalis* 'about Charde...the flowers grew very thyck together as they were writhen about the stalke') are probably of this period; but the story of the Glastonbury thorn, 'about six myles from Welles in the parke of Glassenberry there is a hawthorn which is grene all the wynter as all they that dwell there about do stedfastly holde' (*Herbal*, II, p. 73*b*),[1] does not claim to be from first-hand knowledge. Nor does the report that 'our Alexander [*Smyrnium olusatrum*] groweth...in a certayn Ilande betwene the far parte of Sommerset shere and Wales' (*Herbal*, II, p. 68). He was obviously attracted by the countryside; 'I never saw in all my lyfe more plenty of thys sorte of bulles trees [*Prunus insititia*] then in Sommerset shyre' (*Herbal*, II, p. 104): but was critical of the misuse of it; 'Flax...should grow as plentuously in the South part if men regarded not more theyr privat lucre then the Kynges Lawes and the comen profit of the hole realm. I havę sene flax or lynt growyng wilde in Sommerset shyre wythin a myle of Welles, but it hath fewer bowles in the top then the sowen flax hath, and a greate dele a longer stalk, whiche thynges are a sure token that flax would grow there if men would take the payn to sow it' (*Herbal*, II, p. 39*b*).

It is possible that before his second exile he paid a visit to the Isle of Wight. This is mentioned in the account of Madder (*Rubia peregrina*), 'The wilde kinde groweth plenteously both in Germanye in woddes and also in Englande, and in the most that ever I sawe, is in the Ile of Wyght. But the farest and greatest that ever I saw, groweth in the lane besyde Wynchester, in the way to South hampton' (*Herbal*, II, p. 118). If this refers to a visit in 1551, it would link up with a similar reference to the Isle of Wight in his letter on Fishes.

From this letter it must have been during this time but after the first *Herbal* had gone to the press that he made the visit to Portland which enabled him to clear up the matter of Soldanella. Moreover, if the plant formerly confused with Sea Bindweed was really Scurvy-grass—and this seems plain from the record in *Herbal*, I, pp. 90–1—it is also evident from a passage in De l'Obel's *Adversaria*, p. 92, that at Portland he again found *Crambe maritima*. Writing under the heading 'Brassica marina sylvestris multiflora monospermos', De l'Obel states that it grows on the sea coast of England at the isle of Portland and says: 'Its seeds were given to us long ago by an English doctor, very skilled in this subject, Turner, who

1 It grew on the south side of Weary-all Hill (Werrall Park): of its two trunks one was cut down in Elizabeth's reign, the other during the Civil War. P. Mundy described it as 'ready to fall down for age' when he saw it in 1639: *Travels*, IV, p. 5.

wished it to be named one-stalked and one-seeded from its single seed. For in the elegant clusters of pale white flowers which form a dense umbel, on many branches, there are short thickened pods each holding a single seed, a little smaller than a pea.' His description of the leaves, 'thicker and broader than purple cabbage', and his picture of the plant make it clear that this is the species which in fact Turner had already discovered at Dover.

In 1551 the first instalment[1] of *A New Herball*, dedicated (but less magniloquently than the *Names*) to Edward Duke of Somerset, 'imprinted at London by Steven Mierdman [or in the colophon Myerdman][2] and they are to be soolde in Paules Churchyarde at the sygne of the Sprede Egle by John Gybken', was published. The dedication from which much has already been quoted ends with an answer to two criticisms. 'Some will saye why is thys fyrst tome so lyttle': to it after explaining his difficulties he urges that he wishes before setting out any more to explore England more fully. 'Others will thinke it unwysely done, and agaynst the honor of my art that I professe, and agaynst the common profit, to set out so muche knowledge of Physick in Englyshe, for now (say they) every man...nay every old wyfe will presume not without the mordre of many, to practise Physick.' He answers by asking 'how many surgians and apothecaries are there in England which can understande Plini in latin or Galene and Dioscorides where as they wryte ether in greke or translated into latin', and then argues that at present they know no more of the identity of the herbs than the 'old wyves' or 'the grossers' who sell them. An English Herbal will at least give them an opportunity of knowledge; and the danger of murder is no greater now than it was when Dioscorides and Galen wrote in Greek for a Greek-speaking world. 'If they gave no occasion unto every olde wyfe to practise Physick, then gyve I none. If they gave no occasion of murther, then gyve I none. If they were no hynderers from the study of lyberall sciences, then am I no hynderer, wryting unto the English my countremen an Englysh herball.' After the Table or index at the end of the book is the note: 'Here endeth the Table of the firste parte. God wyllynge the nexte yeare ye shall have the seconde.'

The book is a small folio containing 196 pages numbered A to P. It is divided into alphabetical sections, the headings being with few exceptions in English, though the plants are arranged according to the initial of their Latin names from Absinthium to Faba. It contains 169 woodcuts

1 That a second was expected speedily is clear from Gesner's reference in the list of authors prefixed to Bock, *De Stirpium* (Strasburg, 1552). He wrote, 'part of his work we have seen lately published, the rest we await very eagerly'.

2 He came to England in 1549, set up a press in Billingsgate and lived there till after Cranmer's arrest.

taken in the main from the octavo edition of Fuchs's *De Historia Stirpium*, published at Basel in 1546. Presumably Turner had obtained this book and had the pictures copied: they are of course reversed. The results are, it must be confessed, generally disappointing. Fuchs's originals have a Düreresque quality combining accuracy with grace of line and beauty of design. The copies except in plants of very simple structure are coarse of outline and obviously twice removed from life. But even so they serve admirably the purpose of identification. Anyone, who knew the British flora well, ought with their help to have been able to name most of the species represented without serious difficulty.

On occasion Turner is himself mistaken. Thus under Astragalus, a plant which he reported from Coombe Park and Richmond Heath (*Herbal*, p. E. v*b*) and which Britten (*Turner's Names*, p. 100) and Clarke (*First Records*, p. 44) no doubt rightly identify as *Lathyrus macrorrhizus* Wimm., he noted that 'Fuchsius do contend that this should be Apios', questioned this, and yet printed Fuchs's picture of Apios (*Historia Stirpium*, p. 131) which is obviously *Lathyrus tuberosus*.[1] One picture, that of Smallage (*Apium graveolens*), is repeated, once under Apium or 'Persely' (p. D. iiii*b*) and again under 'Elioselinum' (p. P. i). In several places he has a number of figures, although no attempt is made to describe them. Thus under 'Chikewede' of which he says 'this herbe is so well knowen in all contrees that I nede not largelyer to describe it' (p. B. vii) he has three pictures, one which may be *Stellaria media*, but is much more like *Arenaria serpyllifolia*, and the others certainly *Veronica hederaefolia* and *V. agrestis*; as he adds, 'the Pothecaries call it morsum gallinae' which is usually *Lamium amplexicaule*, the identity of Chickweed can hardly be described as fixed; in fact the confusion thus engendered lasted until Ray's time.[2] So, too, under Germander, by which he apparently means in the text *Teucrium chamaedrys*, the page of four pictures contains two Labiates (*Teucrium chamaedrys* and *T. botrys*) and two Veronicas.

Turner's hopes for the speedy publication of more of his Herbal and for the exploration of England were destroyed by the death of Edward VI in 1553. Soon after Gardiner's release from the Tower on Mary's accession he appointed a Commissioner, 'the greatest Wolfe in Welles' who already 'hath one benefice in Holdernes, and two in Somersetshire, he is Residentiari both in Bristow and Welles and hath iii prebendes one in Welles, one in Bristow and one in Salisbury' (so Turner, *Romyshe Wolfe*, p. D. viii). This was Roger Edgeworth of Oriel College whom Turner names in *Spirituall Physic*, p. 88*b* and who proceeded to make himself Chancellor of Wells. No doubt Turner left almost at once. We know

1 See above, p. 72: Turner interpreted Apios as the Umbellifer, *Conopodium denudatum*, cf. G. Bauhin, *Pinax*, p. 162.

2 See *Catalogus Cantab.* under Alsine.

nothing of the circumstances of his second flight into exile: but as soon as Mary's ascendency was secured his life was obviously in danger.[1] With a wife and three children still young he could only seek refuge by returning to the Continent. According to Miss Garrett[2] he left Wells in September.

He seems to have gone straight to Cologne, to have settled down there and to have stayed long enough to get a well-stocked garden. The reference in *Herbal*, II, p. 71*b*, to Orobanche 'betwene Colon and Rodekirch' probably comes from this period; for there is no mention of the Broomrape, or Chokewede as he calls it, as German in *Names* or the *Herbal* of 1551. But here as in several other places it is difficult to separate the earlier from the later records. Bonn and Mainz, for example, which Jackson includes among places visited in his second exile, were also visited as we have seen in his first.

This time he seems to have moved on first to Worms; for the references to plants there and in the neighbourhood are too frequent to suggest mere passage through it. Thus of *Iris foetidissima*, 'I have sene...hole cartes full in Germany besyde Wormes in the middowes not far from the Rhene' (*Herbal*, II, p. 23); of 'Spourge gyant' (*Euphorbia lathyrus*), 'I have sene in diverse places in Germany, first a litle benethe Colen, by the Rhene syde, and afterward besyde Wormes in high Germany' (*Herbal*, II, p. 93*b*); of *Bupleurum rotundifolium*, which he was the first to name 'Throw waxe', 'in great plenty in a corne field on the Northsyde of the cytie of Worms in Germany' (*Herbal*, III, p. 56); of *Allium ampeloprasum*, 'I never saw farer wilde lekes in all my lyfe then I saw in the sedes about Wormes in high Germany. For they were much larger in the leves and greater heded then they were that I saw about Bon' (*Herbal*, II, p. 101); of *Thlaspi arvense*, 'besyde Wormes growing besyde diches and at Francfort about the walls of the city' (*Herbal*, II, p. 152*b*); of *Medicago sativa*, 'I have sene it growyng wylde in Germany within an half myle of Wormes in the hygh way towarde Spyer' (*Herbal*, II, p. 52); and of *Gratiola officinalis*, 'about Wormes in the close that is hard by the water side beyond the bridge, where as my servantes gathered an hole wallat full at one tyme' (*Herbal*, III, p. 33). A note of this period was added to his account of 'Ami' (*Ammi majus*) in *Herbal*, p. C. iii, 'The othere kind is of late found in Italy....The sede cometh nowe to Frankeford to be sold under the name of Amomi, but some call it verum Ami, namely the grossers of Norinberge [?Neuenburg]' (*Herbal*, I, p. 39).

1 Cf. *Romyshe Wolfe*, pp. B. iii, E. vi*b*.

2 *The Marian Exiles*, p. 314: the evidence is that on 5 September 'Thomas Leigh servant to Turnour Deane of Welles' is named in *Acts of Privy Council*, 1552–4, p. 341. It should be noted that Turner's case tells heavily against Miss Garrett's interpretation of the migration.

Around Speyer he reported that Madder was 'set and planted in greatest plenty that I know' (*Herbal*, II, p. 118); and 'a litle from ye toun of Bathe [Baden] in hyghe Germany' he found his Northumbrian friend *Meum athamanticum*.

Finally, he moved down to Weissenburg where he settled and had a garden and a practice. Jacob Detter, the Apothecary there, was evidently a friend who helped him with the identification of plants (*Herbal*, II, p. 71); and the longest record at Weissenburg has to do with him. It deals with the structure of Spikenard which Turner regarded as a grass and of which he had till then only seen the ear.[1] He was sceptical of Mattioli's claim that the plant had no stalk at all since after long search in the shops of Venice only ears could be found.[2] Then 'although in Germany there is not suche choyse of simples in every place as is in Venis, yet in this yere of our Lorde 1557 I found in the shop of Jacob Diter the Apothecari of Wiseburg on pece of Nardus whiche hath a stalk a fynger long, holow, and of the bygnes of a metely byg straw, which I have to shew at thys present daye' (*Herbal*, II, p. 63). To his dwelling in Weissenburg there is an allusion in *Herbal*, II, p. 112; and to his garden in *Herbal*, I, p. 197—of the Cytisus which he had grown at Cologne 'now I have it growing here in my gardin at Wisenburg'—a reference the more interesting because it proves that during his stay he was revising the first part of the *Herbal*.

From Weissenburg certain other localities were apparently visited, and he noted the Black Poplar (*Populus nigra*) 'by the Renesyde hard by the city called Lauterburgh' (*Herbal*, II, p. 99); Radish (*Raphanus raphanistrum*) 'more common aboute Strasbourgk in highe Germany than in an other place that ever I came in' (*Herbal*, II, p. 111*b*), and Angelica (*A. sylvestris*) 'not far from Friburge in the wood called Nigra sylva, or ellis Swartwald in Duch, where as is the beginninge of Hircinii sylvae'[3] (*Herbal*, III, p. 5).

It is to be remembered that at this time 'High Germany' from Frankfort and Strasburg to Basel and Zürich was the refuge of a considerable body of English protestant refugees.[4] Richard Cox, who was certainly known to Turner, had escaped in 1554, and made his way to Frankfort in time for the 'Troubles' which the exiles brought in their train: was it an interest in the strife between Cox and Knox that brought the botanist

1 It is in fact the rhizome of a Valerianaceous plant of Indian origin, *Nardostachys jatamansi*: but this with the withered stalks and ribs of leaves attaching to it looks very like a dried ear of corn.

2 Cf. Mattioli, *Comm. in Diosc.* pp. 30–1 and Amatus Lusit., *Enarr.* p. 17, who both affirm that the 'spica' is a root.

3 The great forest so named by Julius Caesar and Tacitus that stretched from the Erzgebirge to the Rhine and north to the Harz.

4 For the 'census of exiles', cf. C. H. Garrett, *The Marian Exiles*, pp. 67–349. She estimates the number at 678.

to the city? Cox afterwards moved to Strasburg and to Worms; so the two men must have been in fairly constant contact—a fact which was to have its sequel when after Turner's death his widow married the then Bishop of Ely as his second wife in 1570.[1]

Of other literary activities during his second exile there is not a great deal that need be said. The last of his dialogues dealing with religion, *The Huntyng of the Romyshe. Wolfe*, is ascribed by Jackson to the date *c.* 1554; but the large folded plate of the Bishops, Gardiner, Bonner and Tunstall, with wolves' heads assisting at the death of the Lamb over the dead sheep Bradford, Rogers, Hooper, Ridley, Latimer and Cranmer, must have been inserted at a period later than their martyrdoms in 1555 and 1556. In a copy of the book in the Cambridge University Library the plate is small and shows only one wolf and sheep, and this may well have been the original and earlier picture. It is plain from definite allusions in the text that the work was finished within a very short time after Edward VI's death ('our yonge maister whiche hathe departed from us of late', p. A. iiii) and Gardiner's release from the Tower (he was taken there five years ago and has stayed until 'these fewe monethes', p. B. i*b*—his imprisonment being from 1548). The difficulty against a late date is that the dialogue represents three men, the Forester, the Dean and the Hunter, going up for a session of Parliament. They discuss the persecution under Henry VIII and name some of those done to death in it. But the assumption is that these old and bad days are over; and much of the work is devoted to the advocacy of measures to combat the ignorance and superstition of the people. Turner's own policy is put into the Hunter's mouth. First, he denounces the spoiling of the monasteries by King Henry VIII 'with his covetous counsell'; the divorce of 'the vertuous Lady Anne of Cleve' and the similar divorces obtained by four lords who put away their wives 'because they liked hores better' (p. D. vi); the taking by the Crown of the first-fruits and tithes which properly belong to the Church; and the acquiescence of Gardiner and others in these robberies (p. D. vi*b*). Then after further protest against pluralities and clerical pride he turns to the necessary reforms. 'First ye must provide that there be mo scoles in Englande, and that there be better provision for the universities that the realme may have enough scolers that are learned to make shepherdes of. And ye must provide thorowe out all Englande that every parson and vicare have an honeste living, to bye him bokes and to finde him and his household withal' (pp. E. vi*b* and vii). This leads on to an insistence that the vicarages and other Church property, lately secularised, must be restored, and patronage and the right of choosing incumbents must be surrendered to the Church. Finally,

1 Richard Hilles who gives this news to Bullinger (*Zürich Letters*, II, p. 181) describes her as a 'young widow'—the epithet is perhaps flattering.

'in every lytle shire in England there should be at leste iiii bishoppes, I meane no mitred nor lordlye but suche as should be chosen out of the rest of the clergie every yere and not for ever which should be honest learned men, preachers and graduates...these should examin and admit al the Elders...admit lawful ministers, examin heretikes and appoint ponishment' (p. F. *ib*).

Clearly this must have been written in Edward's reign when there was still a prospect of such reform being accomplished, although the book was not actually finished until after the king's death. Probably the script was made ready for publication while Turner was still living at Wells; for it is dedicated to the 'lordes and yonge gentylmen' of Somerset and other western counties. If F. S. Isaac is correct in assigning it to the press of Egidius van der Erve at Emden,[1] it would appear that Turner sent the script to him in London owing to his connection with a Lasco before he had himself left England, and that Erve when he fled with Myerdman and a Lasco took it with him.

The fact is that the book was certainly printed in various places. There is a version of it *The Hunting of the Fox and the Wolfe* with a preface 'To al my faithful brethren' which is a sermon exhorting to strive 'against Romish reliques' and has been attributed to John Knox. In 1555 Turner's books were again prohibited by Royal Proclamation,[2] and the Wardens of every Company were ordered to take steps for their surrender and destruction. It is small wonder that so few copies have survived, or that bibliographical details are hard to collect.

More characteristic of his outlook at this time is the last of his controversial works, *A Newe Booke of Spirituall Physic*, dated 20 February 1555, and stated no doubt in jest to have been 'imprinted at Rome by the Vaticane churche, by Marcus Antonius Constantius,[3] otherwyse called, Thraso Miles gloriosus',[4] or as the list of corrections 'To the Reader' puts it 'agaynst Marcus Antonius Constantius otherwyse called Thraso or gloriosus Pape miles'. This is much the most illuminating of Turner's religious works; and carries forward the ideas already outlined—though now with a sense of their impracticability. It is in fact a fine and fearless attack not only upon the superstition and tyranny of the Catholics but

1 Cf. *The Library*, 4th series, XII, pp. 339, 343 and C. H. Garrett, *The Marian Exiles*, p. 315: but Miss Garrett has not disentangled Turner's travels: cf. above, p. 96.

2 Printed in Foxe, *Actes and Monuments*, VII, pp. 127–8.

3 Marcus Antonius Constantius is the name applied to Stephen Gardiner, e.g. by Cranmer in the letter printed in *An Aunswere unto a Cavillation* (London, Day, 1580), p. 427, and used by Gardiner in his treatise of July 1552: cf. Turner, *Sp. Physic*, p. 38. Gardiner was still alive when Turner's book was sent to the press.

4 Is it a coincidence that Stephen Gardiner played in the *Miles Gloriosus* along with William Paget and Thomas Wriothesley (cf. *Letters of S.G.* p. 186)—presumably at Cambridge when all three were at Trinity Hall?

upon the greed and profanity of those who in the name of religion have spoiled the Church of its resources, degraded its ministers, and robbed the people of all opportunities for improvement. It is a slashing piece of satire, full of the quaint irony which dictated the colophon, but full also of an honest man's indignation that the good work of purifying religion from Romish errors had been disgraced by upstart rapacity and fraud against God. Nor is his denunciation merely general. He quotes case after case of the abuse of Church patronage, the spoliation of vicarages and churchyards, the sale of preferment, the enclosure of commons and the closing of hospitals by the 'crowish stert uppes' of the new nobility or the 'lordlie bishoppes' of the old.

The book is so complete an answer to the charges of self-seeking and of timidity brought against him and so excellent an example of his vigour of thought and phrase that it deserves some fuller comment. It is dedicated to the Dukes of Norfolk and Suffolk and the Earls of Arundel, Derby, Shrewsbury, Huntingdon, Cumberland, Westmorland, Pembroke and Warwick in a letter which explains that though the whole body politic is sick it is the head, the nobility and gentry that must first be treated and cured. 'I have wrytten', he declares, 'in thys booke wyth reason and scripture' (p. A. 2*b*). There follows a résumé of the contents: an examination of the character and function of a nobleman which is not 'faire houses, riche apparell, dauncyng, lutyng, dycyng nor cardyng, haukyng nor huntyng'; a statement that the nobility is diseased; a diagnosis of the particular ills; a statement of the remedies.

The subject is opened with the pleasant tale of John of Low the King of Scots' fool. 'Is it not possible that one mule may at this day cary another: Master Latemer thought so when Syr Martin Mydas rode to heare Master Myles preache and sayd when the sermon was done that hys mule was as well absolved as he' (p. 9*b*).[1] Then after many scriptural examples of kingliness and good nobles comes the tale of John Ward the painter of Cambridgeshire who put a picture of St Christopher in his pew to remind him to be a Christopher, but when his fellows set candles before it said that it was as lawful to set a candle before his hat as before the painting since 'the hatter is as holye a man as I am and hys handworke deserveth as well to have candelles as myne'—for which remark he was cast into prison (p. 20*b*). Gentlemen, he continues, must deal fairly with their tenants; and with that end must be well educated and know the law; and so must study the scriptures (pp. 21–31). He then illustrates the lack of this partly from the attitude of the bishops towards Henry's divorce, partly by the story of the Commissioners at Calais.[2]

1 Cf. p. 26: 'a witless gentleman that cannot write his own name is a beast sitting on a beast.'

2 See above, p. 75.

He then deals with the chief disease of the nobles, their dropsy and insatiable thirst for estates, benefices and every kind of property (pp. 48–50), a thirst specially characteristic of the 'newe gentlemen'. 'Well,' he concludes, 'they that cam from the donghyl muste go thether agayne.... Howbeit some of the olde are growen out of kynde. There was an erle... got one of the greatest deaneryes of Englande...' (p. 50*b*). Then after illustrating the meanness of the gentry by personal examples he adds, 'The common sort of gentlemen thynke themselves very lyberall yf they bestowe upon dogges xx li in the yeare and the rest of theyr goodes upon gentlemen though they spend not iii farthinges all theyr lyfe upon one poore man' (p. 67); and later, 'God kepeth open houshold at all tymes and as well provideth for flyes, sparowes, pisemiers and yonge ravenes as he dothe for olde goshaukes, herones, bustardes, swannes, egles and great oxen' (p. 70*b*).

Finally, he prescribes his remedies for the two sorts of 'stertuppes', the money-grubbers and the flatterers. Send the young to school: education will shape them rightly. Send the older 'to the grenelande' or somewhere else far away, where they will have to work hard. Restore the lands that 'Syr Matthew Mukforke' has stolen. Guard against the ambition and covetousness of the clergy.

The best known product of his stay at Weissenburg is the remarkable letter which he sent from there on 1 November 1557 to Conrad Gesner and which is printed in the fourth volume of his *Historia Animalium.* This as his final sentence declares contains 'all that for the present, seeing I am deprived of all the help of my notebooks and must rely on memory alone, I have been able to gather about the History of Fishes'. It is in fact a very notable list of the English names of such fishes as he can identify with brief notes of their distribution. It is, of course, written in Latin.

Obviously Turner had been interested in fishing and in fishes from his boyhood; and he has done his best to gather the local names and to assign them to their proper owners. Thus of 'Apua' he records that it is called in Cambridge a Spirlyng, by Londoners when fresh a Sprat, when smoked a Red Sprot or a tryeg (? dried) Sprot, and that it is a fish *sui generis*, not a young Herring; of Gadus that it is called by Northerners a Keling, by Southerners a Cod, by Westerners a Welwel, and that its head if large enough is bought as a delicacy by the rich; that 'Allerfanghe' is a trout that lives in pools under Alders, and Bulltrout is like Salmon but nicer; that Aristotle's Gobio is the Smelt, though Gobio may also be the Gudgeon; that gelu or gelatine is got from the Ling. Two or three notes make it clear that he had travelled more widely in England than is usually supposed. Thus he records that the Cumbrians have a fish called 'Grundlin' (Groundling), rather like a Loach,[1] and found by boys under the stones

1 Groundling is usually a name for one of the fresh-water fishes, Loach or Gudgeon.

on the shore—probably one of the Rocklings, *Motella mustela*; that 'Pour Cuttel' (*Octopus vulgaris*) is found in the Isle of Wight, 'Sore Mullet' (*Mullus surmuletus*) at Portland, and 'Cookefishe' (probably *Labrus mixtus*) in the West; and that at Portland the Small Scallop (*Pecten varius*) is called a 'Cok'.

The interesting problem arises as to the sources of his knowledge. Evidently he knew the Latin equivalents of many of our fishes, and must have learned them by the same process of study and comparison of authorities as he had used for birds and plants. He had also received a note-book or pamphlet from Gesner in which Oppian's *Halieutica* and Pliny's list of fishes were explained and the German and English names added. It is upon this that his letter is a comment. He also refers though only briefly to Belon; and therefore presumably had seen his two volumes, *De Aquatilibus*, published in Paris by Stephanus in 1553. From an allusion in his *Herbal*, II, p. 37, he seems also to have known 'Massarius Venetus hys boke that he writeth of fishes', that is Francesco Massario's *Castigationes et Annotationes* on the 9th book of Pliny's *Natural History*, published by Frobenius at Basel in 1537. He evidently did not know the great works published by Guillaume Rondelet and by Hippolyto Salviani in 1554.

There is, however, further evidence of his continued interest in the subject. Among the paralipomena or latest additions to Gesner's great work, in *Historia Animalium*, IV, Paral. p. 25, there is a picture of the Lump-sucker (*Cyclopterus lumpus*) sent to him by his friend Johann Kentmann of Dresden. Above it is a reference to the section of the *Historia* dealing with the 'Orbis Rondeletii' (p. 633) where he surmises that this may be the 'Seeham' of the Germans and the 'Lumpe' of the English. Below is the following: 'When I had sent this picture to the excellent doctor William Turner anxious to know his opinion on it, he replied, "The picture that you sent is plainly of our Lump: the two front fins are redundant: there were none on the fish which I saw recently caught, and ate [so he had evidently returned by then to England]. It lacks also the round hole under the chin which I saw and touched in my specimen. Otherwise it is accurate enough."' Gesner then adds that he also sent the picture to Geronimo Massario of Vicenza a doctor and presumably the son of Francesco who had written on fishes; and that he confirmed the identification saying that the Lump was often taken in Cornwall. Gesner continues: 'Again in another letter to me Turner says "The Lump has its name from the shapeless mass which we call a lump. It may be Pliny's Orchis...but cannot be Belon's Orb, for our Lump is edible, scaleless, and cartilagineous. It has no true bones or spines inside but cartilages. It tastes like a Skate. It has only two fins below the gills. It has on each side three rows of spines from head to tail, and on the back a single row. These are not continuous but set at intervals. They resemble the thorns of a bramble. A navel of huge size protrudes on its belly. Under its chin is a round hollow like a wheel: by this it seems sometimes to fix itself to the ground or the rocks. It has a frog-like head and small eyes. Its skin is rough to the touch. Some of our folk call it a 'See oul' or as the Scots call it a 'Paddel'."' This last sentence is interpreted in Ray's *Historia Piscium*, p. 208, where we read 'Lump or Sea-Owl, *Scotis* Cock-Paddle'.

In the same year 1557 and on the 10th of March he dated from Basel the prefatory dedication of his treatise on baths. This was printed in 1568 with the second edition[1] of the treatise and the completed *Herbal* by Arnold Birckman of Cologne, but was obviously written during this exile. It is addressed 'unto his wellbeloved neighboures of Bath, Bristow, Wellis, Winsam and Charde', and begins with a pleasant fancy, 'The most part of al flocking birdes, of the whiche nombre are Linnettes, Goldfinches, Sparrowes and Twyes, if they chance upon any good plenty of meat, they cease not locking and calling, if they heare any of their kindes, be it never so far of, until they have broughte them unto that meate which they have founde...then how unkind were I beyng a reasonable creature...if that after that I had travayled thorow Italy and Germany...I should not at the lest offer unto you suche good thinges as I have founde.' Mainly he is concerned with 'the bath of Baeth in England', but he has also included details of German and Italian baths. He had paid a visit with a patient, 'one Myles somtyme one of my Lorde of Summersettes players', to 'the bath of Baeth which is in the strete besyde S. Johns hospitale'; had gone into it himself; had brought out of it 'slyme, mudde, bones and stones' and found them smelling of brimstone; and so concluded that the water was heavily impregnated with sulphur.[2] So, too, he describes the baths at Ems and Baden in Germany, both of which he had visited, and gives full directions for their proper use; and adds similar testimony from his friends to a number of other places. Apparently no one previous to Turner had ever written of our English spa; he apologises for the mistakes liable to occur in a pioneer piece of work. But his enthusiasm for the healing value of baths, his insistence that they must not be tried casually but for a long spell of regular treatment, and his warning that diet and exercise must be watched during the course make his work doubly valuable. Considering the state of chemical knowledge at the time his treatise is very sane and enlightened.

1 In 1562 a different dedication, to Lord Edward Seymour, dated 'At London the xv of Feb.' 1560, is inserted. With the exception of this dedication and the title-page the editions of 1562 and 1568 are identical.

2 A conclusion endorsed by J. Bauhin in his treatise on the Fons Bollensis (Montbéliard, 1598), p. 263.

On Mary's death in November 1558 Turner and his family came back to England to take part in the outbreak of joy which greeted the passing of the Spanish dominion. He must again have visited Antwerp, and spent some time with Coudenberg;[1] and on his way through Brabant he collected 'two rootes or thre' of 'Gratiola' (*Gratiola officinalis*) which he 'gave unto Maister Riche and Maister Morgan Apotecaries in London' (*Herbal*, III, p. 33). He must have gone there on his arrival, and have stayed for some time at the house in Crutched Friars, a street lying between Mark Lane and Aldgate, which he kept until his death ten years later; for he preached to a great congregation in St Paul's on 10 September 1559 as Strype has recorded (*Annals*, I, i, p. 199). On 4 October he went down into Kent to preach at the funeral of the young Lady Cobham, wife of William who had lately been made Warden of the Cinque Ports, and whose garden, as we have seen, Turner already knew. Henry Machin, who records this in his *Diary* (p. 214),[2] clearly distinguishes him from his namesake of Canterbury (cf. l.c. pp. 210, 279) and describes in detail the magnificence of the occasion and the great dinner that followed it. With his friend, Sir William Cecil, in power, with the influence of the Marian persecutors broken, with his son and daughters growing up, his future was at last freed from fear. Nevertheless, he had to bring an action in order to secure his reinstatement at Wells. The case was heard 'in the house of Richard Goodricke near Fleet Street'; a verdict in his favour was given on 18 June 1560; Goodman's appeal was dismissed and Turner's dispensation from residence for preaching confirmed on 20 July: but it was not until 21 March 1561 that a writ ordering Goodman to vacate the deanery made it possible for him to recover his home.[3] In 1563 the Rectory of Wedmore was added to it, and his income, according to Phelps's *History of Somerset*, II, p. 48, then amounted to the opulent figure of £151. 6*s*. 8*d*.; and in spite of the protests and intrigues of the enemies of whom he speaks in his last Preface he was maintained by royal decision in possession till his death.

But as these dates show, there was a considerable interval between his return from exile and his restoration to Wells; and this was almost certainly devoted to botanical and other similar studies in London. Now he was no longer a courtier or concerned with political or theological controversies. He wished to return to his *Herbal* and then to carry out the work on birds and fishes which he had promised. In consequence we

1 Most, if not all, of the references to Coudenberg may well belong to this visit; for they occur in the later sections of the *Herbal*, and the famous gardens were very new in 1547.

2 Camden Soc. 1848.

3 For details, cf. Historical MSS. Commission, Dean and Chapter of Wells, II, pp. 282, 289.

find him in close contact with the herbalists and gardeners of the metropolis; and his later work is full of allusions to them. Thus in the revision of the first part of the *Herbal*, whereas in 1551 on Dittany he wrote 'I have not sene it growynge in Englande' (p. N. iiii), he altered his words to 'I have sene it growynge in England in Maister Riches gardin naturally' (I, p. 203); and in the other parts gives frequent notes such as that on Mistletoe, 'I never saw more plentye of righte oke miscel then Hugh Morgan shewed me in London. It was sente to hyme oute of Essex' (*Herbal*, II, p. 165); and that on Angelica, 'The rootes are now condited in Danske, for a frende of myne in London called maister Alene, a marchant man who hath ventered over to Danske sent me a litle vessel of these well condited with very excellent good honey' (*Herbal*, III, p. 5). His will proves that he had a nephew William Turner, grocer of London, through whom he may have made 'Mr Allen's' acquaintance.

There was in fact at this time an interesting group of pharmacists, students of herbs and owners of herb-gardens in London, with which Turner now came into close touch. Hugh Morgan, the best known of them, had indeed been censured by the College of Physicians in 1556 under their rule which forbade surgeons and perfumers to deal in medicines, and again in 1559 for selling pills without a doctor's permission,[1] but this does not seem to have affected his position. He afterwards figures very freely in the *Adversaria* or Botanical Commentaries of Mathias de l'Obel, the Flemish botanist who came to London in 1569. He is mentioned over twenty times (pp. 74, 84, 89, 101, 124, 145, 172, 176, 217, 251, 276, 337, 343, 371, 376, 389, 394, 400, 409, 426, 441, 452), generally in connection with plants growing in his garden, though later also in regard to his deep interest in West Indian plants and his traffic with sea-captains and even with merchants from Venice. De l'Obel speaks of him as a man learned and liberal-minded with a keen interest in study and discovery as well as in the medical qualities of herbs. He evidently came into touch with the great Continental pharmacists, Francesco Calzolari and Andrea Bellicocco of Verona, Alberto and his brother the Syrian doctor Cechino Martinello of Venice, Jacques Raynaudet of Marseilles, Jacques Farges of Montpellier, Valerand Dourez[2] the Fleming of Lille and Lyons, Pieter Coudenberg and Wilhelm Driesch of Antwerp; and this contact, with the resultant exchange of seeds and specimens, must have been of the greatest value for the progress of botany. He remained in London until the close of the century; for Gerard in 1597 speaks of a great 'Lote or Nettle Tree' (*Celtis australis*) 'in a garden neere Colman

1 Cf. Caius, *Annals* of the College, pp. 43, 52.

2 For this spelling I follow L. Legré. It is to be noted that G. Bauhin, whose brother Jean claimed kinship with him and whose *Pinax* is usually accurate, gives his name in the Explicatio as Donreus; and this spelling is frequent elsewhere.

streete [Coleman Street running north from Lothbury to London Wall near Moorgate] in London, being the garden of the Queenes Apothecary called M. Hugh Morgan, a curious conserver of rare simples' (*Herball*, p. 1308). Later he crossed the river to Battersea, where according to Parkinson (*Paradisus*, p. 437) he had another garden. He died there in 1613 at the age of a hundred and three.

Next to him, and even more freely mentioned by Turner, was John Rich—the 'Maister Riche apothecary' whose garden 'where I saw many other good and strange herbes which I never saw anywhere elles in all England' appears in the *Herbal*, I, p. 203, II, pp. 69, 147, 152*b*, 158, 161, III, p. 33. He also was well known both to De l'Obel, who calls him Johannes Riccius (e.g. *Adversaria*, pp. 347, 351, 450), and to De l'Ecluse, who writes Johannes Richaeus (*Rar. Plant. Hist.* pp. cxl, ccxl), and calls him and Morgan the Royal Pharmacists. Of Rich's later history there seems to be no trace unless he retired and became the 'Mr Rich of Lee in Essex' of whom Gerard speaks in connection with Pinks, *Herball*, p. 473.

In 1560 while Turner was still in London he wrote another dedication to his treatise on Baths. This was to 'Lorde Edward Senar [evidently Seymour] Erle of Herford' and explained that while he was in the service of Seymour's father the Duke of Somerset he got license to go to Bath, and when he became Dean of Wells 'tried the same bathes a litle further'. This edition of the book was published by Birckman of Cologne in 1562.

During the eighteen months before his return to Wells he may perhaps have made the visit to Colchester which has left its mark in one or two places on his *Herbal*. Thus of Linden (*Tilia cordata*), 'It groweth very plentuously in Essekes in a parke within two miles from Colchester in the possession of one maister Bogges' (*Herbal*, II, p. 153*b*); and of the 'Tyme Saxifrage' (a plant which Britten fails to identify),[1] 'I have sene of this kinde growinge in Essexe by the seasyde' (*Herbal*, III, p. 67). He may also have done some botanising in Kent; for in the revision of *Herbal*, I, p. 132, under *Ajuga chamaepitys*, in addition to the note in *Herbal*, p. J. vi, 'in diverse places in England', he added, 'it groweth in good plenty in Kent', and this can hardly have been recorded on the occasion of Lady Cobham's funeral in October.

The last part of the *Herbal* which must have been now in preparation records several of his latest West country finds: *Gentiana amarella* 'in Dorsetshyre upon the playne of Salisberrye' (III, p. 25); *Bupleurum rotundifolium* 'in Somersetshire betwene Summerton and Martock' (p. 56); *Ribes rubrum* 'by a waters side at Clover [Clewer] in Somersetshyre in the

1 Turner, l.c., says he found it also 'aboute Colon in sandye groundes'. Probably the Essex plant was *Glaux maritima*.

possession of Maister Horner' (p. 63).[1] He describes two or three pleasant dietetic details. Of Spinach he writes, 'Spinage or spinech is an herbe lately found and not long in use, but it is so wel knowen amongest al men in al countrees, that it needeth no description....I knowe not wherefore it is good savinge to fill the belly and louse it a litle. But with those profittes it hurteth the stomach and bredeth winde' (pp. 71-2). Of 'Sampere' (*Crithmum maritimum*), after repeating Dioscorides's way of keeping it in brine mentioned in *Herbal*, p. M. iii*b*, he adds, 'but dwelling in the farther of Summersetshyre not far from the sea syde, where as I had good plentye of Sampere, I found an other way whiche lyketh me and all them that have prove it, muche better than the other. I sethe the sampere in whyte wine till it be metely tender: then I put it into so much whyte vinegre as will cover it' (*Herbal*, I, p. 173). Of 'the Fen shrub or bushe called Gall' (*Myrica gale*), after explaining that apothecaries have no business to use it instead of Myrtle he adds, 'It is tried by experience that it is good to be put in beare, both by me and by diverse other in Summersetshyre' (III, p. 48). We can picture him making the experiment.

In 1562 the second part of his *Herbal*, which begins with Fagus and carries on the work to the end of the alphabet, was published by Arnold Birckman at Cologne along with his treatise on Baths, and with the 'Homish Apothecarye translated out the Almaine speche into English by Jhon Hollybush' and dated 1561. It was dedicated to 'the right honorable and his verye good Lorde Syr Thomas Wentworth Knight, Lorde Wentworth', the son of his patron who had died in 1551; and the dedication is not without interest. He stresses that along with the identifying of herbs he desires to 'confute (for the quantite of the learning that God hath geven me) the errours of them which have writen of late, and have notably erred, and stifly have defended their errours'. This will bring criticism upon him from those who regard such disputations as needless and from 'other dark doctores, not bullati, but boletarii, which soddenly, lyke todestolles stert up Phisiciones within two or thre yeares study', who 'in all theyr lives never have found agayn the knowledge of one lost herb, tre, fishe, byrde, beaste, or metall' but who 'will with great arrogance and lofty lookes dispise other mennis short bookes'. Others will complain that he has not written in Latin—though he has only done 'what Tragus, Rembertus and Matthiolus, honest and learned

1 For this locality there is a more unlikely plant (*Antirrhinum asarina*) in De l'Obel, MS. *Stirpium Illustrationes*, II, p. 35 (the MS. being in Magdalen College, Oxford), 'Asarina comitatus Somerseti in a narrow lane by which one goes to Clower the seat of Thomas Horner towards Mendiep'. Presumably Horner was known to both the botanists. He was the father of Sir John, knighted 1584, cf. *Visitation of Somerset* (Harleian Soc. xi), p. 57.

men, have already done'. So he pays his tribute to Wentworth 'whose father with his yearly exhibition did helpe me, beyng student in Cambridge of Physik and philosophy'.

This preface makes it probable that although much of the book had no doubt been written before he left England in 1553, much is subsequent to the publication of Mattioli's *Commentarii* in Latin in 1554 and much was almost certainly penned at Weissenburg. But tempting as it is to suggest that he left the script in Cologne on his way back, the facts make this unlikely. There is a note under Oak on p. 109, 'It was told me by a learned man, a frende of myne, that in the year of our Lorde MDLVII there was a great plenty of galles found upon oke leves in the North countre of England, and namely about Hallyfax', which can hardly have been supplied before his return to his own country. The quality of the printing (two and a quarter pages of 'Fautes' including a picture labelled Sabina, inserted as Ruscus, and representing neither, are listed at the end of the book) suggests that it was set up by someone ignorant of English, and not closely supervised.[1]

Since 1551 not only had he gained a much wider experience of German plants but several important additions had been made to the literature of the subject. Thus in *Herbal*, II, p. 3, he quotes from 'a Christian physician Hieronymus Tragus', 'out of his Herbal'—probably a reference to the Latin version of Jerome Bock's *Kreuter Buch*, published at Strasburg in 1552; and from this book seems to be derived the curious picture of the 'Tribulus aquaticus' (*Trapa natans*), *Herbal*, II, p. 156. So, too, on p. 120, he mentions his friend Conrad Gesner's book '*In Mounte Fracto*' as he calls it—the little treatise on the plants of Mount Pilatus, published in 1552. There had also appeared in 1554 the Latin version of Pierandrea Mattioli's work, the *Commentarii de medica materia in Dioscoridem*, published in Venice. If Turner had known Mattioli before, it was only by hearsay and through Luca Ghini: now he had read him with care 'in the second edition' as he calls it in *Herbal*, II, p. 134*b*—since the book was originally published in 1544 in Italian. Though he speaks highly of him as of a friend of Luca Ghini and in general pays serious attention to his opinion, he is often critical and sometimes caustic in his attitude to it and rarely accepts it without reservations. 'As to the figures of Matthiolus [often charming bits of landscape or bouquets of flowers] I must nedes confesse that they are fayre' (*Herbal*, II, p. 54): yet 'I am far deceyved, except the figure [of Sium] that Matthiolus setteth out, be

1 Arnold Birckman had, as we learn from J. Denucé (*Correspondance de C. Plantin*, IV, p. 34 note), a branch house 'La Poule grasse' in Antwerp directed by Arnold Mylius: it is possible that Turner's arrangements were made with this nearer establishment. It was to Silvius, another Antwerp printer, that he sent his edition of William of Newburgh.

not lyke the monstre that Horace maketh mention of, whych hath a mannis head, set upon a horse neck' (II, p. 138).[1] The fourth new book is the *Enarrationes in Dioscoridis libros* by Amatus Lusitanus[2] published at Venice in 1553. This, as we have already indicated, Turner regarded with contempt: indeed, he never mentions Amatus except to castigate him, as in *Herbal*, II, p. 83*b*: 'Petasites Fuchsii which is the true Petasites Dioscoridis Ruellii and Rembertes Petasites, and myne is not the second Arcion, for all the gaynsaying of Matthiolus the Italian and Amatus the Spanyarde who wold face out learned men with stout checkes without any sufficient proof.' Probably his opinion of the two men was much like that of his friend Gesner who said of Mattioli 'I agree with him and forgive his faults: he accepts my friendship so long as I don't criticise: he is surely too ambitious';[3] and of Amatus 'he is reckless and ignorant'.[4] Of a fifth, Rembert Dodoens, he never explicitly quotes any writing and it is probable that he had not seen any: but he alludes on occasion to his opinion—'the opinion of Ruellius which Fuchsius, Rembertus and I hold alltogether' (*Herbal*, II, p. 80*b*). The great Flemings belong in reality to the next generation.

Another side of his activities during these last years at Wells is mentioned by Cooper in his *Athenae Cantabrigienses*, I, p. 259: he says that Turner, besides a collation of the translation of the Bible 'with the Hebrew, Greek and Latin versions making many emendations', 'also prepared for the press William of Neuburgh's Historia rerum Anglicanarum from a MS. in the library at Wells and sent it to Antwerp to be printed by W. Sylvius. It appeared there 1567 but certain chapters sent by Turner were omitted, as was his preface, for which Sylvius substituted one of his own. A copy of this edition annotated by Turner was formerly in the possession of William Fulman of Corpus Christi College, Oxford.' Cooper no doubt drew his knowledge of this from Thomas Hearne, who gave an account of the work in his *Heming's Cartulary*, II, pp. 669–71, based upon a copy of Silvius's edition corrected by Turner who intended to publish a full text with a new title and preface. His statement is certainly correct in so far that the first printed edition of the Chronicle of William of Newburgh (the Augustinian priory near Coxwold in Yorkshire), the *Rerum Anglicarum libri quinque*, was published at Antwerp by Gulielmus Silvius, printer to the King, in 1567, in a thick duodecimo volume with a dedicatory preface addressed to Queen Elizabeth by him-

1 Of this picture Gesner, *Epist. Medicinal* (Zürich, 1577), p. 146, writes, 'we laughed at this ridiculous and monstrous Sion'.

2 Fallopius speaks of him as 'a physician of good note although a Jew' (cf. Parkinson, *Theatrum*, p. 174). But Mattioli resented his work and attacked it in his *Apologia adversus Amathum* (Venice, 1558), attached to which is a long catalogue of his mistakes.

3 *Ep. Med.* p. 7*b*.

4 L.c. p. 105.

self, and that according to R. Howlett, who edited the *Chronicles* for the Rolls Series, Silvius's edition 'omits many chapters and contains a marvellous number of mistakes'. He adds that it was printed from a manuscript 'apparently not now extant' (Preface, p. lv).

Turner's connection with this work is borne out by the statement in the dedication of his *Herbal* of 1568 where he speaks of his suffering at the hands of 'a craftie covetous and Popishe printer who suppressing my name and levinge oute my Preface set oute a booke (that I set out of Welles and had corrected not without some laboure and coste) with his Preface as though the booke had bene his owne'. If the story is true it is an interesting addition to our picture of Turner; for William of Newburgh is the best of the early English chroniclers, a man with a real insight into the meaning of history and a critical attitude towards the marvellous; and that Turner should have detected and arranged for the production of his work is testimony not only to his love of antiquity but to his energy and versatility. That a foreign printer, capable of such piracy, should have done his work badly, is hardly surprising. It does not reflect upon Turner's own share in the production.

During these years he seems to have been seriously troubled by the stone which had already caused him anxiety ten years earlier. It is notable that many of the plants included in Part III of the *Herbal* are medicines for this complaint—a curious collection of remedies most of them innocuous but revealing the quality of the traditional pharmacy. His book *Of the natures and properties of all wines that are commonlye used here in England*, dedicated to Cecil 'sometime his co-student in the University of Cambridge', is specially directed against those 'that holde that Rhennish and other small white wines ought not to be drunken of them that either have or are in daunger of the stone'. This was bound up with a similar book on *The Properties of the three most renouned Triacles*. The two were printed in London by William Seres and issued in 1568. They may probably be the product of Turner's last years when he had given up his residence at Wells and was living in London, an 'old and sickly' man.

For in spite of his botanising and explorations Wells does not seem to have been wholly congenial. He was a north-country man, outspoken in his comments upon men and things, contemptuous of ritual and clerical clothing, and defiant of ecclesiastical control. His bishop, Gilbert Berkeley, was well qualified to illustrate his strictures and provoke his jests; and the episcopal letter, quoted by Jackson, shows that the dean, and his dog, had found an easy victim, and that the disciple of Latimer had not ceased to 'speak unsemely of all estates'. In 1564, when orders as to the vesture of the clergy were issued, he was suspended for refusal to conform, though no other action was taken against him, and he was left in possession of his stipend. He gave up residence in his deanery, came to his house in Crutched Friars, and spent his last days there.

There is a considerable amount of detailed evidence as to the so-called Vestiarian controversy in which during these last years he was engaged. In addition to his bishop's letter we have several stories of his rather ribald mockery of the square cap and the surplice which as the peculiar badges of the clergy he strongly disliked. We have, in Robert Crowley's *Briefe Discourse against the outwarde Apparell of the popishe Church*, a protest drawn up in 1566 in the name of the London ministers, in which no doubt he had a share. We have in the Zürich letters evidence of the discussion between the English reformers like Laurence Humphrey, Thomas Sampson and Thomas Lever, who shared Turner's views, and the Continental leaders, Henry Bullinger and others. And we have Turner's own letter of 23 July 1566 (*Zürich Letters*, II, pp. 124–6), expostulating with Bullinger for his attitude of comparative tolerance. The matter is not of great importance except as illustrating Turner's own character, in which independence of outlook and what his friends would call firmness and his critics obstinacy were strongly marked. The description attached to his signature, 'A physician delighting in the study of sacred literature', indicates that his deanery was never more than a secondary interest; and this is borne out by the account of his death sent by John Parkhurst Bishop of Norwich to Gualter and Bullinger: 'on the 13th[1] July [1568] Dr William Turner, a good physician and an excellent man, died at London: Lever[2] preached at his funeral' (*Zürich Letters*, I, p. 206).

He was buried in his parish church, St Olave's, Hart Street; and John Stow in his *Survey of London*[3] has left a record of him and of his monument.

William Turner was an ancient Gospeller, Contemporary, Fellow-Collegian and Friend to Bishop Ridley the Martyr. He was Doctor of Physic in King Edward the Sixth's days and domestic Physician to the Duke of Somerset, Protector to that King: He was also a Divine and Preacher, and wrote several books against the errors of Rome; and was preferred by King Edward to be Dean of Wells: and, being an exile under Queen Mary the First, returned home upon her death and enjoyed his Deanery again. He was the first that, by great labour and travel into Germany, Italy, and other foreign parts, put forth an Herbal in English, anno 1568, the groundwork of Gerard's Herbal, and then lived in Crutched-Friars, from whence he dated his Epistle Dedicatory of that Book to the Queen.

1 Or 7th as on Memorial, see below.

2 No doubt Thomas, the eldest of the three exiled brothers: he had been Master of St John's College, Cambridge, 1551–3.

3 Ed. Strype, 1754, I, p. 379: an English translation of the epitaph, slightly different, is in B. Brook, *Lives of the Puritans*, I, pp. 131–2, being taken from J. Ward, *Lives of the Gresham Professors*, p. 130: cf. also C. H. Cooper, *Ath. Cant.* I, p. 257.

In the south-east wall, a Stone engraven, without any plate, bearing this inscription:

'Gulielmo Turnero, Medico ac Theologo peritissimo, Decano Wellens. Per annos triginta in utraque scientia exercitatissimus, Ecclesiae et Reipublicae profuit, et contra utriusque pernitiosissimos hostes; maxime vero Romanum Antichristum fortissimus Jesu Christi miles acerrime dimicavit, ac tandem corpus senio et laboribus confectum, in spem beatissim. resurrectionis hic deposuit; devictis Christi virtute mundi carnisque civibus cap. [Cooper reads 'copiis'] triumphat in aeternum.

'Magnus Apollinea quondam Turnerus in arte,
Magnus et in vera religione fuit:
Mors tamen obrepens majorem reddidit illum.
Civis enim Coeli regna superna tenet.
Obiit 7: die Julii, An. Dom. 1568.'

CHAPTER VII. WILLIAM TURNER, SCIENTIST

Only a few months before his death Turner had written the dedication to the 'most noble and learned Princesse in all kindes of good lerning Quene Elizabeth' of the complete edition of his *Herbal*. This 'preface', besides its mention of his Latin conversation with the queen and of her favours to him, states with some dignity his claim to have been among the very first to produce a herbal 'above thyrtye yeares ago' 'beyng yet felow of Pembroke hall in Cambridge', to have been able to correct Fuchs and others, to have known from Ghini much that was first published by Mattioli, to have 'taught the truthe of certeyne plantes' unknown to or wrongly described by others, and to have done much work in the field at first hand. He concludes it by saying that if he can be protected 'from my enemies, whiche have more then these eight yeares continuallye troubled me verye muche, and holden me from my Booke' and from sickness he will 'set out a booke of the names and natures of fishes'.

Next to the Preface is a table of the Names of Herbs, very incomplete, and a list of authors, eighteen of the ancient world from Damocrates to Simeon Sethi, mostly familiar, and eighteen, beginning with Ermolao Barbaro, from the modern, some of whom are unfamiliar. Among them is Geraldus Delwicus (Gerard Veltwick) already mentioned; Gabriel Gabrielis who must surely be Gaspare de Gabrieli, the nobleman of Padua to whom Amatus makes frequent allusion in his *Enarrationes* (pp. 9, 18, 290, etc.) and of whom Gesner (*Horti Germaniae*, p. 239*b*) says that he devoted all his powers to the care of his garden; and Matthias Curtius, who may be either the Matteo Corti of Mattioli, *Epist.* p. 439, who lectured on medicine at Bologna, 1538–40, and whose *Dosandi Methodus* had been published in 1536, or else the merchant of Libau whose famous garden had been visited by Gesner and was reported by him to be full of rare plants (l.c. p. 238). Gabrieli and Curtius both figure largely in Gesner's catalogue of garden-plants, and were perhaps known on that account to Turner.

The rest of the work consists of a fully revised edition of the *Herbal* of 1551, a reprint without the dedication of that of 1562, and the new third part. It, too, is printed, on the whole very badly, by Birckman of Cologne. The long list of Errata in Part II still appears: but Part III is, if possible, even more inaccurate. It is bound up with the treatise on Baths with its original dedication and the 'Homish Apothecarye'.

The revised edition of the first part of the *Herbal* represents a large amount of new work. This consists of the insertion not only of several long and scholarly discussions as to identification, like the very full treatment of Wormwood at the beginning[1] and of Bean at the end of the book, but of additional details, medical and observational, such as the record of a case of Aconite poisoning in Antwerp with its added note of warning to Londoners (*Herbal*, I, p. 21). Thus the short section on 'Acorus' (*Herbal*, B. ii and ii*b*) is expanded not only by a new description of the plant quoting Ruel and a description from Pliny, but by the expansion of the former paragraph by references to the views of Antonio Musa, of 'Actuarius' and Serapio, of Nicolaus Myrepsus whose work had been translated by Fuchs at Tübingen in 1547, and of Mattioli in opposition to Musa. We have already alluded to the interesting alteration that he makes under 'Brassica marina' (*Herbal*, I, pp. 90–1) by separating the white-flowered fleshy-leaved 'Scurby wede' (*Cochlearia officinalis*) of Friesland and Purbeck from the 'Soldanella' 'which a far of loketh like a Mallowe both in floures and leaves, but when a man commeth nere it loketh so lyke a Withwinde [*Convolvulus sepium*] that a man wold saye that it were no thing elles but a Sea withwinde'. A similar alteration distinguishes between the two forms of 'Cepea', one *Veronica beccabunga* and the other *Atriplex portulacoides* (*Herbal*, I, pp. 123–4)—and in this case he speaks of the additions as made at Weissenburg. Most of the additions, at least twenty, some mere paragraphs, but others as under Calamint (*C. acinos*), 'Berefote' (*Helleborus viridis*) and Cucumber substantial in length, are obviously due to Mattioli. Though he calls him 'a man wel sene in Simples and as some judge best learned in them of al other

1 Pp. 1–16. By a printer's error the picture of 'Absinthium Ponticum' (in *Herbal*, p. A. iiii) has been replaced by a duplicated copy of that of 'A. marinum' (*Herbal*, p. A. iiii*b* and I, p. 10). This mistake is pointed out in J. Bauhin, *De Plantis Absinthii*, pp. 51–2 and *Hist. Plant. Univ.* III, i, pp. 170–1.

new writers' (*Herbal*, I, p. 128) he is still usually critical, often strongly so; but particularly when Amatus can be brought under the same condemnation.[1] Here is an example in *Herbal*, I, p. 6: 'A certayne Spanyard sometyme called Joannes Rodericus[2] and afterward I cannot tell by what chaunge named Amatus Lusitanus, a verye ape unto Matthiolus, but muche behinde him in learninge who semeth to have taken a great parte of his boke out of the Italian Commentaries of Matthiolus.'[3] In one place on the Box (*Buxus sempervirens*) he finds the two in conflict; and then he supports the Italian. Occasionally short passages in the first version are cut out—an indecisive section under Chrysanthemum, p. K. i*b*, and a sentence under Cornell, p. M. ii. At the end of the book there are transfers, of 'Elatine' (*Polygonum convolvulus*),[4] pp. O. vi*b* and P. i, and 'Chokewede' (*Orobanche*), p. P. v, and excisions, of 'Orchall otherwyse called Corck', p. P. i*b*, and 'Epimedium', p. P. v*b*. Two sections, 'Daucus' and 'Eryngium', are almost entirely rewritten.

The third Part of the *Herbal* was dedicated to 'the right worshipfull Felowship and Companye of Surgiones of the citye of London' and dated at Wells, 24 June 1564.[5] In the opening sentences he explains that, whereas the previous parts have dealt with plants mentioned by old writers, there are others found since their time which have value for medicine. These are the subject of this final volume. It is offered to the surgeons partly because many of its contents 'belong unto Surgery' and partly because when he was in London Master Wright, late surgeon to the Queen and others of his acquaintance, laid the task upon him. After the dedication comes a series of lists of plants that are hot in the first, second, third and fourth degrees, and then similarly of those that are cold, moist or dry; and then an index. The contents consist of a number of English plants, some very common: they are grouped together, following the example set by Brunfels and already followed in *Names*, because they are

1 Criticism of these two is found also in Johnson's Preface to his edition of Gerard's *Herball*: he writes, 'Amatus was an author of the honesty of Mattiolus: for as the one deceived the world with counterfeit figures, so the other by feined cures to strengthen his opinion': and Johnson had not seen this work of Turner.

2 Amatus speaks of his commentaries as having been first known 'sub nomine Joannis Roderici Lusitani', *Enarrat.* pp. 6, 168; in the latter passage he defends the right of authors to change their names: in the former he declares that his commentaries on Dioscorides were written fifteen years ago.

3 Mattioli, *Apol.* p. 14=*Epistolae*, p. 153, in a long and savage attack, asserts that 'Amathus' has pirated his *Enarrationes* from his own writings. From *Herbal*, I, p. 100, it appears that Turner knew this attack.

4 This is clearly the right identification of Turner's plant—so Cordus but not Bauhin, *Pinax*, p. 252, who identifies it with *Linaria elatine*—Turner recognises that his plant differs from the Elatine of Dioscorides.

5 R. Potts in his account of Turner prefixed to the 1851 reprint of the *Huntyng of the Romishe Fox* states that this part was first published separately in 1566.

not found in the ancient authorities. In consequence the identification is less laborious and the 'vertues' less expanded. For these he quotes Bock (Tragus), 'the practitioners of Germany', 'the herbaries', and his own experience. Of some, novelties like Spinach and 'Rosa Solis' (*Drosera rotundifolia*), he is openly contemptuous. Of some, and curiously enough *Digitalis* is among them,[1] he knows of no uses. Several including 'Herbe oneberrye' (*Paris quadrifolia*) are plants previously mentioned on which he makes additions and corrections; it is proof of his greatness that he accepts a correction from Mattioli (p. 35). There are included a small number of foreign drugs, and some of these come from interesting places. Thus Guiacum is from 'Taprobana [Ceylon] and Java and the Ilandes of Inde' (p. 34); Nutmeg (*Myristica officinalis*) from an 'Ilande of Inde called Badon' (p. 40); the best Rhubarb[2] from 'Tanguth [Kan-su], that is in Sinarum regione, throw the lande Cataia [Cathay], into the land of the Perses, Whereof the Sophia is the ruler, and from thence it is sent to Egipt, and so to Italy' (p. 64); and Cassia from 'the Weste newe found Ilandes, out of Hispaniola, Cuba and Paria' (p. 20).

Yet even here it is disappointing to find how little he really appreciated the novelties which exploration was bringing to light or the vast changes in human thought and life to which it was the prelude. Only one page in all his writings has any reference to it—*Herbal*, II, p. 90*b*. There, under pepper, he remarks, 'There ar many thynges found out of late yeares by the sayling of the Portugalles and diverse other adventurus travalers in far countres', and quotes 'out of Lewes Bartomanni [Ludovico di Varthema—usually quoted in England at this time as Lewes Wertomannus or Vartomannus] fift boke of the thynges that he sawe in Inde' as to the existence of pepper in Calecut and in Taprabona. But apart from a mere mention on the same page of 'Mappheus the noble physiciane', presumably Giovanni Pietro Maffei the Jesuit and author of *Historiarum Indicarum libri XVI*, there is nothing more.

The pictures are similar to those in the previous parts: indeed two, the Valerians of III, p. 76, are identical with those called Phu in II, p. 86. One printed under 'Duch Pimpinell' on p. 10 reappears as 'Saxifragia alba' on p. 68. Probably this book had less supervision in proof than its predecessors; and the number of printers' errors is as Peter Turner afterwards admitted improperly large. It is regrettable that the attempt which Peter obviously made at one time to produce a revised edition of his father's book was never carried into effect. He and his friend Thomas Penny might then have given us work of outstanding importance.

1 Digitalis was first depicted and named from 'Fingerhut' by Fuchs, *De Hist. Stirp.* p. 892, in 1542.

2 For a discussion of this trade, cf. Laufer, *Sino-Iranica* (Chicago, 1919), pp. 547–51.

William Turner's death was in any case a serious loss to English science, culture and religion. He had expressed the intention of doing further work on birds and fishes, and had in hand material for a fuller treatment of both fields. Freed as he became in the last years of his life from anxiety about his own safety and the welfare of his family, reasonably secure in regard to money, living in London and within easy reach of the Dutch and Flemish centres, and with a rapidly increasing knowledge both of English students and of foreign experts, he ought to have had another ten years of useful research and writing. That he was exhausted prematurely by the tension of current events, by his own long exiles and by lack of skilled assistance, and that if he had confined himself to medicine and cared nothing for social and religious reform, he would have done more and suffered less, is no doubt true—though it is not clear that his value to the world would have been enlarged. Scrappy as it is, enough remains of his learning in many departments for us to be able to appreciate his achievement and acclaim it.

Of his work as a naturalist, indeed as the true pioneer of natural history in England, it is not easy to speak too highly. He began at a time when almost nothing was known of the ancient and scientific studies which Graeco-Roman culture had bequeathed to Europe. The English names of plants show a folk-lore full of charming fancies and quaint beliefs, but wholly unscientific in its attitude; and in itself affording no sort of basis for the development of knowledge. Englishmen had obviously enjoyed their countryside and found delight and occasional usefulness in its herbs and flowers, its beasts and birds: but they had not begun to study it or even to describe and catalogue its denizens. The fact that for the first records of the British flora it is useless to go behind Turner, and that he has left (according to W. A. Clarke) recognisable notices of 238 native species, is in itself sufficient proof of his achievement.

The passion to give accurate names to the flora and fauna was evidently ingrained in him. His *Libellus*, even if it mentions Ruel and Brunfels, shows almost no acquaintance with them: it is an original attempt to discover the English equivalents of Greek and Latin plants, and on the whole a very successful venture. The *Avium praecipuarum*, longer and more explicit, reveals the method—a comparison of the Greek (in the original and in translation) and the Latin authorities and then an examination of the native species to which they seem to be referring. That he is more successful with the flowers than with the birds is due to the medical importance of plants and the fuller literature of herbalism. But even there the difficulties as we have constantly seen were immense. Nothing is more admirable in him than the patience and persistence with which he wrestles with them.

Of his care in identification the clearest evidence can be found by a comparison of the first and the later edition of Part I of the *Herbal.* The rejection of a few species, for example *Epimedium*, presumably on the ground of uncertainty as to its true identity; the alteration of others as in the case of 'Brassica marina'—a case which has even eluded the vigilance of W. A. Clarke; and above all the labour and ingenuity with which the meaning and arguments of the different authorities are examined as in the case of the masterly treatise on 'Absinthium'; these demonstrate the care with which he undertook his task. No doubt he is mistaken in ascribing too great accuracy to the ancient authors and in his consequent insistence that their words must be given an almost biblical infallibility, and, somehow, be accepted as reconcilable with the facts. No doubt he is prejudiced: he plainly disliked Mattioli and carried his hatred of Spain into a vendetta against Amatus Lusitanus.[1] But for the most part his independence of outlook is linked with a sound judgment and a strong desire for truth; and the result is a list which 'may be considered the foundation of our British Flora' (Clarke, *First Records*, p. 187).

His main work was inevitably with identification; and it might well be urged that the compilation of a druggist's list of accurately named simples was his chief concern. If so, his claim to be a naturalist would be arguable. We need not deny that his object in most of his publications was practical, that he was deeply interested in the 'uses and vertues' of herbs, and that the living plant was hardly more important to him than its dried leaves and blossoms. Nevertheless he had done a large amount of field work, not only in the herb-gardens of western Europe and the parks of his patrons, but 'abrode in the feldes', on the Alps, along the Rhine, in Friesland and on the heaths and beaches of England. Nor was his interest solely in the search for new species: he was fascinated by the structure and growth of what he saw, and had a real gift for its description. Here for example is a passage that illustrates his quality—as well as incidentally adding a new 'first record' to his list. It occurs in his treatment of 'Orobanche' in Part II, p. 71*b*—and may be compared with the similar but much shorter account of 'Chokewede' in *Herbal*, p. P. v. We will set them out for comparison. In 1551 he wrote: 'Orobanche is a litle stalke, somethynge red, aboute twoo spannes longe sometyme more, rough, tendre, without leves: the floure is somewhat whyte, turnynge towarde yellow... I have marked my selfe, that thys herbe groweth muche aboute the rootes of broome, whyche it claspeth aboute wyth certayne lytle rootes on every syde lyke a dogge holdyng a bone in hys mouth: notwythstandyng I have not seen any broome choked wyth thys herbe: howe be it I have seen the herbe called thre leved grasse or claver utterly strangled, al the naturall juice clene drawen oute by thys herbe.' And here is the version of 1568.

1 Possibly Amatus's dislike of Turner's friend Fuchs (cf. *Enarr.* p. 160, etc.) or his sojourn at Ferrara may be partly at least responsible.

'The herbe is comenly a fout long and oft longer; I have marked it many yeres, but I colde never se any lefe upon it. But I have sene the floures in diverse places of diverse colores, and for the moste parte where so ever I saw them, they were redishe or turnyng to a purple color in som places, but in figure they were lyke unto the floures of Clare [*Salvia verbenaca*] with a thyng in them representyng a cockis hede. The roote is round and much after the fasshon of a grete lekis hede, and there grow out of it certayn long thynges lyke strynges, which have in them in certayn places sharp thynges lyke tethe, where with it claspeth and holdeth the roote that it strangleth. I have found it oft tymes claspyng and holdyng mervellously soft the rootes of broum, so that they looked as they had ben bound foulden oft about with small wyre. And ones I found thys herbe growyng besyde the comon claver or medow trifoly [*Trifolium pratense*], which was all wethered, and when I had degged up the roote of the trifoly to se what should be the cause that all other clavers or trifolies about wer grene and freshe, that that trifoly should be dede, I found the rootes of Orobanche fast clasped about the rootes of the claver, which as I did playnly perceyve draw out all the natural moysture from the herbe that it should have lyved with all, and so killed it as yvi [*Hedera helix*] and dodder [*Cuscuta europaea*] in continuance of tyme do with the trees and herbes that they fould and wynde themselves about.' Verbose perhaps, but admirably descriptive. The man who wrote and expanded that record was a great naturalist.

His observations of birds are, on occasion, equally accurate. We have quoted his boyhood's account of a Robin's nest: here is a more mature record based evidently upon observations during his first exile and dealing with the Great Grey Shrike (*Lanius excubitor*). 'The bill is black and moderately short and hooked at the tip: it is the stoutest and strongest of all, so much so that once it wounded my hand even though I was protected by a double glove; it very quickly breaks and crushes the bones and skulls of birds....It has short wings and flies as if it were jumping up and down. It lives on beetles, butterflies and the larger insects, but not only on these: like a hawk it lives on birds. It kills Goldcrests and Finches and (as I once observed) Thrushes. Bird-catchers even report that it sometimes slays certain woodland Pies and drives away Crows. It does not fly down the birds that it kills and strike them with its claws like Hawks, but ambushes them and attacks them and (as I have often noted) aims at the throat, and squeezes and breaks the skull with its beak. It fractures and crushes the bones and then devours them, and when it is hungry crams into its throat such gobbets of flesh as the narrow gape can contain. Moreover unlike other birds when it has abundance of prey it stores up some of it against future shortage. It impales and hangs big flies and insects that it has caught on the thorns and spines of bushes.

It is of all birds the most easily tamed; and when used to the hand is fed on meat. If this is dry or bloodless, it requires drink. I have seen it in England not more than twice; in Germany very frequently.' It is of course only a winter-visitor with us but nests freely in the Rhineland.

A later record occurs in his parable of the Kite (*Milvus milvus*) which looks like a hawk but behaves like a scavenger (*Spirituall Physic*, p. 54). 'In the tyme that he byldeth hys neste he caryeth al that he can catche and snatche unto it, ragges, cloutes, napkins, kerches, boyes cappes, and some tyme purses as I have herde saye. And all the hole yeare thorowe there is no pray that cometh amysse unto him, he eateth upon al kynde of carion....He is so bold some tyme in England (I never saw it so nether in Italy nether in any parte of Germany where as I have bene) that he dare take butter and bread out of boys handes in the stretes.'[1]

None the less he was a man of his time; and as such not naturally disposed to set up the results of observation against the authority of tradition or ready to challenge the ideas of his day. The most familiar instance of this is his acceptance of the legend of the Barnacle Goose on pp. 24–5. But here it is the weight of evidence (as he estimates) that overcomes his scepticism. 'No one', he writes, 'has seen the nest or egg of the Barnacle:[2] and this is not surprising, for without the aid of a parent Barnacles have spontaneous generation in this way. When at a definite time the last of a ship, or its planks or yards of pine, have rotted in the sea, from them there break out at first what seem like fungi [*Lepas anatifera*]: in these after a time, the obvious shapes of birds become visible, then these are clad in feathers, then they come to life and fly. This, lest anyone suppose it a fable, is not only the common testimony of all the coastal folk of England, Ireland and Scotland but Gyraldus [Giraldus Cambrensis] the famous historian who wrote the history of Ireland far more happily than could be expected at his date[3] testifies that this is the exact generation of the Barnacles. Yet it seemed hardly safe to trust common report, and for the rarity of the thing I did not wholly believe Gyraldus: so while I was thinking over what I am now writing, I asked a man whose obvious integrity won my confidence, a man by profession a theologian, by race an Irishman, by name Octavian, whether

1 An expansion of the note in *Avium*, p. 95: cf. above, p. 88.

2 This is not in fact true: for Albert the Great had recorded that Barnacle Geese had laid in captivity, *De Animalibus*, XXIII, p. 186; so Max Müller, *Lectures on the Science of Language*, II, p. 599. Albert's statement is quoted in *Ortus Sanitatis*, De Avibus, ch. 20, under 'Carbates'. But Turner's view was widely held by his contemporaries, e.g. by William Bullein, *Booke of Simples*, p. 77, 'Barnacles...never lay egs as the people of the north parts of Scotland knoweth': Bullein adds, 'least it shoulde seeme incredible to many I wyl geve none occasion to any either to mock or to marvel', although in fact he has told the story on p. 12 of the same book.

3 Cf. *Topographia Hibernica*, Dist. I, ch. 15 (ed. Dimock, pp. 47–8).

he thought Gyraldus reliable in this respect. He took oath by the very gospel that he professed, and replied that it was the perfect truth that Gyraldus had recorded about the generation of this bird; that he had himself seen with his own eyes and handled with his own hands birds still half-formed; and that if I stayed a month or two in London he would see to it that some half-formed birds were soon brought to me. This generation of the Barnacle will not seem so very marvellous to those who remember Aristotle's account of the winged Ephemerus....' Thus he has done his best to collect evidence, and realises that the event is, at least, highly peculiar. Nevertheless, that he was convinced of its truth is emphasised in the letter sent to Gesner after the publication of his book.[1] In days when spontaneous generation was universally believed (for was it not both classical and biblical?) he can hardly be blamed.

That he accepted and defended with vigour the belief in such generation is clear from the long introductory section of *The Huntyng of the Romyshe Wolfe*. On the assertion that 'it is not against nature that one beste should go out of kind into another' he writes, 'Pismiers go out of kind in Flies, Case wormes [Caddises] or Cod wormes become Flies, Cole wormes [Cabbage caterpillars] growe into Butterflies, Serpents into Dragons. Aristotle writeth in his boke *de natura animalium* that a Redbrest is turned into a Redtaile...Apuleius was turned into an Asse...Nebuchodonosor (as it appereth by the fourth of Daniel) was chaunged into a beast' (p. B. i). This by itself might be merely metamorphosis—the insect changes are true: he had himself denied Aristotle's bird-case: he cannot have taken the Golden Ass seriously. But he follows it by arguments that are plainly sincere—cases of the appearance of 'Bansticles' or 'Stikhelbeges' (*Gasterosteus aculeatus*) in newly dug ditches, of Eels in isolated ponds, of Rats in fresh-built ships (p. A. vii). These arguments are made by the Hunter, and though to the modern mind the Dean's replies are the more convincing, Turner certainly put his own views in the Hunter's mouth.

Nor does this case stand alone. Turner was a man of his own time; and by temperament averse from novelty. But it is with something of a shock that a modern reader finds him adding to his account of the 'Blewbottel' (*Centaurea cyanus*) the naive remark, 'It groweth muche amonge Rye, wherefore I thynke that good Ry, in an evell and unseasonable yere, doth go out of kynde into thys wede' (*Herbal*, p. N. iiii*b* and I, p. 189), or of Rye, 'thys I do know, that in a countre where as I have ben, wythin the Dukedom of the Duke of Cleve, called Sourlant [Saarland], that wheat if it be sowen in that sourlande, as it is truely called, the fyrste yeare it will bring furth wheat, and in the second yeare, if the wheat that grew there be sowen in the same place agayne, that it turneth into rye,

1 *H.A.* III, p. 96.

and that the same rye sowen in the same ground, within two yeares goeth out of kinde into darnell and suche other naughty wedes' (*Herbal*, II, p. 129).[1] Similarly in the *Avium* on p. 32 he tells a curious tale of anglers smearing their baits with Osprey's fat because of the legend, which he solemnly repeats, 'that when the Osprey hovers, the fishes below him, compelled as is believed by his Eagle-nature, turn on their backs and show their white bellies so that he may choose whichever he likes'—a legend which reappears almost word for word in Drayton's *Poly-olbion*, XXV, ll. 134–8, and is alluded to by Shakespeare in *Coriolanus*, Act IV, Scene 7.

With his respect for the past he is apt to be sceptical, as we have constantly seen, in regard to innovations whether of interpretation or of practice. In *Herbal*, III, p. 79, there is a little note on 'Rosa [*sic*] solis' (*Drosera rotundifolia*) which is characteristic. Of its 'vertues' he writes, 'our English men now adayes set very muche by it, and holde that it is good for consumptions and swouning, and faintnes of the harte, but I have no sure operience of this, nether have I red of anye olde writer what vertues it hath: wherefore I dare promise nothing of it'. Without authorities or experience to guide him he was wise to be cautious: but his preference for the old over the new could hardly be made more plain.

But if in him there is little of that passion for new knowledge which, as we assume, should be characteristic of a pioneer, and still less of that divine discontent and sense of expectation which should belong to one who stands on the threshold of a new age, there were in his reverence for the truth of the past a massive integrity and a hatred of sham which here as in his religious writings find free expression when he detects fraud. The Mandrake (*Mandragora officinarum*) had long been the subject of legend and superstition. This is how Turner deals with those who exploit it: 'The rootes whiche are conterfited and made like litle puppettes and mammettes which come to be sold in England in boxes with heir and such forme as a man hath, are nothyng elles but folishe feined trifles and not naturall. For they are so trymmed of crafty theves to mocke the poore people with all, and to rob them both of theyr wit and theyr money' (*Herbal*, II, p. 46). In spite of his candour, the fraud continued, as the 'Catalogue of Natural Rarities' in Robert Hubert's exhibition in 1664 makes plain.[2]

With this there are all the practicality and commonsense of a man who has lived hard and followed a profession of service to human welfare.

1 So in *Names*, p. E. ii, he had written 'Corne goeth out of kynde into Darnel'. Cf. F. Bacon, *Sylva Sylvarum*, Cent. VI, p. 525: 'It is certain that in very steril years corn sown will grow to another kind'; he devotes the next eight paragraphs to experiments in the 'transmigration of species'.

2 *Catalogue*, p. 41; cf. Peter Mundy's comment, *Travels*, IV, p. 46.

A résumé of his very interesting discussion of Medica or 'Medic Fother'[1] (Lucerne or Alfalfa, *Medicago sativa*) will aptly illustrate his greatness: it fills three pages, *Herbal*, II, pp. 51*b*–53. He begins with descriptions from Dioscorides and from Pliny; and adds: 'Besyde these markes...I have marked, that it hath a yelow flour, and that the lefe, which standeth in the myddes betwene the ii other leves that alwayes grow about it, hath a longer footstalk or stele then the rest have. And the same lefe from the goyng down of the son untill it ryse agayn, foldeth itself inwarde, and then goeth abrode agayn when the son ryseth agayn. After that the yelow flour falleth away, there groweth a litle thyng to conteyn the sede in, which at the first is lyke the end of a writhen gymlet, but after that it beginneth to be rype, it draweth himself together, and is made lyke a litle water snayle or a crooked rammis horne....' Then follow the 'vertues' out of Dioscorides, Pliny, Palladius and Columella—the two latter substantial accounts of its value and treatment as a crop. He ends: 'Thus far have I writen to you the myndes and experience of old autores.... Now it that I have proved my selfe, I will not refuse to shew unto you, my countremen. I have sowen iii kyndes of medic fother, the leste kynde, the grete smoth kynde, and the great rough kynde. The lest kind do I alow leste of all other, because the leves and stalkes are al very litle, and therfore in fedyng of cattel can do but litle service. The grete smoth kynde as I have proved, groweth into a mervelous greate bushe. As for the greate roughe kynde, how greate it will be, I have not as yet proved, for I never sowed it before thys summer. But by all tokens that I can se as yet, it is lyke to be as good and greate as the greate smoth kynde. If ye have but a bushe or ii of Medic, and would fayne have much sede rype before the commyng of wynter, because the medic bushe is very thyck, and therefore hath many floures and sed vesselles that the son can not come to, it is best to take the moste parte of every bushe at the joynt of the herbe...and then ye must set the braunches that ye have plucked of, depe in the grounde, and water them twyse on the day, and they shall bryng furth sede as well as them that are sowen, and muche better then they that are overshaddowed in the bushe and want the help of the son. Thys have I proved diverse tymes, wherefore I dar be bolde to write it.' So he introduced the growing of Lucerne as a fodder-crop to English agriculture.

It is indeed clear that practical devotion to the welfare of his country is a powerful element in his character. Not only did he abandon the project of publishing his *Herbal* in the Latin in which he first wrote it and which would have made it available to the learned world, but he delayed publication until he could more fully explore and describe the

1 The ποία Μηδική of Aristophanes, *Knights*, 606. For its Persian origin and subsequent distribution, cf. Laufer, *Sino-Iranica*, pp. 208–19.

herbal wealth of England so that he might both serve the health and increase the renown of his native land. Though he was evidently appreciated and happy abroad, especially perhaps among the Frisians who even to-day have much in common with the folk of our north-east coasts, he had the kind of patriotism which could not stay happily in exile except when circumstances compelled it.

In the exploration of England for the discovery of its flora and fauna he seems to have been definitely a pioneer. He plainly had the love of nature for itself which as we have seen is lacking in the science of the Middle Ages or even in Wotton. He does not go to plants or birds for their utility, for the fables and moralisings that can be drawn from them, or even for the sake of their literary importance. Unlike Wotton he is plainly attracted by 'the thing itself', its structure and growth, its habit and behaviour. Like the artists who designed the charming woodcuts which he plagiarised from his friend Fuchs he loved what he studied and wished to share with others its attraction.

It is this attitude of objective interest in nature, an attitude at once aesthetic and inquisitive and indeed strongly imbued with specifically religious emotion, which is the source and impetus of the scientific movement. Turner as we have seen shared to the full the literary and archaeological interests of renaissance scholarship. He examined and collated his authorities; he scrutinised his species in the light of them; he laboured with nomenclature and synonyms: it might seem that catalogues were his chief concern. But identification is for him, as for every true scientist, never an end in itself, but a means to appreciation, to the discernment of characteristics and the study of form and function, to that knowledge of the nature of things which is the preliminary to the right use and ordering of man's environment. Medievalism was in the main indifferent to nature, though many a boss and capital, roof-beam and bench-end testify to the craftsman's delight in foliage or figure: it did not regard the exploration of nature or research into it as of any value: indeed it rejected them as an interference with man's real and supernatural end. But with Turner as with many of the Elizabethans came a new spirit of adventure such as had already led to the exploration of the solar universe and the discovery of the New World.

The extent and limitation of Turner's influence abroad and at home are best seen from the evidence of subsequent students. Mathias de l'Obel, the third of the great Flemish herbalists, who was actually resident for most of his later life in this country, refers directly in his *Adversaria*, first published in 1570 in London, to Turner on only four occasions. These are worth noting. The first on p. 92 (edition of 1604) has been already quoted: it is in the account of *Crambe maritima* at Portland, and runs, 'some time ago the English doctor most expert in this subject, Turner, had given me seeds of it and wished it to be called "single-seeded" from its single seed'. This implies that the two men had corresponded, but not met; and this seems probable; for l'Obel was in Italy during the last years of Turner's life. On p. 244 is the note, 'the learned Turner said that the purple primrose reported by the herbalists of Antwerp as growing in England was pulmonaria'—an allusion wholly obscure. On p. 348, 'our very learned Turner told us that this plant [*Meum athamanticum*] is abundant on the Alps of Cheviot on the Scottish border'—a not very exact reference to *Names*, p. E. v, and *Herbal*, II, p. 57. Finally, there is a more general reference when after a stricture upon Dodoens for not testing the medicinal qualities of plants mentioned by Dioscorides and so causing deaths he adds, 'so the pious and learned Turner in his Alphabetical English Herbals'.

Plainly Turner was respected but not well-known—perhaps because De l'Obel's English was at that period not even so good as his Latin. If Turner had published his *Herbal* in the more learned language his influence might have been larger, at least on the Continent. Nevertheless, here is indication that a man who had followed Turner at Ferrara and Bologna, and knew his friends the druggists of Venice and Antwerp and London—Andrea Martinello and Pieter Coudenberg and Hugh Morgan—had a high regard for his work.

More disappointing is the silence of De l'Ecluse. The two men could hardly have met, unless possibly during the last three years of Turner's second exile: for De l'Ecluse never visited England until 1571 when he went to Bristol and met De l'Obel there.[1] But no botanist is more generous than De l'Ecluse in expressing obligation; and we should expect that Turner's name would figure regularly if not frequently among his authorities. In the *Rariorum Plantarum Historia* there seems to be one clear allusion—on p. lxix where he maintains the identification of 'Bistorta' with the Britannica Dioscoridis which Turner had made in *Herbal*, III, p. 13 (see above, p. 89)—an agreement the more remarkable because Dodoens who seems to have no knowledge of Turner rejects the identification (*Pemptades*, p. 333): but there is no mention of Turner nor any direct quotation from his work. This may well be due to the fact that though De l'Ecluse was reputed to know almost all the languages of

1 De l'Obel must have gone to Bristol almost immediately on his arrival in England. His *Adversaria* contains few references to British plants (the 'Small White Waterlily' on the road from London to Oxford and Bristol (p. 257) and the 'Anglica Saxifraga' seen 'on the dry chalk hill [presumably Silbury] piled up by military art between Chipnam and Marlburum' on the road from London to Bristol (p. 183), and two from near Coventry) that do not come from Bristol, where he got *Colchicum autumnale* (p. 54); *Cochlearia anglica* (p. 122); *Hypericum androsaemum* (p. 173); *Blackstonia perfoliata* (p. 173); *Anagallis lutea* (p. 194); *Alchemilla arvensis* (p. 324); *Conopodium denudatum* (p. 325); *Trinia glauca* (p. 331); and *Phillitis scolopendrium* (p. 359).

ancient and modern Europe English was apparently not one of them. Later on, by his visits in 1579 and 1581, the first a brief trip in the autumn, the second a stay of at least six months till July or August, he came to know Rich and Morgan, the Royal Druggists as he calls them; and became intimate with James Garret the younger, the perfumer, whom he describes as 'my dear friend, a man of honour, greatly delighting in the study of herbarism' (l.c. p. cix): but though as we shall see he was intimate with Thomas Penny, and appreciated Richard Garth, Edward Points, Lord Cobham and others, these belonged in the main to a younger generation.

Jean Bauhin, in his very learned monograph, *De Plantis Absinthii* (Montbéliard, 1593), pp. 19, 38, 42, 51, 65, mentions and briefly discusses Turner's identifications (*Herbal*, I, pp. 2, 7 and 9). In his great posthumous book, edited by Chabrey and published at Yverdun in 1650–1, the *Historia Plantarum Universalis*, he constantly quotes and comments upon the *Herbal*, using the edition of 1568 and obviously knowing its principal contents. Thus, for example, he cites Turner on 'Laureola' in I, i, p. 568; on Broom (both the description in *Herbal*, II, p. 7 and the picture l.c. p. 144) in I, ii, p. 396; on 'Cistus' in II, p. 2; on the 'Large Convolvulus' in II, p. 155; on 'Cantabrica Plinii', that much-debated species, in II, p. 159; on the 'Rose of Jericho' in II, p. 209; on 'Moschatellina' at Cologne in III, i, p. 207; and on 'Calfe Snowte' in III, ii, p. 463: he translates a long passage, the whole discussion of 'Brassica marina' (*Herbal*, I, pp. 90–2),[1] in II, pp. 167–8: evidently he appreciated the value and solidity of his work. Yet in general he shows little sign of acquaintance with Turner: nor is Turner ever mentioned in the many letters which he received from Gesner in the years 1560–5. He was too young a man to have seen the exiles of Mary's reign.

Ulisse Aldrovandi regularly quotes Turner's *Avium Praecipuarum*, for example as to the nesting of Cormorants (*Ornithologia*, III, p. 81), but hardly ever mentions his name except when he condemns him (e.g. II, p. 30). He cites him as 'a certain Englishman' and probably knew him only through Gesner's *Historia Animalium*.

Turner's *Names*, though perhaps not his *Herbal*, supplied the English names and a few other details to the *Llysieulyfr Meddyginiaethol* or Welsh Herbal written in the late seventeenth century and probably by William Salusbury, the first translator of the New Testament into Welsh. This Herbal, edited in 1916 by E. Stanton Roberts, is based upon Fuchs, and contains in addition to borrowings from Turner a few, a very few, Welsh localities round the Vale of Clwyd.

William Bullein the contemporary physician and author not only puts Turner at the head of his list of medical authorities in his *Dialogue betweene*

1 Cf. above, pp. 102–3.

Soarenes and Chirurgi, pp. 4, 5, but speaks enthusiastically of him in his *Booke of Simples*, p. 63*b*, 'which doctor is a Jewell among us Englysh men as well as among the Germaynes as Conradus Gesnerus reporteth of hym for hys synguler learning, knowledge and judgement'.

John Gerard in his *Herball* has a few references, couched in respectful language, but of no special importance, pp. 282, 329, 473, 896 (where he records that both Turner and De l'Obel had found *Trinia glauca* at St Vincent's rocks); and on at least two other occasions he lifts notes without acknowledgment, one being Turner's remark about feeding small caged birds on Chickweed (Gerard, *Herball*, p. 491), and the second his local name 'Lucken gollande' (*Names*, C. ii) which appears in Gerard (l.c. pp. 809–10) as 'Locker goulons' and 'Lockron gowlans'. These, for what they are worth, are clearly Gerard's own plagiarisms, for the *Pemptades* of Dodoens on which his *Herball* is based, though it lists Turner as one of its authorities, never afterwards mentions him.

Thomas Johnson, in his Preface to his edition of Gerard, mentions Turner as the first of English students, gives a brief account of his *Names* which he also mentions and quotes later, on p. 1236, in reference to *Lathyrus macrorrhizus*, and adds that he set forth his *Herbal* in 1551, but 'did not treat of many plants'. Evidently he knew nothing of the complete work.

Parkinson, though he mentions Turner some ten times as an authority for nomenclature, nowhere gives any certain proof of first-hand knowledge of his work. He may have known *Names* but almost certainly did not know the *Herbal*. Considering his interest in nomenclature it is a pity that he did not realise the value of his great predecessor's discussions of the subject.

It might indeed seem that Turner's labours had been largely wasted. That he wrote in English greatly limited his usefulness on the Continent; that his final *Herbal* was published in Germany no doubt told against its circulation in England. There is little evidence that even botanists of the learning of John Ray[1] who spoke of him with warm commendation knew his books well, though he certainly used his work on birds, probably at second hand through Gesner and Aldrovandi. Was Turner a pioneer with none to follow?

So far as his writings go this may perhaps be the case, though to have influenced Jean Bauhin is no small achievement. But his influence through his son Peter, and particularly through Peter's friendship with Thomas Penny, was very much wider and more lasting than references in books would suggest. In his own day he stood alone as a scientific botanist in England: it is largely due to him that no one has ever been in that position again.

1 Pennant in his *Tour in Scotland*, speaking of Morpeth, paid an equally warm tribute but in terms which hardly imply personal knowledge of Turner's works.

CHAPTER VIII. JOHN CAIUS AND THE *RARIORUM ANIMALIUM*

Turner's contemporary John Caius shares with him the distinction of being the pioneer of animal study in Britain[1]—though in his case the volume of his work on natural history is relatively small. It has been reprinted in the edition of his works edited by E. S. Roberts, then Master of his College, in commemoration of the quatercentenary of his birth; and to this volume is prefixed a biography by John Venn. Born in Norwich on 6 October 1510, educated in that city, and admitted to Gonville Hall, Cambridge, in 1529, Caius took the usual studies which would normally lead to ordination and theological work. But he was attracted by the new interest in Greek, and became so proficient in it that he became a fellow of his College in 1533. Unlike Turner he was uninfluenced by the religious movement of the time, and although at one period accused of atheism seems to have been always a Catholic. At Cambridge his closest friend was his old school-fellow William Framingham who was first at Pembroke and then gained a fellowship at Queens' in 1530. Caius in the pamphlet on his own books speaks of him with warm affection as 'by calling a philosopher, by degree a Master of Arts—though men spoke of him as Doctor; always a young man of comely eloquence, equipped in all that pertains to liberal studies especially music and rhetoric in which he gained distinction, in all that makes for knowledge of poets, historians and theologians particularly in Greek as well as Latin' (*De Lib. Prop.* p. 2*b*). He died in 1537 being then only twenty-five; and left the eight books that he had already written to Caius. These were put in store when Caius went abroad; and he was never able to recover them.[2] It is perhaps an indication of the source of Caius's own austerity that the two first of these books were on Continence: certainly the description of his character is an indication of the scope of Caius's own interests.

In 1539 he went to Padua where he spent four years in medical studies,

1 This was partially true of their contemporary reputation: cf. William Bullein, *Dialogue betweene Soarenes and Chirurgi*, p. 4: 'By and through the learned doctor Mr John Kaius, revealing unto this fraternity the hidden iewels and precious treasures of Cl. Galenus shewing himself to be the 2 Linacer whose steps he foloweth. Who shal forget the most worthy Doctor William Turner? whose learned actes I leave to the witty commendations and immortall prayse of Conradus Gesnerus. Yet his Booke of Herbes wil alwayes grow greene and never wither as long as Dioscorides is had in mynde among us mortall wightes'—these two being put at the head of a list of some 120 names.

2 He gives an account of their loss in his *Counseill against the Sweat*, p. 6.

attending the classes of Johannes Baptista Montanus the eminent physician and of Andreas Vesalius the still more eminent anatomist. With the latter he lived for eight months and during this time the *De Fabrica Corporis Humani*, published in 1543, was in preparation: but Caius does not seem to have shared Vesalius's somewhat critical attitude towards Galen. After receiving his M.D. in 1541 and serving as lecturer on Aristotle in Greek for the next two years, he toured through Italy, visiting many of the northern cities and spending part of the winter in Rome, whence he obtained the picture of 'Pliny's deer' which he saw carved on a pillar there (cf. Gesner, *Hist. Anim.* I, p. 306). Returning through Zürich where he met Conrad Gesner, Basel where he published his *De Methodo Medendi*, and along the Rhine where he made acquaintance with Melanchthon the Reformer, with Joachim Camerarius the Greek scholar who was at this time at Leipsic, and with Sebastian Munster, he reached England in 1545 and began the lectures on anatomy to surgeons in London which he claims to have continued for twenty years. In the exercise of his profession he seems to have travelled widely and become the leading doctor of the time. In 1551 he took a house in the Close of St Bartholomew near Smithfield and rented it till his death; and in the same year was certainly at Shrewsbury[1] during the outbreak of the Sweating Sickness, against which he published his famous *Counseill* in 1552. In 1556 he published, at Louvain, his *Opera aliquot et Versiones* dedicating it to Wotton's friend Sir John Mason. In 1557 he became the co-founder of Gonville Hall; in 1558 he revisited Cambridge—and found it disappointing; in 1559 he was chosen Master of his College and for the next fourteen years divided his time between it and the Royal College of Physicians in London of which he was President from 1555–60, and again in 1562 and 1563 and finally in 1571.

At this time we have a vivid suggestion of his way of life from John Parkhurst,[2] afterwards Bishop of Norwich, to Gesner in the *Zürich Letters*, I, p. 31. 'Hail again and again, most illustrious and very dear Gesner. As soon as I came to London, I sought out your friend Caius that I might give him your letter; and as he was from home I delivered it to his maid servant; for he has no wife nor ever had one. Not a week passes in which I do not go to his house two or three times. I knock at the door, a girl answers the knock, but without opening the door, and, peeping through a crevice, asks me what I want. I ask in reply where is her master?

1 Cf. *De Ephemera Brit.* (*Works*, ed. Roberts), p. 67. Venn (l.c. p. 7) is surely right in rejecting Cooper's statement that he practised in Cambridge, Norwich and Shrewsbury before settling in London.

2 His name occurs in the list of learned men to whom Gesner expresses obligation in *Hist. Animalium*, IV; and he contributed some verses prefixed to it: but there is no evidence that he had any interest in nature.

Whether he is ever at home, or means to be. She always denies him to be in the house. He seems to be everywhere and nowhere, and is now abroad, so that I do not know what to write about him. I shall certainly tell him something to his face, whenever I have the chance to meet him; and he shall know what kind of a man he has to deal with.' This has naturally been quoted to demonstrate Caius's eccentricity or even misanthropy: but it is probable that as the letter is dated 21 May 1559, four months after his election as Master of his College, he was then settling into his Lodge at Cambridge and had left his London home in the charge of a damsel too shy or too scared to be communicative. In any case the affection between him and Gesner is plainly proved—as it is, more clearly, by his lament over Gesner's death in *De Libris Propriis*, p. 21*b*.

In December of this same year, 1559, occurred another incident very characteristic of the times and of the man. As President of the College of Physicians he was in the Chair when an Oxford M.D., John Geynes, was charged with stating that Galen had made mistakes. This charge, laid by the veteran Thomas Wendy, was investigated a month later. Caius then decided that unless Geynes explicitly recanted, he should be put into gaol as a charlatan.[1] This attachment to the infallible utterances of the ancients was eminently typical of Caius; and in this case it coincided with his special interest: for his chief literary work was the recovery, study and translation of the Greek text of Galen on which he published a number of books between 1544 and 1549. Of the preparation and writing of these he gives a very interesting account in the little book *De Libris Propriis* printed by Seres in 1570 and addressed to Thomas Hatcher, the antiquary; and incidentally pays tribute to John Clement and Thomas Linacre, to Henry Knolles, George Owen, Christopher Mount (Mont), John Gilpin and Thomas Wendy, to the learned ladies Anne Bacon and those of the family of Sir Thomas More, and to the books of John Claymond. But his most famous work was his treatise on the Sweating Sickness printed in 1552 after he had been dealing with the terrible outbreak of the disease at Shrewsbury. In spite of his medical eminence he seems to have taken little interest in pharmacology or botany; and his contributions to natural history are almost all in the field of zoology. Apart from a record of two white Ravens from the same nest trained like Falcons to come to the hand, which he saw in 'Cumbria' in August 1548 (*De Rar. An.* p. 23*b*); and of a fish 'Variata', which his pupil Mouffet renders 'Bleak' but which was evidently one of the Wrasses, caught when he was visiting the submerged town off Selsey in a boat in July 1555 (l.c. pp. 25, 25*b*), none of his observations refer to any work in the field.

1 The story is told in Munk, *Roll of R. Coll. of Physicians*, I, p. 62, and is quoted in Harveian Orations by Sir George Baker and Francis Hawkins.

They are contained in a collected work, *De Rariorum Animalium atque Stirpium Historia*,[1] printed by William Seres in London in 1570; and most of them had originally been sent in letters to Conrad Gesner and incorporated by him in his *Historia Animalium*. Most of the animals are beasts that he saw and drew in London, two of the most interesting in the royal menagerie in the Tower, the 'Ounce' and the Lynx. Of the 'Ounce' a pair had been brought to England from Mauretania—and Caius described them at length, spoke fully of their intractable wildness, so that when they had to be moved the keepers had first 'to strike them so hard on the head that they would lie half dead' (p. 4), told how one of them killed and tore open a 'great Mastiff dog', and drew an admirable picture of 'this most cruel beast' lying as quiet as a tabby cat. From picture and description it seems evident that the beast was a Hunting-Leopard or Cheetah (*Cynaelurus jubatus*). The Lynx (*Felis lynx*) he also described from a specimen kept in a cage big enough to allow rapid movement and some climbing. So restless was it that it seemed hopeless to get a picture until a countryman who wanted to see the lions came up carrying a woodpecker in a basket. When the lynx saw the bird it came to a halt, and the drawing was begun. But the countryman moved off; the lynx resumed its movements; and it was only when Caius sent his servant to buy the bird that 'the beast stood still until its portrait was completed' (p. 5*b*).

Most of the other animals that he describes and depicts came like the Hunting Leopard from North Africa. The first in Gesner's book is a Getulan Hound, 'long-legged, almost tailless, with a face like a hedgehog', that had come to him in 1554. The next is the Civet Cat or Zibeth which he had bought from an African merchant 'for eight pounds, that is 24 French crowns or 48 Rhenish florins' (p. 6*b*): it was tame and tractable, but when angry growled like a dog 'doubling the letter R'; he had it pictured 'as like as one egg to another' and described it carefully with special attention to the civet-gland. Then he describes the 'Tragelaphus or Goathart', evidently the Barbary Sheep (*Ammotragus lervia*) pictured in 1561: of this he knew both sexes, and states that they bear only one lamb at a birth (pp. 76–9). Next is another 'Mauretanian' from the desert, 'in shape and appearance half-way between a hind and a heifer, I call it a Moschelaphus or Bucula cervina'—an elaborate description of head and horns makes it clear that it was a female Hartebeest (*Alcephalus buselaphus*): he comments upon its grace and agility, and its tameness and hunger: it died very emaciated three days later (pp. 9–10). Long-tailed and fat-tailed Sheep from Arabia, the latter of which he identifies with that mentioned by Herodotus, are his next subjects. Then he describes the Getulan Squirrel, evidently

1 Included in *The Works of John Caius*, edited by E. S. Roberts, Cambridge, 1912. The page-references quoted here are those of 1570.

Xerus getulus, comparing it with our Red Squirrel and commenting on its slothfulness, timidity and dislike of the cold: this also he had obtained from a merchant. Finally, he mentions the Barbary Ape as another species sent from 'Mauretania'. It is interesting to find such clear evidence of traffic with North Africa at this period.

Another of his prizes is the 'Saguina', 'a little animal, the size of a Sorex, not much bigger than a baby rabbit', and evidently from his description a Marmoset (*Hapale jacchus*) (p. 15): of this he (or Gesner) states that it is very much smaller than the one whose picture Pieter Coudenberg the famous apothecary of Antwerp had sent to Gesner (cf. *Hist. Anim.* I, p. 869).[1] He also discusses at length the identity of the 'Strepsiceros' whose horns were used for lyres by the Greeks and which some identify with the Cretan sheep. This Caius criticises; and therefore sends to Gesner the picture of a lyre so that the contrast with the Cretan horns may be evident.

Two more of his pictures are of horns preserved in the halls of his noble patients—those of the Bonasus or Wild Bull he saw in 1552 in Warwick Castle along with the armour of the legendary Guy of Warwick (a shoulder-blade of this beast is said by Caius to hang on the north gate of the city of Coventry, and a rib in Guy's chapel where it was mistaken for a bone of the legendary boar slain by him) (pp. 13*b*–14*b*), and of 'Cervus palmatus' the Reindeer 'at Killyngworth [Kenilworth] in Warwick' (p. 7*b*).

But the most interesting of his records is probably that of the 'Hippelaphus', a beast brought out of Norway over which he shows both his accuracy and his handicap (pp. 10–11*b*). The picture is well drawn and obviously represents a young female Elk; and the description bears this out beyond question: it is long, accurate and interesting, much of it being derived from the Norwegian keeper: among other details is the statement that 'it drinketh water and also English Ale in great plenty, yet without drunkenness'. He then states, 'in Norway they call it an Elke or Elend, but in this they are plainly mistaken; for it has not the legs of an Elk since they never bend'. He knew and implicitly believed the strange account in Caesar's *Gallic War*, Bk. VI, Ch. 27, of how the Elk having no joints in its legs cannot lie down[2] and so sleeps leaning up against a tree, and how the hunter saws through the tree so that the beast falls over and is caught. Caius, like Turner, but much more emphatically, accepted the ancient lore; and for him its verdict was final.

The Chamaeleon of which a long description is given is the only reptile noted (pp. 15*b*–17): but among the birds, though the species described

1 This is not in the 1570 edition.

2 Is this a variant of the story told about Elephants in Aristotle, *H.A.* II, i, 498, but rejected by him?

are few, several points of real interest occur. First, there is a good picture and description of the 'Meleagris, or Numidian or Moorish Fowl', our familiar Guinea-fowl, which Gesner inserts in a chapter to itself (*H.A.* III, pp. 424–5). His own first species is the Haliaëtus, of which Gesner printed his description as an Appendix (*H.A.* III, p. 73) with his picture of it.[1] 'Nostri an Osprey vocant'; Caius first describes it very accurately; then rejects the account of the Aurifrisius as given by Giraldus Cambrensis in his book on Ireland (*Top. Hib.* Dist. I, ch. 16) and by Neckam—both of whom assign to it one taloned and one webbed foot; then narrates the common British belief that fish when seen by it turn themselves on to their backs and offer themselves as its prey—a legend perpetuated by Shakespeare in *Coriolanus*, Act IV, Sc. 7;[2] and then after saying that it is common round our coasts and in the Isle of Wight tells of one that he kept for seven days without food, at the end of which time it apparently died; for he remarks that 'its flesh was black'. The belief about the fish turning over for the Osprey is also mentioned by Turner (*Avium Praec.* p. 32)[3] and it seems probable that Caius derived it from him (pp. 17–18). Another bird which he kept for much longer ('I kept one in my house for eight months') is the Puffin (*Fratercula arctica*). Of this he sent a picture, recognisable but not one of his best, to Gesner (*H.A.* III, p. 657).[4] He describes it very clearly, including its nesting in rabbit burrows, its refusing to fly unless within sight of the sea, its rejection of cooked meat and of water except for washing, and its cry of 'pupin, pupin'[5] when hungry. He speaks of it as hibernating 'like the Cuckoo and the Swallow'. He concludes by saying that 'it cheerfully bit those who gave it food or touched it, but in kindly and harmless fashion. It is satisfied with a small meal and is not voracious like our Corvorant which our people corruptly call Cormorant' (pp. 216–22).

This quotation introduces us to Caius's favourite enthusiasm, the derivation of words. With his interest in language and in symbolism such

1 In the same year, 1555, a picture of the Haliaëtus or 'Orfraye' with two clawed feet was published in Pierre Belon, *De la Nature des Oyseaux*.

2 'I think he'll be to Rome
As is the osprey to the fish, who takes it
By sovereignty of nature.'

A rationalisation of this legend is given by Aldrovandi, *Ornith.* I, p. 108, who reports that the Osprey hanging over the water drops a fatty secretion from its tail which has a stupefying effect upon the fish.

3 Cf. above, p. 132.

4 De l'Ecluse, *Exoticarum*, V, p. 104, commenting on this picture and description sent by Caius to Gesner says that owing to its badness he had himself had another drawn from a specimen received by Pieter Pauw from Hendrik Hoier: hence the Puffin became known as the 'Anas arctica Clusii'.

5 More successful as an explanation of its name than as a rendering of its cry—which is a deep and growling sequence of three notes.

as is expressed in the gates of his College, we should expect him to be concerned with philology; and as compared with his predecessors he is if not more interested certainly more interesting. The best example of this is his account of the 'Anser Brendinus' which 'the common folk of the British sea-board name from its varied colour a Brendgose, and the London poulterers a Bernacle for which I suspect they ought to say Berndclac or Brendclac' (p. 18). He gives as his reason that 'the ancient Britains and Scots called all geese of sea or marsh or lake Clakes just as to-day we speak of Fenlakes and Fenlagges when we ought to say Fenclakes' (does this throw light on the hotly debated origin of Grey-lag—of which Skeat's explanation Grey Lag-goose, 'the grey goose who lagged behind', is surely the feeblest ever perpetrated by a great philologist?).[1] 'They likewise called a mottled colour Brend or by metathesis Bernd, so that a mouse-coloured creature spotted with white like this goose is called bernded or brended.' Some of this Caius may probably have got from the description 'of the nature of claik geis' in Hector Boece's[2] *Hystory and Chroniclis of Scotland* published in a translation by John Bellenden, Archdean of Moray, in Edinburgh about 1536, wherein Boece describes 'the procreatioun of thir clakis' in a manner similar to Turner (cf. *Description of Scotland*, Ch. XI, Holinshed's edition, p. 17). Of Caius's etymology Max Müller, in his long account of the legend of the Barnacle Goose (*Lectures on the Science of Language*, II, pp. 585–604), takes respectful notice though it does not agree with his own. It is at least a great advance upon the usual guesswork of the age. In his description of the bird as sent to Gesner (*H.A.* III, p. 96) Caius states that the black-faced Brent (*Branta bernicla*) is the young of the white-faced or Barnacle (*Branta leucopsis*), but this does not appear in his later version of his letter.

So, too, in treating of the 'Spermologus' (*Corvus frugilegus*) which only differs from the Crow by its harmlessness, its pouch under the beak in which it carries food to its young, and in the white patch of skin between beak and head, he argues that since there are many Latin and Greek words in English this bird is called Rouke from the Latin *rauce* and its harsh cry. He adds a note still relevant and arguable, 'Our country folk doubt whether it is more useful to man while devouring worms that damage the crops or useless while feeding upon grain' (p. 22).

He also describes the 'Sea Pie' (*Haematopus ostralegus*) (p. 196) and the Morinellus or Dotterel (*Eudromias morinellus*) (p. 21), the latter's

1 A century later in 1639 P. Mundy retailing the barnacle story says that 'these fowl are by the country people called Clawgeese', *Travels*, IV, p. 51: elsewhere he spells this claegeese—which is certainly Clakis or Clagis, the Scottish name: claegoose is surely lag-goose.

2 First Principal of the newly founded University College of Aberdeen. His *History*, with its grand reputation for credulity, was published in Latin in 1527.

picture and description having been sent to Gesner and inserted in *H.A.* III, p. 554. He describes it being caught by candle-light when it imitates the actions of the fowler; 'for if he stretches out an arm, the bird spreads a wing; if he lifts a leg, it does the same; in short, whatever the fowler does, the bird imitates'.[1] Finally, he describes a large Parrot from Brazil, apparently the Red and Blue Macaw (*Ara macao*) whose picture he sent to Gesner—probably that reproduced in *H.A.* III, p. 653.

His notes on fish which were made in the main too late for sending to Gesner chiefly consist of records of large stranded species which found their way into the London market in the last few years of his life. The 'Maculo' (of which the picture sent by him is in Gesner, *H.A.* IV, p. 216, and which seems from it to be neither a cetacean nor a Shark),[2] 60 feet long stranded at Lynn in 1555 (p. 24*b*); the Cherke (Shark, probably *Carcharias glaucus*), 'like a Dolphin', taken off Dover on 16 June 1569 and seen by Caius in London on 22 June (p. 26); the Dolphin[3] from Shoreham in Sussex, seen in London in December 1569 and sent to Thomas, Duke of Norfolk, (of this Caius says that Belon's picture is the best—so he must have known the *De Aquatilibus*) (pp. 27, 27*b*); and the 'Ceruchus', apparently a Sturgeon (*Acipenser sturio*) which came ashore between 'Lastofte' and Pakefield in February 1570 and whose head was brought to him in London in May—'we saw it and had it drawn' (pp. 28–9); these are his principal records. He also mentions the Swordfish (*Xiphias gladius*), the Ruff (*Acerina vulgaris*) at Yarmouth of which Gesner mentions that he had sent a picture (*H.A.* IV, p. 702),[4] the 'Horsefish' (apparently *Hippocampus hippocampus*), the Acus or Hornbeak (*Belone vulgaris*) at Boston, the Chrysophrys or Dory (*Zeus faber*) and as already mentioned a Wrasse at Selsey. In the index at the end of the book he adds 'the Phoca which we call Seele a contraction of See Vele that is See-calf' of which he had written in his *De Canibus*, p. 5*b*.

Of plants the only interesting notes are those of the famous Peas (*Lathyrus maritimus*) on the beach near Orford in 1555 (p. 29*b*) which Camden reported as a miracle (*Britannia*, p. 339), and on the Ilex in the

1 Drayton, *Poly-olbion*, XXV, ll. 345–8, writes:

> 'The Dotterell which we think a very dainty dish
> Whose taking makes such sport a man no more can wish;
> For as you creep or cow'r or lie or stoop or go,
> So marking you with care the apish bird doth do.'

So, too, J. Childrey, *Britannia Baconica* (London, 1661), p. 120.

2 Experts consulted cannot identify this more precisely.

3 He seems to have missed the Dolphins and 'the three great fishes called Whirlepooles taken at Gravesend and drawen up to the King's bridge at Westminster' which John Stow reported from the Thames in 1552: cf. *Annales*, p. 1026.

4 For this he is quoted by Aldrovandi, *De Pisc.* p. 243.

royal garden at Westminster, to which De l'Ecluse also paid tribute when he visited England in 1581.

We have dealt with these notes both as they appeared in Gesner's great *History* and as Caius published them some ten years later. But before leaving them it will be of interest to look at the one long letter which has been preserved addressed by Gesner to their author with his comments upon some of them in August 1561.[1] From this it appears that he had previously received and bound up with pictures of his own a number of portraits of animals in their proper colours sent by Caius; and that the gift now being acknowledged consisted of some shells, some skulls or jaws of fishes, and above all a set of 'new and very beautiful pictures and descriptions', including the Barbary Sheep, the Hartebeest and the Fat-tailed Sheep, and accounts of the Chamaeleon, various Monkeys, the Rook and the Turkish Duck. Of all these Gesner has something to say—a request for the horns of the Barbary Sheep, if it should die, in exchange for those of the Ibex, a comment upon the first 's' in the name Buselaphus ('forgive this grammatical trifling!'), and a further request for pictures of Rook and Duck. He goes on to explain his treatment of Caius's pictures of Lynx and Civet-cat, to comment upon the Bison and the Elk, discoursing at length upon the identity of 'Alces' and 'Machlis' and stating that no one nowadays believes that the legs of Elk or Elephant cannot be bent, and to thank Caius especially for the gift of the Osprey's foot. The letter contains three references to Turner, the first acknowledging an affectionate letter from him; the second alluding to two fishes, one like a Gurnard, the other a Blenny; the third saying that he has planted Turner's Onion bulbs which he has received along with other gifts. There is further an allusion to Parkhurst, Bishop of Norwich, who has promised to send him a Globe-fish, to Claymond's notes on Pliny which Gesner thinks that he has seen, and to letters of George Bullock, lately Master of St John's College, now, since Elizabeth's visitation of Cambridge in 1559, an exile, which Caius had sent to him apparently by way of introduction if Bullock should come to Zürich. Bullock being a staunch Catholic, Gesner's response to this dossier is a trifle lukewarm. The rest of the letter deals with the treatises on baths and spas which Caius and Gesner were both proposing to write and with a discussion of the plant called Elleborine of which he says that the learned think it a sort of Orchid ('alisma' or 'damasonium'—both then Orchid names), but Anguillara and Bock disagree, identifying it with a shrub whose twig he encloses.[2] He sends a copy of his edition of the Works of Valerius Cordus just published at Strasburg and of the prescription

1 In *Epist. Medic.* pp. 133*b*–6*b* and above, p. 83.

2 J. Bauhin, *Hist. Pl. Univ.* III, ii, p. 378, records of this letter that Gesner identified the Elleborine of the ancients with Herniaria.

for his special medicine Oxymelitis on which he afterwards printed a tract.

The letter is an interesting sample of the means by which learned men kept in touch—a good illustration of the change which better postal arrangements had made in the learned world. It indicates the close contact between Caius and his friend. It is in fact longer than most of the 209, almost all written in the last five years of his life, which Wolf collected and published. But it is very similar in character to many of them.

A later letter addressed to Theodor Zwinger of the medical school at Basel on 8 April 1565 also refers to Caius. Gesner reports that Caius has just written to say that he is sending some of his books to Frobenius and Episcopius for printing, and wishes them if not accepted to be sent to him: 'in one of them addressed to me he treats of British dogs.' Gesner asks Zwinger to find out whether and if so when the printers can undertake this; and adds that as the 'Dogs' lacks pictures and Froschover his own publisher has some that would be suitable, it might perhaps be possible for Froschover to publish it. It is well known that this famous work, written for Gesner, actually arrived too late for the great naturalist to use, and was eventually printed by Seres in London in 1570.[1]

The publication of these notes was one of the last of Caius's literary efforts. For the last year or two of his life he was concerned with his purely mythical History of the University of Cambridge. His relationship with his colleagues in the College and the officials of the University became embittered: he was high-handed and they were jealous and ungrateful. It seems certain that their hostility shortened his life. He died in July 1573 and was buried in the College Chapel with the shortest of all epitaphs, *Fui Caius*.

An interesting sidelight upon his last years of life is preserved in the strange book on dietetics and cookery, called *Health's Improvement*,[2] published posthumously in 1655, but written sixty years earlier by Thomas Mouffet, best known for his share in the *Theatrum Insectorum*, who came from Trinity to Gonville and Caius in 1572. Mouffet, in his chapters on fish, recorded that Dr Caius called 'Bleaks of the Sea' Variatae 'because they are never of one colour but change with every light and object, like to changeable silk' (p. 148); supposed 'Currs to be all one with our Gurnard' (p. 151), in which Mouffet thought him mistaken, although Turner in his letter to Gesner had said that Gurnards were called Curres in East Anglia; and imagined that 'Minoes [Minnows, *Phoxinus laevis*] Aliniatae Caii were so called because their fins be of so lively a red as if they were died with the true Cinnabre-lake called Minium' (p. 183), an

1 An English version by Abraham Fleming was published in 1576.

2 For this, cf. below, pp. 177–9.

etymological absurdity very characteristic of Caius but in this case based upon an error of fact. He also gives us another reminiscence—'I remember that Dr Caius (whose learning I reverence) was wont to call Tenches good plaisters but bad nourishers: for indeed being outwardly laid to the soles of one's feet, they oftentimes draw away the ague; but inwardly taken they engender palsies, stop the lungs, putrifie in the stomach, and bring a man that much eats them to infinite diseases'—a verdict which if correctly reported hardly enhances Caius's reputation in medicine: no doubt he derived it from the superstition which we have seen in the *Treatyse on Fysshyng* that the Tench had a curative power for other fishes which rubbed themselves against it.

One further record of him is preserved in the *Hortus medicus et philosophicus* by Joachim Camerarius the younger, the doctor and botanist of Nuremberg, son of the great German classical scholar. The book, published at Frankfort in 1588, has a quotation from a letter of 'that very learned man Caius the Briton' to Gesner dealing with the Sundew, *Drosera rotundifolia*. Caius writes that it is called by the common name of Sun-rose[1] and is not so well thought of by doctors as by 'empirics' and herb-women, since sheep feeding on it in England tend to get diseases of the liver and lungs: some doctors, however, use a decoction of it for the treatment of consumption.

Apart from his friendship for Gesner and the fact that he had worked at Padua there is little to suggest that Caius knew or influenced many of his contemporaries in the field of natural history. He mentions Doctor Turner in the preamble to his treatise upon dogs, but knew him rather as a friend of the great Zürich savant than personally. The two men must have met and had many common friends, Thomas Wendy and others of the College of Physicians. But they differed in religion, and very evidently in temperament; for Caius was a solitary and rather disagreeable bachelor, critical even where he was generous, stern and secretive and autocratic, and wholly devoid of geniality and of humour: and their interests were not closely allied. Turner was a botanist, herbalist and physician, a disciple of Luca Ghini and the school of Bologna; Caius was an anatomist, an enthusiast for Galen rather than Dioscorides, and trained at Padua by Montanus and Vesalius. Moreover, though Turner practised in Germany, he was hardly known as a doctor in England, and was certainly more of a divine than a physician, and perhaps more of a field naturalist than either, whereas Caius was a medical man first and foremost whose scientific interests were mainly in the study of his profession and whose zoology

1 *Rosam solis*, which is perhaps merely a printer's or copyist's error, for Camerarius calls it *Ros solis* and Sundew has always been its English name. Yet the confusion lasted till at least a century later: cf. J. Childrey, *Britannia Baconica*, p. 10, '*Rosa solis* more properly called *Ros solis*'.

seems to have been a rather casual hobby pursued to gratify a friend. But it may well be argued that in his presidencies of the College of Physicians, in his benefactions to Cambridge and above all in his zeal for the recovery of the best ancient medical wisdom, for anatomical and physiological knowledge, and for clinical practice and observation, he did more for the real advance of science than he would have done if he had given more time to the cataloguing of flora and fauna. Both tasks, the removal of imposture and of the folklore and white magic which did duty for pharmacy, and the exploration and identification of nature, were necessary. Sound therapeutic method was perhaps the most effective means of preparing for the recognition of inductive procedure, of observation and experiment, in place of blind reliance upon tradition. Caius was not by temperament or intention a pioneer: but by his years of service to propagating the knowledge of anatomy he did in fact play an important part in the development of biological science in England. It was of real value that during the upheavals of reformation and counter-reformation he was able to maintain unbroken his weekly lectures to the surgeons and his regular correspondence with continental savants. In the long run such influence counted heavily, even if his obstinate conservatism and subservience to authority delayed its effects.

How necessary it was may be seen from a brief consideration of the contemporary of Caius and Turner whom we have already mentioned, William Bullein.[1] Born in 'the isle of Ely, my native countrey' (*Booke of Simples*, p. 21), claimed, quite without evidence, by Wood and by Cooper for their respective 'Athenae',[2] in 1550 vicar of Blaxhall[3] 'by Orford in Suffolk' (l.c. p. 20*b*), then a practising doctor at Alston and Tynemouth, and finally settled in London, he wrote several short treatises on medical subjects including the small Herbal already quoted. Almost nothing is known of his life except the sensational incident which he describes twice over in the *Booke of Simples* (Letter to the Reader and p. 79*b*), how when he was in Durham his patron Sir Thomas Baron Hilton[4] had died of a malignant fever and he had been accused of murder by William Hilton the dead man's brother, how he had taken ship for London, been wrecked and lost his library and a completed manuscript, and how though acquitted of murder an attempt had been made on his life by a man named Bellises whom he had cured,[5] and he had been thrown into prison by

1 Cf. *Visitation of Yorkshire* 1563, pp. 42–3.

2 *Ath. Oxon.* I, c. 538; *Ath. Cantab.* I, p. 343.

3 'I was neere kinsman unto the chiefest house of that town' (*Government of Health*, p. 59*b*).

4 Described in dedication of *Government of Health* as 'Knight, baron of Hilton and Captaine of the King and Queen's majesties castel of Tinmouth'—plainly therefore this part was finished before Mary's death in 1558.

5 Cf. *Booke of Simples*, p. 37*b*.

William for debt—in spite of which he married Sir Thomas's widow (Agnes, his third wife, herself widow of Matthew Baxster) and wrote his largest book in gaol! He died in 1576.[1]

Bullein's work, *The Government of Health*, dedicated to Sir Thomas and published in 1558, is typical of the outlook of the traditional medicine of the time. Like all his books it is in form a dialogue—a series of short questions and long but varied expository answers. Early in the catechism comes a question on the varieties of temperament, and the answer is an offer to sing 'one small song of the foure complections'. It begins:

> The bodies where heat and moysture dwel
> Be sanguine folkes as Galen tell
> With visage faire and cheeks rose ruddy:
> The sleepes is much and dreames be bluddy.[2]

And so on through 'Flegmaticke', 'Cholericke' and 'Melancholie' (cf. pp. 6*b*, 7). A few pages later is a plate or 'Anatomie' illustrating the parts and internal organs of the body, each labelled with the appropriate planet or sign of the zodiac to which it belongs (p. 17). The early part of the book contains general rules for health and the diagnosis of ailments. Then follow questions about herbs and foods—a large number of plants being mentioned roughly in the order of their Latin names but without any attempt to identify them or give localities. The book concludes with a eulogy of Mithridatic treacle, the most famous of all traditional 'compounds'.

Four years later when in prison he revised this work, or rather the enlarged version of it called *Healthfull Medicines* which had been lost at sea, and issued it in 1562 as *The Bulwarke of Defence against all Sicknesse* in four separate treatises, *The Booke of Simples*, *A Dialogue betweene Soarenes and Chirurgi*, *The Booke of Compounds*, and *The Booke of the Use of Sicke Men and Medicines*. The whole was dedicated to Henry Carey Baron Hunsdon, cousin of Queen Elizabeth, and published in London as a small folio.

The first part, a Herbal and Dietary, is very close to his previous book, indeed many of the plants are identically treated. But there is evidence of fuller reading and larger experience. Instead of references only to Fuchs and Gesner (as in *Government*, pp. 41, 42 and 61) he now freely cites Mattioli (e.g. pp. 2*b*, 5, 24, 49) and Ruel (pp. 38, 55*b*) and under Tamarisk (p. 63*b*) pays a great tribute to 'the famous learned man Wylliam Turner Phisicion... which Doctor is a Jewell among us Englysh men as well as among the Germaynes as Conradus Gesnerus reporteth of

1 For his literary work, cf. A. H. Bullen, *Elizabethans*, pp. 155–81.

2 He inserts also long poems on the inspection of urine and excrement—subjects of great importance in early medicine (pp. 35–7).

hym for hys synguler learning knowledge and judgement'. In spite of this he shows no sign of any acquaintance with Turner's works and never elsewhere either cites or alludes to him. The book ends with minerals and animals, and with three pages of small woodcuts of plants, twenty-four in all, similar to those in Turner's *Herbal* but copied not from him but from Fuchs.

Unlike the corresponding sections in *Government* the accounts of herbs here are enlivened by occasional references to cases like that of Sir Richard Alie of Berwick whom he purged with a concoction of Senna (p. 38), or of Thomas Colby of Suffolk whom he helped by cresses (p. 40*b*), or of Cuthbert Blunt almost done to death at Newcastle by 'the doltish impericke' Edwards (p. 64*b*); by references to other practitioners, 'Mayster Jhon Sharman' of Norfolk (p. 21*b*), 'Maister Luke' (p. 26); and by a ferocious diatribe against witches, 'M. Line in a town of Suffolk called Perham [Parham]', and 'M. Didge in a town called Kelshall [Kelsale]', 'more hurtful in this realm than either quarten, pox or pestilence...the fyre take them all' (pp. 56*b*, 57). He also gives a few localities for plants, notably for Eryngium (*E. maritimum*), 'I have seene mutch thereof growing betwene Lestoffe [Lowestoft] rode and Orforde neere by the shore side' (p. 44) and for Mistletoe (*Viscum album*), 'I have seene great plenty growing upon Thornes Peretrees and Okes...in these most auncient parkes of Framingham [Framlingham], Kelshal [Kelsale], Nettlestede, Lethringham, Parham, Some [Monk Soham], Heningham [Heveningham], Westwood, Huntingfeeld, Henham [? Rendham], Little Glenham [Glemham], and Benhal. These Parkes be old neighbors'[1]—a list which gives him an opportunity for a eulogy of 'the right gentlemen', whose qualities he describes with enthusiasm (p. 50*b*). Among such passages is a pleasant note on the Chameleon appended to the section on the Thistle so named, 'This beast doth never eate nor drynke nor winke but liveth by the ayre only which litle beasts be yerely brought from Barbarie by the ships of London into this realme' (p. 44*b*). So, too, on p. 76*b*, he subscribes to the belief that 'Ospraies oile is good to put in water to gather fish ready to be taken'. Yet like Turner he condemns the shaping of Mandrakes and denies that they ever resemble human beings (p. 41*b*). He was a man of his time; although he gives references to later books, his science is that of Bartholomew and based upon Dioscorides, Pliny, Galen and the tradition. He has almost nothing to say of the physic-gardens and new plants[2] which meant so much to Turner; and in spite of his lively interest in the reformed religion and the public weal belongs to the old world.

1 They are all in East Suffolk within reach of Blaxhall.

2 It is of interest that these, e.g. Rhubarb, are still only known as imports from semi-mythical Eastern climes.

That this world was passing is at least suggested by his very interesting list of 'the men of credite both olde and new' who have written on surgical or medical subjects—a list included in the *Dialogue betweene Soarenes and Chirurgi*, pp. 4, 5. This begins as we have already noted with John Caius and William Turner; continues with Sir Thomas Elyot and his *Castel of Helth*, with Thomas Faire (Phaer) of *The Regiment of Life*, Andrew Boorde of *A Compendious Regyment* and *The Breviary of Healthe*, Thomas 'Paguinellus or Pannel' (Paynell) translator of 'good bookes of Physicke',[1] William Keningham physician, astrologer and engraver, and Robert Recorde of *The Urinal of Physicke*; and ends with an alphabetical list of 120 names ranging from Aesculapius, Machaon and St Luke to his own friends and contemporaries. Of doctors in England besides those named already he mentions Augustine de Angustinis the Venetian to whom Elyot refers with gratitude for treatment,[2] 'Bartleus' (possibly Richard Bartlot), Sir William Butts, Nicholas Carre one of the original fellows of Trinity College and Regius Professor of Greek, Cambridge, 1547, John Clement, John Chambre an original member of the College of Physicians, Robert Huicke, President, 1551–2, 1564, Thomas Linacre, Richard Master, President, 1561; doctors of Oxford, Richard Caldwell, President, 1570, John Geynes the famous 'heretic' who was cited before the College in 1559 for impugning the infallibility of Galen,[3] and Simon Ludford one of the two Oxford graduates whose medical degrees were condemned by the College under Dr Caius;[4] doctors of Cambridge, 'Bunus',[5] David Edwards, M.D. 1548–9, author of *De Indiciis* and *Introductio ad Anatomicen*, Cambridge, 1532, John Hatcher, Regius Professor of Physic, *c.* 1554, John Frere (Fryer), President, 1549–50, Christopher Langton, expelled in 1548 for immoral conduct but on this list as author of *Introduction into Physicke*, etc., Thomas Lorkin, Regius Professor, 1564, and Thomas Wendy; surgeons, Edmunds of York, Thomas Gale of London, author of *An Institution of Chirurgerie*, 1563, *An Enchiridion of Chirurgerie*, etc., Robert Balthrop and Thomas Vicary of London, author of *A Profitable Treatise of Anatomie*, 1548, and afterwards in 1585 of *The Englishman's Treasure*; and finally one druggist, Thomas Colfe. John Porter of Norwich has no classification. Contrasted with these genuine practitioners are the 'dogge leaches', 'petty Foggers to make mountaynes of molhils', 'counterfeit chirurgions' and other bad men who brought disgrace upon the profession. The doctors were becoming conscious of their professional status and of the necessity for sounder knowledge; medical

1 He translated *Schola Salernitana*, 1528; *De Morbo Gallico*, 1533, etc.

2 *Castel of Helth*, p. 86 (numbered 60).

3 Cf. above, p. 140.

4 Cf. Munk, *Roll of R. Coll. of Physicians*, pp. 64–5.

5 Dr Venn tells me that this may perhaps be William Bunne who incepted in Grammar, 1512–13: but there is no evidence that he was a doctor.

literature was being rapidly produced; and in spite of the tenacity of the tradition, observation and clinical studies would eventually prevail against it; these facts are as obvious as the need for the work which Turner and Caius had initiated if the reforms were not to be unduly delayed.

CHAPTER IX. THOMAS PENNY THE BOTANIST

Cambridge produced another naturalist and doctor in the generation below that of Turner and Caius. Thomas Penny was born about 1530 at Gressingham on the Lune nine miles north-west of Lancaster. His father John Penny seems to have owned a small property there and though according to Dr Willoughby Gardner, whose account of Penny published in the *Annual Reports of the Lancashire and Cheshire Entomological Society* for 1928–30, pp. 31–52, is the fullest available, the name disappears from the Subsidy Rolls in 1546, Thomas's will leaving £10 'to the poore of Gressingham and Eskrigge wheare I was borne' shows that at least a sentimental link remained. He had one brother, Brian, but of the family and his early life we know almost nothing. That he was at school in the ancient Grammar School of Lancaster is a plausible conjecture.

In 1546 young Penny went up as a pensioner to Queens' College, Cambridge, but in 1550 a sizarship at Trinity was awarded to him and he transferred to the larger College. There he graduated in 1551, became a fellow in 1553, Senior Bursar in 1564, and apparently remained, working at divinity and medicine, until 1565. Soon after the accession of Queen Elizabeth he began to preach and was ordained deacon at Ely on 8 June 1561 (Venn, *Al. Cantab.* III, p. 342). In 1560 he had been given the prebend of Newington in St Paul's Cathedral and in 1561 had preached before the University at Cambridge. He was a strong supporter of the Reformation and in 1560 was the author of some Latin verses on the restitution of Bucer and Fagius (Cooper, *Ath. Cantab.* II, p. 78). He was almost certainly a friend of William Turner, for we find him closely associated with Peter Turner, William's son, a few years later (*Zürich Letters*, II, p. 203); and Peter had come up to St John's, Cambridge in 1564.

In 1565 having been the preacher at Easter of the annual Spital sermon at Paul's Cross he was censured by Matthew Parker the Archbishop as being 'ill-affected towards the Establishment'. In consequence he decided to develop his interest in medicine, and having got an introduction to Gesner presumably from Turner set off for Zürich, where both the great scientist and Henry Bullinger, the pastor and protestant divine, assured him of a warm welcome. Nevertheless, though thus 'become a layman and a physician', he seems according to Strype (*Parker*, II, p. 241; *Grindal*,

p. 269) to have continued in possession of his prebend until 1573. In 1577 he was one of the signatories of a letter to Thomas Cartwright the nonconformist leader, and was in consequence deprived (Cooper, *Athenae Cantab.* II, p. 78).

Already as it seems Penny had begun to form a *hortus siccus* and to draw and describe plants. At least it is certain from De l'Obel (*Adv.* pp. 289–90) that he had botanised in the north and probable that the admirable picture and history of *Rubus chamaemorus* which he sent to De l'Ecluse were obtained in his early years. As l'Ecluse had no other record of the plant which he names 'Chamaemorus anglicana' and was a very accurate writer, we may confidently describe the notice as Penny's own writing. It occurs in *Stirpium Pannonicarum Historia*, pp. 117–19, and when translated runs, 'among thornless brambles we must include this elegant plant whose picture and history were communicated to us by the eminent Thomas Penny, the London doctor above mentioned, under the title of Chamaemorus. It consists of stems twelve inches long on which alternately grow three, four or rarely five leaves, rough in texture, not unlike those of a mallow or rather a mulberry [if fresh and from a seedling: I would rather compare them with leaves of common currant] divided into five points and serrated (*per ambitum serrata*) on long pedicels, and springing out of two wings or processes (*apophyses*) embracing the stalk. The top of the stem bears a single flower, standing out of blackish purple bracts. The fruit is very like that of a mulberry [rather of a common blackberry] (hence the name "ground-mulberry") but a little smaller, at first whitish and bitter, then red and sharply sweet. The root is knotted, sending out a few fibres from each knot. It spreads wonderfully and creeps very far, so that it quickly covers a wide area. It flowers in June and early July: the fruit is ripe in August. It loves snowy and open places and the tops of hills, and grows in great plenty among heather on mount Ingleborrow the highest in all England, twelve miles from Lancaster. The English call it knotberries from the knot-like fruit: to the taste it is astringent, drying and cooling.' Here as elsewhere in De l'Ecluse the square brackets represent his own comments; the passage will then be one of the few examples of Penny's descriptive power that have come down to us. It goes far to establish his reputation, as against the slurs which Mouffet subsequently passed upon his style and diction.

Nor were his interests confined to plants. In the *Theatrum Insectorum* in which his work was eventually embodied we read on p. 270 that 'in the year 1563 while Penny was writing these records he was summoned to Mortlake a hamlet on the Thames to two noble ladies'. So evidently he had begun both entomology and medicine before his journey to the Continent.

That he was singularly well equipped for scientific studies is clear from the description just quoted and from the drawing of the plant which accompanies it. There is perhaps no other of the early botanists who has his command of terse and exact phrasing, who employs technicalities so precisely, or who can give so clear and vivid a classification of the chief points in any species. His pictures which must surely have been his own work are worthy to stand alongside the admirable illustrations by professional artists which decorate the Herbals of the period and were collected by the great publishing house of Plantin. We happen to have another description and picture of 'Chamaemorus' by Thomas Hesketh the friend of Gerard preserved in his *Herball*, p. 1368; and the contrast between Penny's accuracy and its crude and defective portraiture is so great that the two were regarded as referring to different species. It is of course likely that these actual records were compiled by Penny later in life: no doubt his contact with the scientists of Montpellier and Heidelberg developed his powers and technique. But when he left England he was evidently a man of sound education and brilliant parts, a scholar and an artist. Travel for such a man was bound to be fruitful.

How long he was at Zürich before Conrad Gesner's death on 13 December 1565 is uncertain, for we do not know the date of his arrival: it cannot have been more than a few months, but the time was evidently well filled.[1] Fortunately there is unexpected proof of his activities in the miscellaneous collection of documents and pictures connected with Gesner published in 1753 at Nuremberg by Seligmann—a volume usually catalogued as Book v of the *Historia Stirpium* of Valerius Cordus, since this series of descriptions of his finds in Italy just before his death in 1544 is the first item in it. All the latter part of the book consists of an elaborately indexed series of plates, the first a sumptuous and coloured copper-plate of *Swertia perennis* and four smaller plants; then 22 woodcuts with 9 pictures on each; and then 20 copper-plates containing 175 pictures in all. These pictures are all reproductions of the drawings which Gesner had collected, which Caspar Wolf sold to Joachim Camerarius, and which having come into the possession of C. J. Trew were worked over and prepared for publication by C. C. Schmiedel.[2] The notes in the list of names preceding the plates give the source of the various pictures and here Penny's name occurs against twenty-two of them, usually with a brief explanation. Two (*Lycopodium clavatum* and *L. selago*) are on the coloured plate with *Swertia*. One, Tab. 3. 27 (*Helianthemum guttatum*), is labelled

1 Gesner, as the letter to his friend A. P. Gasser written ten days before his death (*Ep. Med.* p. 44) proves, kept his vigour and scientific interests until the end. It deals mainly with minerals, but reports that he has had two live vipers sent him from France, that one has escaped—to his peril—and that he hopes to breed from them. In September he had written to Adolf Occo (l.c. p. 65): 'All this summer I have been very busy in getting pictures of plants painted from life.'

2 Part of Gesner's collection has lately been recovered at Erlangen: cf. Arber, *Herbals*, pp. 111–13.

'T. P. Angli ex picta' with a note added to warn the engraver to correct the artist's mistakes; and another (*Sagittaria sagittifolia*), Tab. 8. 72, also from a picture, is stated to be the figure already published by Camerarius in his *Epitome* of Mattioli, p. 875.[1] One, Tab. 11. 91, 'from the fresh plant', is *Calamintha acinos*. Ten, Tab. 1. 6 *Lotus corniculatus*; 5. 42 *Homogyne alpina*; 5. 44 *Erigeron* ? species; 6. 46 ? *Inula*; 6. 52 *Hieracium* ? *villosum*; 8. 71 *Viola* ? species; 12. 106 *Veronica spicata*; 14. 125 *Rapistrum* ? species; 15. 129 *Trinia glauca*; 18. 156 two branched specimens of *Botrychium lunaria*, are from dried specimens evidently taken out of Penny's *hortus hyemalis* and some no doubt originally found in England. The rest have merely his name or initials except Tab. 1. iv, a woodcut of the aquatic *Enteromorpha intestinale* stated to have been found by him *in Lacu Muratensi*, in the Murtensee north of Fribourg; a specimen doubtless collected on his journey from Zürich to Geneva, and an indication that after his visit to the west he came back to Wolf.[2]

It seems tolerably safe to assume that the bulk of these dried plants had been given to Gesner soon after Penny's arrival in the autumn of 1565. At that time, as his letter[3] of 24 November (only three weeks before his death) to his friend Benedictus Aretius, professor of theology at Berne, indicates, Gesner having finished his book on fossils (a trifle of 338 octavo pages) in July, had devoted himself to the more serious business of his *Historia Plantarum* and was wholly occupied in getting 'the innumerable dried specimens received from Jean Bauhin painted as quickly as possible so as to return them to their owner'. As he does not mention Penny or his plants it is probable that these were put aside into his general collection and perhaps only drawn in the last few days of his life or even by Wolf's instructions after his death. In any case it is possible that they were in the nature of an exchange—the sort of exchange of which Gesner's letters contain constant examples. For De l'Ecluse (*Rar. Plant. Hist.* p. ciiii) records that Penny gave him a picture of a mountain *Geum* painted in its proper colours which he, Penny, had obtained from the herbarium of pictures belonging to Gesner:[4] no doubt it was given

1 Published Frankfort, 1586. Many of the Gesner pictures including several of those commented upon by Penny (e.g. *Gentiana punctata*, *Epit.* p. 416) were used by Camerarius for this volume.

2 For help in identifying these pictures I am indebted to my friend H. Gilbert-Carter.

3 *Ep. Med.* p. 121 *b*.

4 Gesner had been collecting pictures of plants for some time and used a few for his edition of Cordus in 1559 (cf. Dedication). He had received a number of specimens from Felix Platter—dry plants which he had pictured (cf. *Ep. Med.* p. 101 *b*)—and from Jean Bauhin (e.g. *H.P.U.* II, p. 4). To these pictures, and to their painter and engraver employed by him, Gesner makes frequent allusion in his letters to Bauhin written between 1560 and 1565, and published as a supplement to *De Plantis a Divis nomen habentibus*, Basel, 1591, and in his *Epistolae Medicinales*, in one of which to Zwinger on 30 November 1565 he notes that his painter has fallen ill (p. 114).

him at the same time as that of *Swertia*. In addition to plants he certainly obtained possession of a number of pictures of insects and their accompanying records, presumably on the understanding that he would work them up for publication. For not only did they form the nucleus of his own collection of material, but when the fifth and posthumous volume of the *Historia Animalium* was eventually issued in 1587 it included no insects except the Scorpion. It is of course possible that these pictures only came to Penny after Gesner's death, when he certainly helped Caspar Wolf, the doctor, who took over all the master's collection. But it seems more probable that he received them as a return for his own botanical records—in his last months Gesner was working whole-heartedly at his History of Plants and may well have realised that entomology was impossible for him.

Penny's plans were upset by Gesner's sudden death from bubonic plague on 13 December at the early age of forty-nine. He moved on across the country with the coming of spring; for there was nothing to keep him in Zürich; the world was before him; and he had the chance of seeing alive the plants whose pictures he had been studying. So he set off by the direct route past the Murtensee to Geneva and did a considerable exploration of the neighbourhood and of the Jura and Savoy. Fortunately the second volume of C. C. Schmiedel's work, the selection from the pictures and notes prepared by Gesner for his History of Plants published at Nuremberg in two parts in 1759 and 1760, contains very many references to and quotations from the notes which Penny attached to it. Schmiedel in his account of the work (*Historia Operis*, p. xliii) says that while the material was in Wolf's hands it was scrutinised by Jean Bauhin and Thomas Penny, and that the latter 'passed over but few of the sheets of pictures without adding some observations and notes, particularly about localities'. These, for the plants selected, have fortunately been printed. They make it clear, as we could have already guessed, that Penny returned to Zürich after his expedition to Geneva[1] and then went through Wolf's papers, commenting upon them and jotting down records of his finds.

In the two 'Fasciculi' of Schmiedel's work there are thirty-three of these notes in all. A small number of them deal with nomenclature: 'Great Maple', 'Wild Clarye', 'Hunny Suckels *vel* Great Woodbynd', 'Wild Straw Berries' are English names added in Penny's hand to the titles of the plates. More important are the following: II, p. 5 of *Pyrus torminalis* 'Penny finally added that it is called Service tree in English, and in southern

1 One note, that on the abundance of Gratiola round Orléans, suggests that his return to Zürich was after his time in France. This is not impossible. But since there is no reference to Montpellier or Majorca, or with this one exception to anything except Geneva, it is more likely that the Gratiola-note represents either a later visit or, more probably, a letter to Wolf when he was with Caperon.

England is common in woods. Unless the fruit is kept for two or three months like Medlars they are not pleasant.' II, p. 55 of *Blackstonia perfoliata* (after a great debate as to its name) 'Penny does not support the change of name suggested above, but urges the retention of Perfoliate or Yellow Centaury: he admits that it has not such bitterness as the Lesser Centaury; nevertheless it is plainly bitter'. II, p. 18 of *Cypripedium calceolus* 'Penny says Dodonaeus "in Coronariis" [can this be his *Florum et coronariarum Historia*, published at Antwerp in 1568?—if so this must be a later communication from Penny]. calls it Damasonium nothum. I prefer to refer it to Helleborine or to call it Satyrion polyrrhizon. It grows in the woods on the slopes of Mont Salève near Geneva. The native people in their dialect call it Braguette d'Alemagne.' Or more interesting still the two notes II, p. 27 and II, p. 31 on *Epilobium angustifolium* and *E. hirsutum* 'Penny adds that he found the plant in this form on Mont Salève but near Geneva found it with six or seven branches full of flowers on the top of the stalk. There are, he says, some who call it Alipon: I with Ghini refer it to a kind of Lysimachia' and 'Penny adds Turner thinks this a kind of Lysimachia, and not without reason [in *Herbal*, II, p. 44 *E. hirsutum* is "Lysimachiae primum genus" and *E. angustifolium* (or *Lythrum salicaria*?) is "Lysimachia III"]. It grows in wet places and near water. In England it is called Milner Flower [a misprint for Miller] as if one were to say Flos Molitoris.'

Of his notes on localities almost all have to do with Mont Salève, with the Jura or with Mont Mole in Savoy. Records are most frequent among the Orchids and Gentians. Of the former he discusses the name of *Spiranthes aestivalis*, II, p. 10; records from Salève *Epipactis latifolia* (p. 14) and *E. palustris* (p. 17); on its summit *Nigritella nigra* (p. 22); and 'here and there near Geneva' *Horminium monorchis* (p. 19). Of the latter he notes *Gentiana verna* 'on Mont Salève flowering in May' (p. 44)—an important notice of the time of his visit; *G. acaulis* on Jura and 'Mola' (p. 45); *G. purpurea* on Mole 'with a root almost the thickness of an arm, two feet long' (p. 52); *G. punctata* which he declares to be a separate species (p. 53); and *G. pneumonanthe* of which he fills a vacant line by saying 'it is Pliny's Calathian Violet' (p. 51). The only other locality mentioned in this source is of *Lonicera xylosteum* 'near Geneva at the river Arve' (I, p. 37).

Gesner's pictures are not the only evidence for this time of vigorous botanising. Some of Penny's results—more interesting indeed than those which merely confirmed Gesner's finds—were passed on to the great Flemish botanist Mathias de l'Obel and reported by him. Thus in his *Adversaria*, p. 41, we read under 'Typha minor' (*T. minima*) that 'the very skilful and accurate student of plants Thomas Penny the Englishman gathered it some years ago and showed it to me' in the neighbourhood of

Geneva at the confluence of the Rhone and the Arve[1]—this specimen Penny preserved 'among the dried plants on sheets of paper' and showed to De l'Ecluse in 1581 (*Stirp. Pann. Hist.* p. 716; *R.P.H.* p. ccxv); on p. 233 under 'Scabiosa montana' (*Cephalaria alpina*) that 'my intimate friend Mr Penny showed it to me from the Jura mountains and Salève'; and on p. 327 under 'Myrrhis altera' (*Chaerophyllum hirsutum*) 'my skilful friend Doctor Penny got it from the mountains near Geneva and Salève, had it sown in London, and showed it to me'.

Others of his finds in the Alps were afterwards shown to Charles de l'Ecluse and published by him in his book *Stirpium Pannonicarum Historia* (Antwerp, 1583) and in the *Rariorum Plantarum Historia*. Among these is his 'third Alpine Buttercup', really *Anemone narcissiflora*, of which he 'received a picture and description by the generosity of Dr Thomas Penny'. Penny had found this in abundance on Mount Jura near Geneva, flowering in June and July and with ripe seed in August (*S.P.H.* pp. 267–9; *R.P.H.* p. 235).[2] Another portrait and description of a plant which Penny 'thinks to be the Securidaca Dioscoridis' were accompanied by a dried specimen. It grew in the neighbourhood of Geneva 'not far from the trembling bridge'[3] and is evidently *Coronilla varia* (*S.P.H.* pp. 749–50; *R.P.H.* p. ccxxxvii). Finally there is the very notable plant which De l'Ecluse named *Gentiana punctata Pennaei* and of which he received picture and description during his visit to London in 1581. Standing a foot high with the leaves of a gentian but with 'branching sprays of three pale blue flowers elegantly marked with black dots and furnished with five golden stamens (*apices*)' it is evidently *Swertia perennis*. Penny recorded it as 'in Monte Bockemuto Suitensium', that is on the Bockmattli Pass in the Canton of Schwys leading from the Hinter-Wäggital to the valley of the Linth and Nüfels[4] (*S.P.H.* pp. 290–3; *R.P.H.* p. 316); and this is the locality in which Gesner had found it on 1 August 1564.[5] It is in connection with this plant that Gerard describes Penny as 'doctor in Physick of famous memorie and a second Dioscorides for his singular knowledge of plants' (*Herball*, p. 352). A further allusion to this period is found in Gerard's *Herball*, p. 564: 'There is a kinde of wilde

1 Still its chief locality, cf. Schinz und Keller, *Flora der Schweiz*, p. 22.

2 So, too, De l'Obel MS. *St. Ill.* II, p. 92.

3 Gerard reporting this represents Penny as having found the 'Hatchet Fetch' in northern England (*Herball*, p. 1055): Johnson, *Gerard*, p. 1233, corrects this and gives 'Pontetremile' as the locality: so does De l'Obel, MS. *St. Ill.* III, p. 242.

4 I owe this identification to the kindness of Colonel G. R. de Beer, F.R.S.

5 As already noted, it is depicted in colour on the first plate in Schmiedel's book of Gesner's pictures, 'e Bokkemato monte Suitensium' being the locality, and Gesner himself the finder. No doubt he had given Penny its picture; for, as is noted by Schmiedel, *Gesn. Hist. Pl. Fasc.* II, p. 56, De l'Ecluse's picture is plainly copied from Gesner's.

Horehound, called Stachys Spuria Flandrorum, which doctor Pennie the physician brought first into England out of the clifts of the mountains Jura and Salana: this herbe beareth brave skie-coloured flowers standing in thrummie, prickley, scaley or shellie crownets.' It is apparently *Sideritis syriaca*.

We have dealt with this scattered evidence at some length partly because it has not previously been collated, but mainly because it discloses very manifestly the eminence of our forgotten fellow-countryman. The fact that his notes on Gesner's work were thought fit to stand alongside of those of Jean Bauhin, and that two hundred years later Schmiedel was ready to print them, is proof of the contemporary and the later verdict on his worth: and, brief as most of them are, they show a mastery both of the literature and of field-work, and a good insight into taxonomy. Moreover, from these details it becomes plain that Penny spent a full summer, May till July or August, in the mountains near Geneva, on the very spots which John Ray visited exactly a century later, when he like Penny was journeying from eastern Switzerland to Montpellier. It is possible that during part of this time he was accompanied by his friend John James, afterwards physician to Queen Elizabeth; for De l'Obel in the manuscript of his *Stirpium Illustrationes*, II, p. 88, in the library of Magdalen College, Oxford, has a note under *Trollius europaeus*, 'John James doctor to the Queen of England collected this on Mont Mole'.

If, as we have suggested, Penny returned to Wolf after his botanising and then went through the sheets of Gesner's material, he certainly departed again before long; and on this occasion seems to have gone straight through to Provence.

At Montpellier assuming him to have arrived in 1566 he must just have missed the great Guillaume Rondelet who died on 20 July in that year and to whom Rabelais had paid his matriculation fee in 1530. His successor, the Regius Professor who presided over the garden, was Laurent Joubert, more a physician than a botanist. Assatius (Jacques Salomon d'Assas)[1] and Stephanus Barallius (Etienne Barral: homunculus 'de plus de valeur que d'apparence' as Edouard Morren puts it)[2] were teaching.[3] Guillaume Pellicier[4] was writing commentaries on Pliny and studying plants. Jacques Farges, the apothecary, had a garden.[5] Pierre Pena was in practice.[6] Mathias de l'Obel who had been at Padua in 1562

1 He had married Catherine, Rondelet's daughter: for Assas, cf. J. E. Planchon, *Rondelet et ses Disciples*, pp. 28–9.

2 *M. de l'Obel: sa Vie et ses Œuvres* (Liège, 1875), p. 24.

3 So De l'Obel, *Adv.* (ed. 1605), p. 114.

4 L.c. p. 237: 'Guillaume Pelisser Bishop of Montpellier.' He was in fact Bishop of the neighbouring see of Maguelone; cf. A. Germain in *Revue Hist.* II, p. 69.

5 L.c. p. 57.

6 So Gesner in letter of August 1565 (in J. Bauhin, *De Plantis a Divis*, p. 163).

and Verona in 1563 was in Montpellier in 1565 and still there in June 1566 when he gathered a Cirsium which he identified as the 'Ghini's Thistle' of Mattioli (this is certainly *Saussurea alpina*: its picture is in Mattioli's *Comment.* p. 572) at the oil-mills on the river. Almost certainly the two men met there[1] and began their friendship; and this was evidently close and regularly maintained by letters; for, when four years later De l'Obel who went to England in the autumn of 1569 published his *Adversaria*[2] in London in 1570–1, Penny's name appears, as we have seen, a number of times in it. Amongst other information he reported the story of the Suffolk peas in 1555—of how on a stormy shore between Orford and 'Alburnus' (? Aldeburgh) in the autumn there appeared a crop large enough to feed a thousand men (*Adversaria*, p. 397)—the story told by Dr Caius and by Camden, and perhaps learnt by Penny from Gesner who had published it along with Caius's report of the stranded 'Maculo' in his *H.A.* IV, p. 216. Another of these records (l.c. pp. 289–90) is important as indicating the friendship between Penny and Turner. It deals with 'Aconitum Pardalianches' which Turner following Fuchs identified with Herb Paris and described as occurring near Morpeth in *Names*, p. A. v*b*; it reports that 'my friend of much correspondence and travel the learned doctor Penny assured me that he had seen and gathered it on the hills of Northumbria'[3] which must mean that he had visited them, surely on Turner's suggestion, before leaving for Zürich.

In Montpellier itself Penny's finds naturally became less notable; for the district was the centre of botanical studies at the time, and had been thoroughly explored. One discovery of his is reported—by Joachim Camerarius in his *Hortus Medicus*, p. 36—'Caryophyllata vulgaris' (*Geum montanum*) with white flowers 'in monte Lupo', on Mont St Loup, the hill north of the town.

It would be tempting to suppose that Gerard (*Herball*, p. 31) under 'Juncus marinus gramineus' (*Imperata arundinacea*) was giving us further news of Penny's work at Montpellier when he writes 'Francis Penny of famous memory, a learned physition and expert herbarist, found this Sea Rush in the coast of the Mediterrane sea in the way as he travelled

1 This seems the obvious inference from *Adv.* p. 25.

2 It is mainly a detailed flora of Montpellier and the Cevennes; and is published under the names of Pierre Pena and De l'Obel. The problem of Pena's share in it and of his history need not be discussed, but cf. Ed. Morren l.c. pp. 10–12 and L. Legré, *P. Pena et M. de Lobel* (Marseilles, 1899).

3 Penny's plant was doubtless Herb Paris, but it must be noted that Gerard, *Herball*, p. 621, says that 'Doronicum radice repente [*D. pardalianches*] hath been gathered in the colde mountaines of Northumberlande by doctor Pennie lately of London deceased, a man of much experience and knowledge in Simples whose death myselfe and many others do greatly bewaile'. Gerard has abandoned Fuchs' identification and is following Dodoens who stated that Aconitum pardalianches is not Herb Paris but Doronicum, cf. *Pemptades*, pp. 437, 444.

to Piscaire [Pescais]'. Johnson alters the position of this record to *Gerard*, p. 44, and gives the locality as 'betwene Aigues Mortes and Pescaire': he adds a reference to De l'Obel. So far this might seem an obvious allusion to Penny; for the misprint of Francis for Thomas is easy. But research into De l'Obel discourages it. In *Adversaria*, p. 42, he says nothing of Penny; in his *Kruydtboeck* (Antwerp, 1581) he says, 'Franchois Pennin' found it 'by Aigues-mortes op den wegh nae Piscaire'; and in the manuscript on Grasses (*Stirp. Illustr.* 1, p. 19) 'my very dear friend of blessed memory Franciscus Penninus of Antwerp a very skilled pharmacist found it on the sandy shore of the Mediterranean near Montpellier near Aigues-Mortes on the road to Piscaire' and adds, 'Its picture I took steps to get drawn at Antwerp in the shop of the aforesaid Pennin from a dried plant mounted on paper.'[1]

But, if we know very little of Penny's stay at Montpellier, it seems certain that at this time he undertook a piece of field-work that was genuine pioneering, an expedition to Majorca. No details have been preserved: we do not know how or when or with whom he went;[2] and only one of his finds has survived. But this is the highly characteristic *Hypericum balearicum* whose picture and description he handed over to the great Fleming, Charles de l'Ecluse, and which that great botanist in his *Stirpium Pannonicarum Historia*, pp. 66–7, named 'Myrtocistus pennaei'.

It was probably after his stay at Montpellier and before going on to Germany that Penny visited Paris and Orléans. On his way he may well have stayed at Lyons; for he certainly made contact with Jean Bauhin who had settled there in 1563 and with Valerand Dourez his kinsman.[3] Of Paris there is little evidence except the bare mention of a visit: but of Orléans we are told that he went there on purpose to study plant-physiology (*Phytognomices studio*) in association with Natalis (Noël) Caperon, an apothecary who is credited by Parkinson (*Paradisus*, p. 44)[4] with being

1 On p. 75*b* of the same MS. he mentions François Penin as a wealthy perfumer of Antwerp who used to do business at the sign of the Rose.

2 Possibly De l'Ecluse's expedition to Spain which he had explored as physician and escort to the two daughters of Anton Fugger may have been his pattern; for this had happened in 1563 (Legré, *Anguillara*, etc. p. 78) and was known to Gesner (*Ep. Medic.* p. 121*b*). But it seems impossible that Penny met De l'Ecluse at this time; for the great Fleming was in Antwerp in 1565–7 (*Rar. Plant. Hist.* p. 35).

3 Cf. below, p. 183 and *Theatrum Ins.* pp. 25, 207; 84, 271.

4 Parkinson no doubt got this from G. Bauhin, *Pinax*, p. 64, where Camerarius is quoted as the authority. The actual authority for the statement is De l'Ecluse, *Stirp. Pann. Hist.* p. 172, who says that an apothecary of Orléans, 'Noël Capperon', sent him the first specimen and soon after gave it the name Fritillary from its checker-board pattern—it being previously called the Variegated Lily (cf. also P. Reneaulme, *Specimen Hist. Plant.* p. 147). Caperon found it abundantly in meadows by the Loire and sent it to De l'Ecluse in 1572.

the first discoverer of the Fritillary (*Fritillaria meleagris*), and who was afterwards massacred as a Huguenot at the St Bartholomew's tide of 1572 when in Orléans some five hundred victims were slaughtered by the Catholics. Here Penny made observations of the Bed-bug (*Cimex lectularius*) which proved that, instead of being spontaneously generated from rotten timber, it laid eggs from which young bugs hatched and grew up (cf. *Theat. Insect.* p. 269). He also reported to Wolf at Zürich, as we have seen, the abundance of *Gratiola officinalis* in all the marshy places near the city (*Gesneri Hist. Plant.* Fasc. II, p. 65).

His travels during these years 1566–9 are uncertain and perhaps cannot now be mapped accurately. But in 1569 he was certainly at Heidelberg, where the great Johann Lange pupil of Niccolo da Lonigo and author of *Epistolae Medicinales* had lately died, with his friend Peter Turner and with William Brewer, both of whom were keen naturalists and were working at the University, having matriculated there in April 1566 and taken the M.A. degree in February and August 1569.[1] The latter Penny afterwards described as 'a man of great learning and my closest friend' (*Theatrum*, p. 110); and the notes supplied by him on a Shrike, 'Lanio avis', feeding on crickets and beetles and setting them on thorns near its nest (i.e. pp. 136, 157); on a 'Scolopendria' or centipede (probably *Geophilus electricus*), brightly luminous (p. 113); and on the history of Glow-worms (p. 110) bear out the claim that he was a real student of nature. He was also a keen fisherman[2] and not only describes the catching of trout by flies on a hook but also quotes, as from Aelian, an account of artificial flies 'made from purple and mixed wool and fitted with two wax-coloured hen-feathers' (p. 73)—this in order to demonstrate the true antiquity of fly-fishing! At Heidelberg there were certainly made a number of observations duly recorded either by De l'Obel or in the *Theatrum*. Of plants there are 'Cyperus graminea' (probably *Scirpus sylvaticus*) 'gathered by the industry and care of Doctor Thomas Penny by the river Neckar near Heidelberg' (*Adversaria*, p. 38);[3] 'Bellis media' (*Bellis sylvestris*)—'Mr Penny gathered this in the wooded mountains of Germany and sent it to me' (p. 199); Bock's 'Passerina' (*Thymelaea arvensis*) 'to be seen in very few places in Germany whence the industrious Doctor Penny obtained it' (p. 358 and *Kruydtboeck*, p. 980). Of insects we are told that 'along with Penny Peter Turner and William Brewer

1 G. Toepke, *Die Matrikel der Univ. Heidelberg*, II, pp. 39, 464: Turner by special permission proceeded M.A. without taking his B.A., Brewer took his B.A. in 1568.

2 Is it an accident that Lange was also deeply interested in fish, cf. *Ep. med.* 59 and 60, and in insects, l.c. 81 (on a plague of caterpillars devastating the pines in the Black Forest for which he recommends fumigation)?

3 In the MS. of his *Stirpium Illustrationes*, I, De l'Obel after transcribing this record says, 'we found it later in wet places in Somerset'.

studied the life and vices [*vitam et vitia*, a typical "elegance" of Mouffet] of the Buprestis in the country round Heidelberg' (*Theatrum*, p. 142), and Penny has given descriptions of two species. So, too, Brewer is credited (pp. 133–4) with having seen a Longicorn beetle resting with its horns clasped round a twig and with comparing it with the Bird of Paradise which having no feet ties itself by its wiry feathers to the boughs and so gets a rest when tired—which familiar legend he had no doubt derived from Cardan!

At this time Penny must have made the acquaintance of Joachim Camerarius, son of the great Classical scholar and himself a doctor learned in botany who had settled in Nuremberg after his work in Italy in 1564. For in his *Hortus medicus* (Frankfort, 1588) Camerarius mentions a number of plants brought to his notice by Penny and on p. 36, the first mention, introduces him as 'the eminent London doctor, very skilled in natural history, my particular friend'. Such terms coming from him are the more significant because although he mentions nearly every known contemporary botanist he is very chary in his praise and does not similarly distinguish any other single person. The two men must certainly have met,[1] and it is probable that Penny visited him at his home where he was already beginning the famous garden of which the *Hortus medicus* is a catalogue. Penny certainly kept up a correspondence with him sending to him plants of *Lactuca virosa* (*H.M.* p. 82), of the double *Matricaria parthenium* (p. 97), of a Spurge which 'would not stand the winter either in a pot or out of doors'[2] (p. 128) and of *Sedum roseum* 'in monte Engleborreno' (p. 139). In return, as we shall see, he received a number of specimens and particulars of insects.

It seems clear that before returning to England Penny went up to Prussia and the Baltic; for the most famous of his records of plants, that of the 'Chamaepericlymenum' (*Cornus suecica*) whose portrait, in seed, and description he gave to De l'Ecluse in 1581 (cf. *Stirp. Pannon. Hist.* pp. 87–9; *Rar. Pl. Hist.* p. 59), refers to this neighbourhood. On this point there has been a widespread mistake: it has been assumed that Penny found it in northern England, as he certainly did the 'Chamaemorus' or Knot-berry (*Rubus chamaemorus*). It has even been stated that its locality was the Cheviot.[3] But in fact appended to the description is a statement that it is well-known 'Prutenis rusticis circa Gedanum vulgo Dantiscum appellatum', the plain meaning of which is 'to the country-folk of Prussia around Gedanum commonly called Dantzig':[4] and a note

1 Cf. *Theatrum*, p. 63.

2 He remarks on p. 144 that the winter of 1586 was peculiarly severe.

3 E.g. by Gunther, *Early British Botanists*, p. 234. Unfortunately Gunther's account of Penny is full of errors both of fact and of conjecture.

4 Cf. Gesner, letter to Pellizer in preface to *Val. Cordi De Halosantho*: 'Dantisci quod priscis Gedanum fuit, nobilissimum hodie totius Germaniae emporium.'

is added that it is also common in Norway[1] and Sweden. Gedanum as a place-name disappeared about this time though the Dantzig press still imprinted their books 'Gedani' till 1580; it still survives in the names Gedansk and Gdynia. Parkinson explicitly states that it 'was found by Doctor Penny as Clusius saith, by Dantzicke who gave him both the figure and the description as it is here expressed' (*Theat. Bot.* p. 1461).[2] As a British 'first record' Penny's find will have to be deleted: it remains as perhaps the sole surviving evidence that he went to Dantzig in August, probably in 1569.

It is possible that in this journey he was accompanied by his friend Peter Turner; for Turner certainly visited Copenhagen about this time and gave to Penny the specimen of *Geranium bohemicum* which was passed on in 1581 to De l'Ecluse and appears in his books.[3] But as Turner took his doctorate at Heidelberg in 1571 it is probable that his Danish tour is later. He must have been back in England by 1575 when he was incorporated M.D. at Cambridge. At this time he was evidently still interested in plants; for Jean Bauhin[4] in his *Historia Plantarum Universalis*, I, ii, p. 225, records that 'thirty years ago an eminent doctor, Peter Turner, presented us with a dried herbarium (*nos donavit herbario sicco*) of English plants and among others this [*Myrica gale*] with the name 'Gale' attached to it. In this very year 1605 I tasted the leaf...'—a note trebly important, since it is perhaps the earliest use of 'herbarium' for 'hortus hyemalis', proves that Peter inherited his father's habit of collecting, and explains how Linnaeus came to make the English name Gale the specific title of the Bog-myrtle. This seems to be his last appearance as a botanist. He became famous as a doctor, sat in two parliaments as member for Bridport[5] and died on 27 May 1614, aged 72. He had married Pascha, sister of Henry Parry, Bishop of Worcester, who erected a memorial 'with half the lively figure of the party it concerneth' in St Olave's, Hart Street.

In 1569 Penny seems to have returned to England with his M.D. degree

1 De l'Obel MS. (Magdalen College Library), *Stirpium Illustrationes*, II, p. 65 *b*, notes Penny's record and adds, 'I found it in abundance on the Norwegian barrens near Flekkefjord', i.e. on his journey with Lord Zouche in 1598: see below, p. 235. Tradescant (Hamel, *Tradescant in Russland*, p. 184) reported it independently from Archangel and sent seeds to Vespasian Robin.

2 Gerard, *Herball*, p. 1113, includes the record 'sent unto Clusius by that learned Doctor in Physicke Thomas Penny' but gives no locality. Johnson, *Gerard*, p. 1296, adds that it grows in the maritime parts of Norway and Sweden. Coles, *Adam in Eden*, p. 158, exactly copies Parkinson's account.

3 For Penny's description of this plant see below, p. 169.

4 A further allusion to this gift is in *H.P.U.* I, ii, p. 358, where a dried specimen of *Erica tetralix* sent by Peter Turner is recorded. For this cf. Ray, *Cat. Angl.* p. 101.

5 So Venn, *Al. Cantab.* IV, p. 276.

and to have settled in London; for in September of that year he presented the letters testimonial of his doctorate to the authorities of the College of Physicians. Next year he was examined by them in January and forbidden to practise: he disobeyed their instruction and was subsequently put into prison for so doing. In August he was again examined, but not approved:[1] but this does not seem to have interfered with his work or damaged his reputation; for in 1572 when he was called in to attend Rudolf Zwingli, grandson of Henry Bullinger, the great Reformer, who was taken ill at the house of Richard Cox, Bishop of Ely, and Peter Turner's stepfather, he was described as 'considered the most skilful physician in all England' (*Zürich Letters*, II, p. 203). Further evidence of his eminence in the profession is found in the urgent calls which were despatched to him on 13 September 1576 by Walter Devereux, Earl of Essex, then Earl Marshal in Ireland, and his friend Edward Waterhouse. The Earl had been taken ill with dysentery in Dublin: two local practitioners failed to cure him: so he proposed to sail at once to Milford and begged Penny to meet him at his house at Lanfey (Lamphey), Pembroke, bringing with him 'all things convenient for this disease'.[2] In fact he was unable to move and died on 22 September: but the letters summoning Penny from London bear out Zwingli's estimate of him.

Penny had now made London his home, and lived in the parish of St Andrew's Undershaft, Leadenhall Street, having a garden in the Moorfields outside Moorgate, for the rest of his life. But he seems to have travelled extensively, perhaps visiting the north and certainly East Anglia. Records relating to Peterborough connect up with the fact that his brother Brian had a farm in the neighbourhood; those from Essex with his marriage in 1573 to Margaret,[3] the daughter of John Lucas of St John's near Colchester who had been Master of Requests to Edward VI, and half-sister of Sir Thomas Lucas whose name was known to Parkinson. Among these records there are the two accounts of observations on the top of Cartmel fell: 'In 1582 I saw two different kinds of Wasp fighting whence I infer that the large ones raid the nests of the small' (p. 45) and a similar undated record of a battle of horse-flies, *bombylii* (p. 74). Of Peterborough there is a dramatic story of how he saw a Hornet pursuing a Sparrow in the public street; the insect stung the bird and killed it, and then to the admiration of a crowd of spectators proceeded to gorge itself with its

1 The evidence for these events is in the Annals of the College written by John Caius and printed in his *Works* (Cambridge, 1912), pp. 64, 66–7: cf. also C. Goodall, *Historical Account of the College's Proceedings against Empiricks*, p. 315.

2 Cf. *Devereux Papers* (Camden Miscellany, XIII), p. 4. Scandal stated that he was poisoned by agents of Lord Leicester who married his widow two years later.

3 Cf. *Visitation of Essex*, 1612, p. 236.

blood (p. 51)[1]—a story the more remarkable because told by Edward Topsell of himself at Duckworth in Huntingdonshire (*History of Serpents*, p. 659).[2] There is also a long description of a 'fly' taken there by him in 1582 (p. 77); and this is so characteristic as to deserve translation. 'Its body is oblong and butterfly-like; its head is small and yellowish; its eyes large, prominent and black; its proboscis [*promuscis* is regularly used of the Elephant: Penny probably derived this use of it from Wotton, *De Diff.* p. 189] with which it sucks dew from flowers is curled up, and of a mullein-yellow colour; its horns (*cornicula*) are two, black, rather long, attached a little above the eyes; its abdomen and back are leaden greyish, the tip of the tail is almost yellow. *It has only four feet*, the hind ones yellowish, the ends of the front ones black. It has four wings (as many wings as feet), the outer ones leaden grey with the outer edge dusky yellow, the inner ones dusky flaxen. The outer wings when joined so as to cover the body are so closely united that you can scarcely, if indeed at all, see the join. It is slow in flight and does not fly far. *It dies three days after birth.* It lives among mallows and nettles.' With the exception of the two phrases in italics—the first an error and the second a rash generalisation—the description is precise, accurate and adequate, fully up to the high standard of Penny's descriptions of plants. It applies exactly to one of the 'Footmen' moths, *Lithosia lurideola*, which is of this colour, folds its wings in this unusual way, and feeds freely on the honeydew on nettles.

Either in London or perhaps in Bristol soon after his return he found his friend De l'Obel who had just published his *Adversaria* and was living in Somerset 'at the house of the generous patron of this history Edward Saintloo [St Loo]' (*Adv.* p. 370). How long the Flemish botanist stayed in England is not certain. He was certainly in Antwerp in October 1574; for Christophe Plantin (*Correspondance*, IV, p. 169) mentions that he was then working hard at the enlarged version of his book, his *Plantarum seu Stirpium Historia* which Plantin published for him in 1576. Soon after, and certainly before 1579 (cf. Plantin, l.c. VI, p. 49), he became physician to William the Silent and was with him until his assassination in 1584. Then he went for a year or more to Antwerp and afterwards came back to England, where he lived almost continually till his death in 1616.

During De l'Obel's earlier visit Penny's interest in botany was certainly still strong, even if insects absorbed most of his attention. Thus De l'Obel

1 If this story seems unworthy of a great naturalist, it must be remembered that it comes to us edited by Mouffet who was proud to have improved Penny's style and given a literary flavour to his records. But the tales of the ferocity and venom of Hornets are ancient and still persist: no doubt the Scriptural references (Ex. xxiii. 28, Deut. vii. 20, Josh. xxiv. 12) are partly responsible.

2 For the problem thus raised see below, p. 224.

in his *Kruydtboeck* published by Plantin at Antwerp in 1581 records that 'the learned English doctor Thomas Penny' had shown him 'Plantago aquatica minor' (*Alisma ranunculoides*) growing in watery places in England (p. 370); and the plate and record are reproduced by Parkinson in *Theatrum Botanicum*, p. 1245. This notice antedates Johnson's record in 1633 given by W. A. Clarke (*First Records of British Flowering Plants*, p. 151).

Further, Parkinson (*Theat.* p. 613) records that it was in Dr Penny's garden in London that De l'Obel found the 'Lesser Water Betony' which he regarded, wrongly, as distinct from *Scrophularia aquatica*. In this garden William Mount noted *Acorus calamus* growing (cf. Gunther, *E.B.B.* pp. 234, 256). De l'Obel also records in the notes of his *Stirpium Illustrationes*, edited by William How and published in 1655, that certain Wild Peas, growing like a Bindweed and with leaves like Lucerne but not serrated, were first observed and communicated to him by 'his dear friends the English doctors Thomas Penny and J. James of the Isle of Wight, physician to the Queen'[1] (p. 162). It must have been at this time that De l'Obel received from 'an illustrious lady Dame Catherine Killigrew', wife of Henry Killigrew Elizabeth's ambassador, and fourth daughter of Sir Anthony Cook,[2] specimens of beans cast up on the coast of Cornwall where no ships had been wrecked—a first record of the effects of the Gulf-stream (*Adv.* p. 395); for she died in 1683.[3]

It may have been in 1571 that Penny first made personal contact with De l'Ecluse who visited England in September[4] of that year, and did some botanising on St Vincent's rocks with De l'Obel as is described in *Rar. Plant. Hist.* pp. 211, ccxiiii. This certainly happened eight years later when the great Flemish savant dashed over from Vienna for a brief visit in the autumn, and reported the occurrence of *Gentiana amarella* in flower and seed at Dover in September (l.c. p. 315); saw Double Feverfew (*Chrysanthemum parthenium*) in a London garden and took it back with him to Antwerp (p. 337); and noted *Salvia verbenaca* 'at Greenwich near the race-course (*Hippodromus*) of the Palace' (p. xxxi). Next year Penny sent him the picture and description of his great Balearic discovery. But in 1581 early in the spring De l'Ecluse came back, stayed until July, meeting Philip Sidney, Francis Drake[5] and other men of eminence, in-

1 For whom see below, p. 172.

2 His other daughters had married Cecil, Sir Nicholas Bacon, and Sir Thomas Hoby respectively: they were famous as learned ladies.

3 Her husband married again in 1590 (cf. Sir Henry Wotton's comment, *Life and Letters*, I, p. 259), was knighted in 1591 and died in 1603.

4 In *R.P.H.* p. clxxx he records having seen *Blackstonia perfoliata* in Essex at that time.

5 So Vorst in the funeral speech appended to De l'Ecluse, *Curae posteriores*; and E. Morren, *C. de l'Ecluse: sa Vie et ses Œuvres*, pp. 31–2. He seems to have visited Windsor, cf. Ray, *Cat. Angl.* p. 102.

spected Penny's herbarium and took from him several pictures and descriptions: *Cornus suecica* as already described; *Carduus heterophyllus* from the foothills of Ingleborough (p. cxlviii); *Trollius europaeus* from 'the northern mountains' (p. 237); and especially the 'small kind of hawk-weed' whose picture, however, he does not reproduce (p. cxlii). The description of this last which De l'Ecluse gives in full is so excellent, and so complete an answer to the sneers at Penny's powers of style that it may be quoted in translation: 'The least Hawkweed is a plant nine inches high or a little taller, with a root single, white, with few fibres; leaves from the root, five, six, seven or even more, like those of a daisy but rather longer and serrated; from the middle of them rises a single stalk, sometimes two or three, entirely without leaves, tubular, reddish near the root, sometimes branched at the top, the branches being reddish at the base; contrary to what happens in other plants, all the stalks swell (*extuberant*) and become thickened towards the summit; each bears a single round flower-head on which rest small flowers level at the top and finally producing pappus (*pappescentes*). The plant is milky and rather bitter. It grows among crops in sandy places and friable soil and is frequent in many places in England; elsewhere I only remember to have seen it in a single place near Heidelberg. Thus Penny.'[1] It is a quite admirable characterisation of *Arnoseris pusilla*, and constitutes a first British record for that species—though both Clarke in *First Records* and Druce in *Comital Flora* have overlooked it.

In addition he gave to him the picture and description of a Cranesbill growing in Denmark in the neighbourhood of Copenhagen (*agro Hafniensi*) which his very learned and very dear friend Turner, an English doctor (obviously Peter Turner), brought and gave to him. This he describes as follows (*S.P.H.* pp. 419–20 and *R.P.H.* p. c): 'The bulbous Geranium has a bulbous root consisting of many small and oblong bulbs, fibrous towards the top; a stem a foot high, with nodes (*geniculatum*), slightly reddish at the root and nodes; at each node two leaves resting on rather long pedicels, cut into five principle partes, each of which is unequally jagged at the edge: from each junction at the foot of the leaves rise five sheathed leaflets: flowers at the top of the branches, joined in pairs and resting on small pedicels, rosy, of five leaves, in colour purplish pink (*ex purpureo rubescentes*) like the blue flowers of the Geranium but rather smaller.' Presumably it is *Geranium bohemicum* described from a specimen with abnormally developed rhizomes: De l'Ecluse himself in *Rar. Plant. Hist.* l.c. notes that it appears very similar to *G. sylvaticum*.

1 This record, with the picture and the note that description and picture were given him in London in 1581, appears also in De l'Ecluse, *Stirp. Pannon. Hist.* pp. 647–50; and in abstract in MS. De l'Obel, *St. Ill.* II, p. 1.

Penny, moreover, gave him some useful information by word of mouth. 'I learnt from him', writes De l'Ecluse (l.c. p. 301), 'that this plant [*Primula farinosa*] sometimes abounds in the moist pastures of northern England: and for this reason the learned and kindly Richard Garth[1] Chief Secretary of the Chancery in London had brought to me from the county of Derby and Wybsey not far from the city of Halifure [Wibsey just outside Bradford, and Halifax] some of these plants in full flower before my departure from London....I sent them to Antwerp to my friends'[2] and on p. 311: 'He told me that this plant both the blue-flowered [*Pinguicula vulgaris*] and the white-flowered [*P. lusitanica*] are found in many parts of England, and in the north where it grows along with the red-flowered primula, is called "Butterwort", that is root of butter from the fat like butter with which men heal the cracks in the udders of cows, and by the southern English "Whytroot" because it kills sheep if for lack of food they eat it.'[3] Garth, who was a keen and competent botanist, had a close connection with Yorkshire through his daughter Jane who was married to John Savile of Methley afterwards Baron of the Exchequer.[4]

De l'Ecluse certainly formed a close acquaintance with Garth whom he had evidently met on his former visit;[5] according to Gerard, *Herball*, p. 757, he sent him from Austria specimens of *Polygonatum officinale* which Garth grew and 'imparted unto me'. He also saw much of James Garret the apothecary of Lime Street, who had a garden near London Wall at Aldgate and who became a regular associate of his through his brother Peter Garret of Amsterdam (cf. *Exotica*, p. 68, etc.); visited the gardens of the 'royal druggists' John Rich and Hugh Morgan (p. cxl); and saw a peculiar Plantain in the garden of Edward Points (p. cix).[6] He has left us two other glimpses of the London of 1581. He found a middle-sized Bulrush [*Typha angustifolia*] 'in a muddy ditch at Tyburn near the cemetery where malefactors hanged there are buried, one mile west of London; a narrow-leaved plant but much larger than the small one that Penny showed me among the dried plants which he kept between sheets of paper and which was published by Pena and L'Obel in *Adversaria*'

1 For Garth see below, p. 243. At this time he lived at Groutes near Croydon. He received a grant of arms in 1564 and is given in the *Visitation of Surrey*, p. 189, as 'one of the 6 Clarkes of the Pettey Bag'.

2 Cf. *Stirp. Pannon. Hist.* p. 342.

3 L.c. p. 362.

4 Their eldest son Henry, afterwards knighted, was born in 1579; for his half-brother John see below, p. 300.

5 L.c. pp. 321, 337, which record plants and seeds sent to him by Garth in 1579 and 1580.

6 He received a picture of this foliated variety of *Plantago lanceolata* and also a picture of an abnormal *P. major* from James Garret in 1591; and Parkinson, *Theat. Bot.* p. 495, records that a figure of an abnormality was sent to him by Garret: but the picture of this var. in De l'Ecluse is *P. major* var. *brachystachya*; that in Parkinson is the very different var. *polystachya*.

(p. ccxv). And a more cheerful one: 'Daffodils are so abundant in the fields near London that in the famous street commonly called "Ceapside" in March countrywomen sell the flowers in the greatest abundance and all the shops are bright with them' (p. 164). And like De l'Obel he enjoyed the palace garden at Westminster where he noted the huge ilex (p. 23) 'yet standing' in 1633 (Johnson, *Gerard*, p. 1343) and 'just against the backe gate that openeth into the High streete over against the Tennis Court' (Parkinson, *Theat.* p. 1395),[1] even as his countryman had noted the flowering pomegranates (*Adv.* p. 419). He sailed from 'the house of Lord de Cobham Knight of the Garter and Warden of the Cinque Ports'[2] in July finding a few plants while waiting for his ship.[3]

For the rest of his life Penny's interest seems to have been devoted almost entirely to insects and the building up of the mass of material now partially preserved to us in the *Theatrum*. He was a rapidly aging man who suffered from asthma. As with many other victims of disease the treatment prescribed can hardly have improved his health: according to Mouffet[4] he treated himself with woodlice crushed in wine until Mouffet who was a 'chymical' doctor persuaded him to inhale fumes of sulphur—which cured him! In 1587 he lost his wife, and the blow hastened his end. His digestion was already impaired: he suffered seriously from headaches: his memory began to fail—all these troubles Mouffet describes in *Health's Improvement*, p. 248, and explains as due to abstinence from salt. Whatever the accuracy of this diagnosis, he never rallied. In January 1588 he died. He was buried 'without any ceremony whatever, no mourning apparel, nor ringing nor singing' in the churchyard of St Andrew's Undershaft. A portrait of him, apparently taken just before his death, by William Rogers is contained in a medallion on the title-page of the manuscript of the *Theatrum* as prepared for publication in 1589, and now preserved in the British Museum (Sloane MSS. 4014). This has been reproduced in Dr Willoughby Gardner's paper. It is a crude engraving, the head and shoulders in profile of a bearded man with a good forehead and a bulbous nose, wearing his hair short and holding a book.

He left a mass of papers, some, perhaps the medical and botanical, apparently to Peter Turner,[5] others, and these the entomological, to his younger friend and colleague Thomas Mouffet. The latter declared that he found them ill-arranged, ill-written, 'disfigured with smears and dubious symbols'; that he preserved and sorted them; and that at great expense of time and money he prepared them for publication.

1 Sir T. Hanmer in 1660 described it as the largest of its kind in England, *Garden Book*, p. 127.

2 William, Lord Cobham, for whom see above, pp. 97, 115. He died in 1596.

3 Cf. Parkinson, *Theat. Bot.* p. 22—the plant found being *Calamintha acinos.*

4 So *Theat. Insect.* p. 204.

5 So Pulteney, *Sketches of Botany*, I, p. 86, quoting Jungermann. If so, they have left no trace.

CHAPTER X. THOMAS MOUFFET AND THE *THEATRUM INSECTORUM*

For fifteen years, as Mouffet tells us, Penny heaped up materials for his History of Insects, 'receiving exceeding great help' from his wide circle of friends at home and abroad. Several of these were men of European reputation. De l'Ecluse was interested in bees and sent several notes and pictures from Vienna (cf. *Theatrum*, pp. 63, 75, 90, 93, 119, 150). Jean Bauhin, 'the learned doctor Johannes Bauhinus' of *Theatrum*, pp. 25, 207, and probably the greatest botanist of the age, sent him a note on the Scorpion which as he and Penny agreed was produced both sexually and from putrefaction: this recorded that he had once hidden a pot of ointment in a wall in Paris and on his return found two scorpions in the hole. Joachim Camerarius, the younger, who came into possession of Gesner's botanical remains and was then settled at Nuremberg and working at his *Hortus medicus* sent him drawings and specimens including a picture of a Goliath Beetle from the Duke of Saxony's museum (p. 152): a letter from Penny dated 18 June 1585 to him enquiring about the identification of certain insects mentioned by Aristotle and Pliny and telling his discovery of the wings of Earwigs has been printed by Casimir Schmiedel (*Vita Gesneri*, Nuremberg, 1754, p. xv, note). But even more useful were the members of the Quickelberg family, Samuel the father a doctor of Amsterdam who published a catalogue of his collection in 1565 and was known to Gesner (cf. p. 257) and his two sons Jakob and Pieter who lived at Antwerp: they sent him many specimens and pictures of exotic insects, a locust from Africa 'whose body we still preserve in our treasury of insects' (p. 136) and beetles from Vienna (p. 142) and elsewhere (pp. 150, 161). The fact that he could obtain such help from continental savants shows how much the interest in science and the status of English students had increased since Turner's day.

At home in addition to Peter Turner, for a few years resident physician at St Bartholomew's hospital, and William Brewer there were several notable helpers: James Garret[1] 'the very painstaking druggist, in the front rank for his knowledge of simples' (p. 255), the friend and correspondent of De l'Ecluse, and well-known in the *Herball* of John Gerard, who reported 'tiny, black, swift-moving worms among violets'—possibly young larvae of one of the Argynnids; John James (whom Dr Gardner, misled by the Latin, calls John Jacob) a fellow of Trinity College, Cambridge, but junior by ten years to Penny, a doctor of great note and

1 Probably James junior, for whom see below, pp. 185, 192.

Physician to the Queen, who attended Sidney after his wound at Zutphen and with whom Penny had collected the Wild Pea; Roger Brown, Fellow of King's College, 1559–66, 'my very learned friend' who sent a warning about a fly (p. 77) and was apparently at this time a Canon of Windsor; Timothy Bright, successor of Turner as Physician at St Bartholomew's, warmly commended by Gerard (*Herball*, p. 313),[1] and the inventor of modern shorthand who gave him the note of a swarm of flies on p. 78 and of an intestinal worm at Cambridge on p. 285; and Edward Elmer (? Aylmer) a surgeon who sent him several insects from Russia (*Theatrum*, pp. 60, 151).

More important for their contributions to the book and for what can be learnt of their interest in natural history are 'Sir Thomas Knivet and his learned brother[2] Edmund' to whom Mouffet expresses particular obligation and the latter of whom sent several pictures 'painted with his own hand' (p. 197). Of one of these, four portraits of the Field Cricket (*Gryllus campestris*), Penny writes that he received it from 'the noble Knight Edmund Knivet, famous above most men by his race, his virtue and his interest in natural objects' (p. 121); another, the 'second Staphylinus' or Devil's Coach-horse as Penny decides it to be, is in fact a very accurate likeness of the strange larva of the Lobster Moth (*Stauropus fagi*) which Knyvet describes as then common in Norfolk; another is of a 'fly' (p. 75) which is in reality the white Plume Moth, *Pterophorus pentadactylus*, the only one of its kind that Penny had seen in England; another is of an Oil Beetle (p. 162). Sir Thomas supplied one observation on the parasites of caterpillars (p. 57).

These two, members of a famous Norfolk family, belonged to the branch of it that had settled at Ashwellthorpe when Edmund Knyvet sergeant-porter to Henry VIII married Joan heiress of John Bourchier, Lord Berners. Their eldest son John died before his mother but left an heir, the famous Sir Thomas Knyvet, immortalised in the ballad of the magic oak.[3] Their youngest son is the Edmund whom Penny praises, and who was evidently a man of rare distinction. They were in fact remarkable people, and made a great collection of books and objects of interest in their museum at Ashwellthorpe. Of the library a double catalogue in manuscript is preserved in the University Library at Cambridge. This consists first of a list of books arranged in subjects—theology, medicine, mathematics, history and philosophy—with sub-headings according to their language, Latin, English, French, Italian and Spanish;

1 'That famous learned phisition nowe living, master doctor Bright.'

2 Actually uncle: see below.

3 For which see Blomefield, *Norfolk*, III, pp. 102–3. Sir Thomas's eldest son, also Sir Thomas (born 1569, knighted 1603, died 1605), married Elizabeth, daughter of Sir Nathaniel Bacon of Stiffkey: he and his father are frequently mentioned in *Stiffkey Papers* (Camden Soc. 3rd Series, XXVI).

and then of a catalogue arranged in classes according to their position on the shelves. The second list was certainly written in or about the year 1638–9 when Thomas Knyvet, grandson and heir of the Knight and author of the Knyvet letters,[1] was head of the family; for at the end of the book there are dated notes of volumes borrowed which for this year are written in the same hand as the catalogue. The first list is certainly older, possibly of Sir Thomas's time. Hardly any of the volumes are later than the sixteenth century.

It is a very interesting collection and astonishingly complete. The medical books in the first list occupy ten folio pages, fifteen to twenty titles to a page. They include Gesner's *Historia Animalium*, the four volumes, and his *De Fossilium Genere*, 1561; Georgius Agricola (Georg Bauer), *De Subterraneis*, 1538; *Ortus Sanitatis*, 1527; Brunfels, *Simplicium Liber* (undated); Fuchs, *De Historia Stirpium*, 1542; Ruel, *De Natura Stirpium*, Basel, 1543; Bock, *Imagines Herbarum*; Mattioli, *Commentarii*, Venice, 1554, another edition Venice, 1565, and apparently a third dated 1560; Lonitzer, *Plantarum Historia*, Frankfort, 1565; Dodoens, *Frumentorum leguminum, etc. Historia*, Antwerp, 1565 and *Florum Historia*, 1569; Pena and De l'Obel, *Stirpium Adversaria*; Carew, *Cornwall*, 1602; and Garcia de Orta, *Simplicium apud Indos*, Antwerp, 1574. In addition there are William Gilbert, *De Magnete*, and two volumes by John Caius; Camerarius, *Symbolorum et emblematum ex animalibus*; several books by Paracelsus; and, almost the latest in the whole library, Sir Walter Raleigh's *History of the World*, 1614. It is noticeable that there are none of Turner's books, and that the later botanists De l'Ecluse, Lyte and Gerard are absent. It looks as if the library had been collected mainly during the middle of the century and that after 1580 little effort had been made to keep it up to date. This at least suggests that it was the work of Edmund, Sir Thomas's youngest uncle, and Penny's chief correspondent.

Many of the actual books belonging to this library are now in the Cambridge University Library, and among them is a volume which testifies vividly to the family's interest in natural history. This has Thomas Knyvet's name on the fly-leaf and is listed in the second MS. catalogue, p. 104*b*. It is entitled *Portraits d'Oyseaux et Animaux etc. par P. Belon de Mans*—a book of pictures illustrating Belon's works,[2] published in 1557 by Guillaume Cavellat of Paris. The names are printed in Greek, Latin, Italian and French above the pictures and a quatrain generally emblematic in character is added in French below. To the pictures of

1 Addressed by him to his wife Katherine and now preserved in the British Museum.

2 At least one of the birds, the Eagle with wings displayed, is from Gesner. The last bird, a Four-legged Duck, has a reference to Gesner's book and may be a redrawn version of his picture—his is apparently a Puffin; this looks like a Coot. The pictures of animals and fishes are almost all from Gesner.

birds—which form the largest part of the book—have been added in ink the appropriate English names: this in a very beautiful script and in the spelling and style of the mid-sixteenth century.

Some of these are full of interest etymologically. Thus he names 'Collurio', which seems to be *Lanius excubitor*, 'The Warriangle'[1] (p. 20*b*), 'Caprimulgus', 'The Night Raune, a night Clapper, a Shrike owle' (p. 28), 'Boscas', 'The Widgenne' (p. 37*b*), 'Vanellus', 'The Lapwinge, a Bastard Plover, a horne Pie' (p. 47), 'Himantopus', 'the Oxe eye' (p. 53), 'Oedicnemus', 'the Stone Curlewe or Field Courlewe' (p. 57), 'Cornix varia' (Hooded Crow), 'the Devonshier Crowe' (p. 69), 'Monedula', 'the Chofe, a Caddow, a Dawe' (p. 69*b*), 'Upupa', 'the Houpper or houping birde' (p. 72), 'Sitta', 'the Nutthacker' (p. 75*b*), 'Ficedula' (probably *Emberiza hortulana*), 'the Bullfinche' (p. 84*b*), 'Ruticilla', 'the Redstert a Prest's Crowne' (p. 87*b*), *Pyrrhula pyrrhula*, 'the Aupe'[2] (p. 92*b*), 'Parus maior, Fringillago', 'the Collmouse, Tydie'[3] (p. 95), 'Parus sylvaticus' (? *Aegithalos caudatus*), 'the Capon Wagtayle' (p. 96*b*), 'Fringilla', 'the Chafefinch, a Jack Baker, a finch, a Sheld Appell, a Spink'[4] (p. 96*b*), 'Apus, Cypselus maior', 'the Martelett' (p. 99).[5] On pp. 23–4 are two pictures of the Phoenix, the first named in ink 'the True Phoenix' being obviously drawn from a dried skin of the Bird of Paradise (*Paradisea minor*).

From these examples and from his identifications in general it is clear that he is an independent student, not copying Turner or Cooper and probably not acquainted with them; that his knowledge is often more exact than theirs; and that his range is limited to England and is fullest for the small birds of the countryside.

Finally, there was of course Thomas Mouffet, Penny's younger colleague and literary heir under whose name the book eventually appeared. Mouffet was the son of a Scottish haberdasher living in the parish of St Mary Colegate in London, and his parents may possibly have been known to Penny before his visit to Zürich in 1565. Thomas, their second son, was born in 1553, educated under Richard Mulcaster at Merchant Taylors' School, and sent to Trinity College, Cambridge, in 1569. Thence he migrated to Caius College where, as we have seen, he gained some acquaintance with the Master, and graduated in 1572. He returned to Trinity, possibly, as Munk, *Roll of the R.C. Physicians*, I, p. 91, states, owing to his anti-Catholic views, and took his M.A. in 1576. Two years

1 Ray, *Ornithology*, p. 88, says that it is called Wierangel in the north of England.

2 Usually 'the Nope' as Drayton, *Poly-olbion*, XIII, l. 74.

3 Cf. Drayton, l.c. l. 79.

4 Cf. Turner, *Avium*, p. 59, 'Fringilla a chaffinche, a sheld appel, a spink'.

5 This book and its English inscriptions are mentioned by Newton, *Dictionary of Birds*, pp. 680, 962.

later he went to Basel and studied medicine under Felix Platter, himself a pupil of Rondelet and maker of an early herbarium,[1] and under Theodor Zwinger, both former correspondents of Gesner. His first book, *De Venis Mesaraicis obstructis*, a thesis submitted to the University and published there in 1578, was dedicated to Penny.[2] He seems to have travelled into Italy and Spain before returning to England. But in June 1580 he was certainly at Nuremberg; for he addressed from there the first of his four letters to Petrus Monavius of Wratislaw printed by his friend Laurenz Scholz in his *Epistolae Medicinales* (Hanover, 1610), cc. 529–33. In September he wrote again, from Frankfort, and said that he was just starting for England. In December of the same year he married Jane Wheeler in London. She seems to have died shortly afterwards; for he married again a year or two later, his second wife being Katherine Brown.[3] In 1582 he was incorporated M.D. at Cambridge and accompanied Peregrine Bertie, Lord Willoughby d'Eresby, on his mission to Denmark.[4]

On his return he settled in London and in 1584 published at Frankfort his dialogue *De Jure et Praestantia Chymicorum Medicamentorum*. In this he defended the doctrines of Paracelsus as against the tradition of Galen. The book may have been responsible for bringing him into touch with the Lady Mary Herbert, wife of Henry Earl of Pembroke, and with her brother Sir Philip Sidney who seems to have been interested in chemistry and in touch with the famous Dr John Dee.[5] Lady Mary was a notable patroness of science and literature, and had quite possibly been acquainted with Henry Lyte and his *Niewe Herball*.[6] In any case the book is on a theme of some importance; for the application of alchemy to the service of medicine, and the consequent development of iatrochemistry, was a significant stage in the growth of science.[7]

Certainly at this period Mouffet was working hard along with Penny both in natural history and in medicine. There is a story in the *Theatrum* of a simpling expedition in an Essex wood[8] by him and Penny in which he rashly inspected a wasps' nest; the whole swarm flew out and pursued

1 Cf. Arber, *Herbals*, p. 141; and Gesner, *Ep. Medicin.* p. 101*b*.

2 A copy in the University Library, Cambridge, contains an inscription by Mouffet to Thomas Larkin or Lorkin, Regius Professor of Physic. This is written in a beautiful print-like script.

3 These facts are given in the brief biography prefixed to the edition of his *Nobilis and Lessus Lugubris* by V. B. Heltzel and H. H. Hudson (San Marino, 1940).

4 So Cooper, *Athenae Cantab.* II, p. 400.

5 Cf. Heltzel and Hudson, *Nobilis*, pp. 119–20.

6 As suggested by Arber, *Herbals*, p. 127.

7 Cf. J. Read, *Prelude to Chemistry*, pp. 29–30; J. R. Partington, *Short History of Chemistry*, pp. 41–64.

8 Mouffet's elder brother John was married at Aldham in Essex—there were oysters at the wedding breakfast! (*Health's Improvement*, p. 161, and Gunther, *E.B.B.* p. 219).

him and his servant, and if they had not been carrying branches of broom which they used for catching insects and so been able to defend themselves, they would have been seriously stung (p. 45). There is also a pleasant reference to bees occupying the space between the beams and sending out two or three swarms annually for thirty years at the house of Anne Seymour, Duchess of Somerset and widow of Turner's patron the Protector, at Hanworth where he and Penny attended her in 1586 and attested her will (pp. 14, 21 and cf. p. 61). Obviously he only came into close touch with Penny in the last phase of the older naturalist's life: but he had certainly known him and perhaps been interested in insects before his sojourn on the Continent, and had studied the silk-worm during it. In this connection he was responsible for the statement that the head of the larva makes the tail of the moth—a statement doubtless due to observing the change to the pupa at the moment when the larval head is being pushed down to the anal segment[1] (p. 182). During this period he worked on a book on diagnosis, the *Nosomantica* published at Frankfort in 1588, and dedicated to Lord Willoughby d'Eresby. Peter Turner contributed Latin verses to its preface.

Of his general ability and equipment as a naturalist there is not much evidence: but what there is seems plainly decisive. We have already referred to his *Health's Improvement*,[2] and the reminiscences of John Caius contained therein. The book consists of a very detailed list of articles of diet, of beasts, of tame birds, of wild fowl both of the land and of the water, and of fish both marine and fresh-water. This is in fact an exhaustive catalogue with more than a hundred birds and very many fishes; and his notes on them, though usually culinary rather than scientific, are often of interest. Dr Gardner says that they prove him to have been a good ornithologist.[3]

A study of the book does not bear out this conclusion. His allusions to Turner's letters to Gesner on the nesting of Cranes in our English fens (p. 91) and on the absence of the Hoopoe (p. 100) and to Caius's knowledge of the Sea-pie (p. 108) at first sight suggest competence, though of Cranes he has nothing to add except a few remarks as to their edibility; of the Hoopoe he says, 'yet I saw Mr Serjeant Goodrous [cf. p. 160; William Gooderus, Master of the Barber Surgeons' Company 1594, was

1 This observation John Ray explicitly contradicted in his first publication (*Cat. Cantab.* p. 87).

2 That it is 'generally ascribed to 1595' and that the part dealing with wild birds is 'Muffett's work entirely' is stated by W. H. Mullens and H. K. Swann, *Bibliography of British Ornithology*, p. 425. The work was not published till 1655 and then was edited by Christopher Bennet. It is largely reproduced by Robert Lovell in the sections of his *Panzoologia* (Oxford, 1661) dealing with birds and fishes.

3 This is also the belief of W. H. Mullens; cf. his article in *British Birds*, vol. v, pp. 262–78.

"chirurgion" to Queen Elizabeth] kill of them at Charingdon Park [probably Clarendon Park is meant], when he did very skilfully and happily cure my Lord of Pembroke at Ivychurch [near Salisbury]';[1] and of the Sea-pie he contradicts Caius by stating, quite mistakenly, that 'they have whole [i.e. full-webbed] feet like water-fowl'. In other passages his dependence upon and mishandling of Turner become obvious. Thus he includes the name 'mire-dromble' (one of Turner's names for the Bittern) with the 'heronshaw' and contrasts it with the 'Byttor or Stork'; calls the Godwit 'Fedoa', Turner's inexplicable name which Gesner also copied; says with Turner that 'the Cuckoe ever lays his egg in the Titling's nest' (p. 105); divides, as Turner does, the Larks into three sorts; recommends Fieldfares when feeding on Juniper in a way that recalls Turner's *Herbal*, but is also found in Gesner; and identifies the Clotbird (*Saxicola oenanthe*) with the Smatch or Arling and both with the Coccothraustes, whereas Turner identifies Clotbird, Smatche and Arlyng with the Coeruleo while Gesner prints Coccothraustes on *De Avibus*, p. 242, and Coeruleo with these synonyms on p. 243. With such evidence the question of his originality becomes soluble.

Add to it his remarks on the 'Branta': 'Barnicles both breed unnaturally by corruption and taste very unsavoury. Poor men eat them, rich men hate them, and wise men reject them when they have other meat' (p. 107); so he evidently accepted the Turner-Gesner story of their origin; and on the Puffin, two paragraphs, pp. 108, 166, as 'birds and no birds, that is to say birds in shew and fish in substance...permitted by Popes to be eaten in Lent' which is not only a reminiscence of Caius but an almost verbal reproduction of Gesner, *De Avibus*, p. 99: 'Angli puffinum avem non avem vel avem piscem faciunt', 'The English make the puffin a bird and no bird, or a bird-fish.'[2]

Similarly every one of Mouffet's birds not found in Turner (and these are very few) is found explicitly in Gesner, for example 'Anas muscaria' in *De Avibus*, p. 103;[3] Merganser as Shell-drake, *Health Impr.* p. 107, *De Av.* p. 119; Fulica as Coot (which is not the identity in Turner), *H.I.* p. 108, *De Av.* p. 345. The plain fact is that the whole of Mouffet's imposing list is a mere abstract from Gesner, the references to Turner or to English birds being lifted almost always without acknowledgment and sometimes without intelligence.

1 Cf. the poems by Abraham Fraunce (London, 1591), *The Countess of Pembroke's Ivychurch*: Henry Earl of Pembroke, Mouffet's patron, leased Ivychurch; and Sir Philip Sidney wrote much of his *Arcadia* there: it adjoined Clarendon Park: cf. R. C. Hoare, *Modern Wilts.* v, Alderbury, p. 187.

2 Cf. Drayton, *Poly-olbion*, xxv, ll. 81–2:

> 'The Puffin we compare, which coming to the dish,
> Nice palates hardly judge, if it be flesh or fish.'

3 This bird is also mentioned in *Theatrum Insect.*, p. 72, as called Muggent at Zürich.

There are, so far as can be judged, only these remarks in all which may perhaps be original—that Bustards in the summer 'lie in a wheatfield fatting themselves as a Deer will do' (p. 91),[1] that 'Snites' (apparently *Capella gallinago*) probe for worms and blow them out of the ground (p. 95), that 'Rock-doves breed upon rocks by the sea-side' (p. 96)—and this may be an inference from a quotation in Gesner; that Godwits are sold at four nobles the dozen and that 'Lincolnshire affordeth great plenty of them' (p. 99); and that 'Curlues feed wholesomly upon cockles, crevisses, muscles and periwinkles' (p. 108). These are hardly enough to commend their author as an ornithologist.

So, too, in his chapters on fishes. Detailed study makes it clear that he had got Gesner's *De Aquatilibus* and taking Turner's letter as his starting-point had then added a few notes—as for example that 'Luces are so rare in Spain', *H.I.* p. 156, *De Aquat.* (quoting from Amatus), p. 502; or that 'Orbes, Lumps, are of two sorts', *H.I.* p. 156, *De Aquat.* p. 631. There are here also a few original records: the two allusions to Sir Francis Drake, pp. 154, 160; the story of the 'Luce' obtained out of Sussex from Mr Huzzy of Cookfield[2] and sent by Mouffet 'to the Mirror of Chivalry (the Lord Willoughby of Eresby)', p. 156; of the 'twenty mussels that almost poisoned him at Cambridg', p. 159; of the Oysters at his brother's marriage, p. 161; of the gentlewoman in Warwick Lane who sent for a pasty of Porpoise 'when the prisoners in Newgate had refused the fellow of it', p. 165; of the Scallops and Cockles at Selsey and Purbeck, p. 167; of the Bleaks which 'troubled with a worm in their stomach' leap into your boat, of an evening, as you row on the Thames (p. 176); and of the 'Old Wife' bought at Putney 'as I came from Mr Secretary Walsingham his house about ten years since' (p. 184). There are several direct but unacknowledged quotations from Turner, for example 'Codlings are taken in great plenty neer to Bedwell [which should be Bednel] in Northumberlandshire' (p. 155), 'Alderling betwixt a Trout and a Grayling lyeth ever in a deep water under some old and great alder' (p. 175): but here as before the vast majority of the material is borrowed from Gesner, and in such a way as to make it evident that Mouffet's interest was almost purely dietetic or literary. He had pleasure in eating and some knowledge of cookery; he wrote easily and with a certain distinction; he was emphatically not a great naturalist or a first-hand student.

Nevertheless, he seems to have thrown himself with eagerness into the

1 He has already stated that 'out of sowing-time' they 'feed upon flesh, livers and young lambs'.

2 Undoubtedly John Hussey Esquire of Cuckfield, son of John Hussey, M.P. for Horsham in 1570. J.H. junior died in 1600, cf. J. Comber, *Sussex Genealogies*, Horsham, p. 187.

business of clearing up Penny's papers and reducing his notes and memoranda to order. No doubt the last months of his friend's life, his wife's death and his own illness, had given him no chance of finishing or even arranging the material: it may have suffered from casual servants and the upheavals of the household. But Mouffet worked hard, and by 3 March 1589, the date attached to the manuscript as prepared for publication, the book was finished. He had according to his own account in the Introduction put the history into order, added to it the literary style which Penny lacked, altered both its 'method' (classification) and its language, had cut out a mass of repetitions and trivialities, and had added many histories and above a hundred and fifty pictures. How far these changes improved the work is, as we shall see, open to question. Penny's style may not be literary; it is admirably scientific: his arrangement may have been defective; it could hardly be worse than Mouffet's.

In any case the facts are that he took out a licence in 1590 to print the script at the Hague; that he prepared an elegant title-page, with portraits of the four authors in medallions, engraved by William Rogers the first Englishman to practise the art of copperplate engraving, and wrote an elaborate dedication to Queen Elizabeth; and that this manuscript is still preserved in the British Museum (Sloane, 4014). A reproduction of the title-page and a brief account of the script by Dr Malcolm Burr were printed in the *Field*, 27 August 1938. We have already mentioned Penny's portrait: that of Mouffet depicts in Dr Burr's words 'a well groomed fashionably dressed, pleasant looking man', one with something of a complacent smirk and foolish expression. His picture, perhaps the only one drawn from life, is the most vital of the four. It seems certain that no Dutch publication ever took place. But Cooper who gives in his *Athenae Cantabrigienses*, II, pp. 400–2, a long list of Mouffet's works states that the *Thĕatrum* had been imperfectly edited by Laurenz Scholz in 1598; and the *Dictionary of National Biography*, XXXVIII, p. 103, adds Frankfort as the place of editing. There does not appear to be any copy of such an edition in existence: but if the book were printed there, this fact might explain how Topsell, who, in his *Historie of Serpents* in 1608, got access to it, certainly used some of its contents (though not precisely in the wording of our *Theatrum*).[1] A critical comparison of Topsell's chapters on Insects and Spiders makes direct borrowing certain, but strongly suggests that he drew from a version different from ours. Thus in the passage quoted from Columella (actually from *De Cultu Hortorum*, the poetical Book X, p. 354, of the Lyons edition of 1541) Topsell, *Serpents*, p. 106, cites it accurately and refers also to Palladius, I, p. 35, a chapter dealing with a similar subject, whereas our *Theatrum* transposes two lines, alters a verb and the sense, and omits all reference to Palladius.

1 Cf. below, pp. 223–4.

So, too, in the quotation from Homer, *Iliad*, XII, 167 (*Serp.* p. 85, *Theat.* p. 42), Topsell is fuller and gives a further quotation. This and a mass of other details[1] support the view that Mouffet's book before its final revision was known to Topsell.

Mouffet was himself too busy to take much further interest in its fate. As we have seen, he had become a fashionable doctor with a large practice; he was well known to Walsingham, Lord Willoughby and the Earl of Essex, and in 1591 went to Normandy as personal physician to the Earl during his campaign. Returning he seems to have left London and joined the Earl of Pembroke's establishment. He was given a home at Bulbridge[2] on the Pembrokes' estate at Wilton near Salisbury and apparently a seat in parliament. In 1593 he produced his *Nobilis: a View of the Life and Death of a Sidney*, printed from the script in the Huntington Library in 1940, a panegyric in a style of affected but not uninteresting euphuism, which discloses evidence of his own contact with Sir Philip Sidney as with Willoughby, Essex and Walsingham. This is dedicated to William Herbert, the son of the Countess of Pembroke. He followed it in 1599 with his poem on *Silkwormes and their Flies*;[3] and wrote largely on medical subjects. The *Theatrum Insectorum* was forgotten.

Willoughby Gardner, in the very valuable paper on Thomas Penny already quoted, suggests that the publication of Aldrovandi's volume, *De Animalibus Insectis* in 1602, was a blow to Mouffet and diminished his chances of securing a publisher for his work. This may be the case. But from the facts it looks as if Aldrovandi's forestalling of him had stimulated Mouffet to a last belated effort. Certainly at this time he did some slight revision of the book,[4] wrote a new dedication to King James who placed his Court at Wilton for a time in 1603, and made some attempt to secure publication. But his death in 1604 put an end to the effort. Eventually the script was sold for his widow to Sir Theodore de Mayerne, doctor to the royal family, a man of letters and a friend of Thomas Johnson the botanist. He succeeded at last, in 1634, in getting it printed, not very satisfactorily, in London, dedicating it to Sir William Paddy, physician to the King, who had befriended him on his first coming to England from France. An English translation by John Rowland, bound up with revised editions of Edward Topsell's two books,[5] was published in 1658. For this the pictures were redrawn, losing somewhat in the process. The translation is tolerably accurate but shows no sign of expert knowledge.

1 E.g. *Serp.* p. 103, *Theat.* p. 191, quoting Monins; *Serp.* p. 101, *Theat.* p. 179, quoting Ovid etc.

2 So J. Aubrey, *Brief Lives* (ed. Clark), II, 90.

3 Cf. Chamberlain's Letters (Camden Society, LXXIX), p. 47: 'The Silkworme is thought to be Dr Muffet's and in mine opinion is no bad piece of poetrie.'

4 Including e.g. a note about fireflies seen in Trinidad by Robert Dudley in 1595: cf. below, p. 185.

5 For these see below, ch. XIII.

It begins with a series of chapters on bees, the longest on the varieties and uses of honey and chiefly remarkable for a recipe for mead and a statement that 'the Irish prepare a distilled drink from honey, wine and herbs which they call uskebache [whisky], suitable for a people that lives on raw or half-cooked meats' (p. 31). In the chapter on wax there is a note that Penny when in Paris was shown how to break open red sealing-wax and refix it without leaving traces, but refuses to describe the method in the interests of public morals (p. 36). In the chapter on drones it is decided that they are not female bees, but their status and use are left uncertain. So, too, with wasps though the book calls the large ones which live through the winter 'matrices' it also calls them 'reges': the smaller ones that die in the autumn are explicitly said to be workers and males. These chapters are mainly a patchwork of quotations, almost certainly part of Gesner's papers; and the true character of the life of the nests and the sex of their inmates are obscured because tradition both classical and scriptural insisted that both bees and wasps were generated spontaneously from putrefying animals. A few particulars of other species of wasps may be from Penny. So probably is much of the chapter on hornets, and the classification (which Mouffet has ignored) of the other winged insects as solitary and either honey-making like the humble bees, or not as flies, butterflies and moths, and the winged glow-worm. To Penny is definitely assigned the description of a sort of wasp 'whose body is all black excepting the back which is red from waist to tail: the end of the tail is black: it has silvery wings of which the front pair is twice as large as the hind: it nests in walls and the edges of ditches and enclosures. I do not know whether they have a sting' (p. 53).

The first chapter on Flies contains several notes definitely ascribed to Penny, and an interesting and accurate observation from Sir Thomas Knyvet which deals with the hatching of flies or of white grubs from caterpillars or half-perfect chrysalids (p. 57). There is also a note of a shower of blood and afterwards a swarm of flies in England 'in 766 B.C. when Rivallus was prefect of Britain'[1] (cf. also p. 78). The second describes many species including *Asilus crabroniformis*, *Hippobosca equina*, and two Gadflies, one sent from Virginia by 'Candidus', John White, artist to the colonists of 1585,[2] and the other from Russia by Edward Elmer 'a careful surgeon' to Penny; another, a green Tabanid, was seen by him at Han-

1 The record is in T. Cooper, *Chronicle*, p. 35 (London, 1565), the second part of Thomas Languet's chronicle, and in Holinshed, *Hist. of England*, I, p. 21.

2 Sailed with Sir Richard Grenville in 1585. Twenty-three of his paintings were engraved by Theodor de Bry for his *Narratio* (Frankfort, 1590). Plate XIII (of natives fishing) contains pictures of a Turtle and of the King-crab (*Limulus polyphemus*) which his friend Thomas Harriott describes under the name of Seekanauk (*Briefe Report*, ed. H. Stevens, p. 44). Some of his water-colour drawings of Virginian subjects are in the British Museum.

worth in August 1586, no doubt on one of his visits to the Duchess of Somerset; two other species in 1582. Penny is also credited with having separated 'Asili' from 'Tabani', with having described the 'Whame and Burrell Fly' (p. 62), various ichneumons including a *Mesostenus* in 1573, and a *Gasteruption* at 'Hinningham' (Heveningham or Hedingham in Essex), once the seat of the Earl of Oxford (Edward de Vere), and another at Greenhithe in Kent (p. 64), various dragonflies including a large one hatched from a nymph 'wholly unfamiliar to him and to me' in Penny's museum (p. 66), and a number of 'Cados Wormes' (p. 69), 'Crane Flies' including *Tipula paludosa* (p. 70), as well as species sent to him by Camerarius, De l'Ecluse and Brewer or seen at Zürich. Among these on p. 63 are a picture and description of the Hymenopteron *Ammophila sabulosa* with a note that 'it runs about quickly as if by jumps, makes its nest in the earth, and satisfies its hunger on flies and small caterpillars' (pp. 63–4). In the next chapter the 'uses' of flies lead to a discussion of the creatures that feed on them, including the following: 'The Chamaeleon of which many have written that it lives only on air, feeds on flies which it suddenly strikes with a tongue six inches long and shot out very swiftly—strikes, draws in, and consumes—as I saw with my own eyes in 1571.' 'The worms of flesh-flies called in English "maggots" and "gentles" are fixed to hooks by fishermen and used for catching roach, perch, carp and other fishes. Trout are allured by the gay path fly[1] and the dung-fly...as we hear from William Brewer' (p. 73).

Of gnats Mouffet ridicules Penny's claim that 'their proboscis is given them by nature not for singing but for sucking up blood through the skin' (p. 81). There is also a good account of English midges, presumably by Penny; of gall-flies from Valerand Dourez the druggist of Lyons, the kinsman of Jean Bauhin and well known to De l'Obel, whom Penny must have known through them in his time at Montpellier[2] (p. 84); and of northern gnats noted by Martin Frobisher at 'portus Nicolai', St Nicholas's Bay, on the opposite side of the Dwina to Archangel.[3]

Chapter 14, 'On Butterflies', contains a larger number of pictures and descriptions than any other, and many of its species can be identified. Unfortunately, nothing is said about the localities or captors of the specimens; they are listed and briefly described; but that is all. They are divided into the night-flying moths which emerge from the chambers of their caterpillars in the earth, coming to light and with the wings closed on the

1 This is called the 'gray path fly' on p. 75.

2 A detailed account of Dourez is given by L. Legré, *La Botanique en Provence au xvi siècle* (Marseilles, 1904), pp. 88–105. Dourez had travelled to Greece in 1564 but had returned and married in Lyons in the following July.

3 For a similar account of the place and its 'mosqueetos', cf. Peter Mundy, *Travels*, IV (Hakluyt Soc. II, vol. LV), p. 134.

back; and the day-flies emerging from suspended chrysalids, with longer and clubbed antennae, and with wings spread out along their sides (p. 88). Of the moths the first is *Acherontia atropos*, and the second *Saturnia pyri*; there follow *Sphinx ligustri*, *Smerinthus ocellatus*, *Arctia caia*, *Lasiocampa quercus* (2), *Cosmotriche potatoria*, *Phalera bucephala*, (one unidentifiable), *Hepialus humuli* ♀, *Zeuzera pyrina*, *Arctia caia*, *Choerocampa elpenor*, ? *Catocala nupta* (this seems to be the species described, but the picture does not fit the description), *Callimorpha hera* (from Clusius) (2). Then he places the 'middle-sized moths', the first of which is a rather poor picture of upper and under wings of *Parnassius Apollo*: there follow it *Saturnia pavonia* (2), *Pieris rapi*, ? *Melanargia galatea*, *Pararge megaera* (2), (two unidentifiable), ? *Triphaena pronuba*, *Abraxas grossulariata* (2), *Spilosoma menthastri*, *Hepialus humuli* ♂, (one unidentifiable), ? *Spilosoma lubricipeda*, *Malacosoma neustria* and an evident Caddis-fly, and finally an unillustrated Silk-worm (*Bombyx mori*). Finally, the small; and first, if lepidopterous at all, *Lithosia lurideola*, *Ino statices*, then a chrysalis of *Pieris brassicae* and an indeterminate geometer, and finally *Hipocrita jacobeae* (2) and *Zygaena filipendulae*.

So begins the second main division, the day-flying butterflies. First, a huge exotic Papilio, probably the American *P. turneri*; then two *Papilio machaon* and a caterpillar, *P. podalirius*, *Vanessa io* which he calls the Queen, *Colias edusa* (2), *Pyrameis atalanta* (2), *Vanessa polychlorus*, *Pyrameis cardui* (2), *Argynnis aglaia* (2), *V. urticae* and chrysalis, *Pararge egeria*, *P. megaera* (underside), *P. egeria* (underside), *Callimorpha dominula* (2). *Gonopteryx rhamni* (2) is the first of the middle-sized: it is well described but said to emerge from a gold-smeared chrysalis, and this as depicted is apparently that of *V. urticae*. The next is not illustrated, and is very obscurely described: there follow two undersides, one possibly *Argynnis paphia*, the second perhaps *V. antiopa*; *Epinephele iurtina* (2), the pictures obvious, the description uncertain; *Pieris rapae* (2); *Grapta c-album*; three, of which neither pictures nor descriptions are recognisable, though No. 10 may be *Melanargia galatea*; then three hawk-moths which by their descriptions seem to be *Metopsilus porcellus*, *Macroglossa stellatarum* and *Hemaris tityus*, but the pictures are not helpful. The last section, small butterflies, is extremely difficult: no doubt because specimens were easily rubbed and hard to preserve. No. 1 reads like one of the Coppers, probably *Chrysophanus phloeas*, no. 2 may be *Celastrina argiolus*, nos. 3 and 4 are probably *Lycaena icarus* ♀ and ♂, no. 5 may be *Lampides boeticus*, and the three pictures below it are surely *Euchloe cardamines*: but none of the other descriptions are identifiable, though no. 9 is possibly *Xanthorhoe sociata* and no. 10 *Coremia ferrugata*. It is delightful to find that the account of this ends with the words 'it displays to us rather the indescribable power of God than any colours that can be described by their proper names' (p. 106). The final section of the chapter records certain migratory swarms of butterflies in the year 1104, on 3 August 1543, and in 1553 just before the death of Maurice of Saxony.

The book next deals with the Glow-worm (*Lampyris noctiluca*), recognising that the male of the European species is winged and if it occurs in England never shines there, whereas the female is sluggish, wingless, and brilliantly luminous. It records the experience of Julius Scaliger, confirmed by Penny's friend William Brewer, of a pair kept in a pillbox until the female laid her eggs and these in twenty hours had hatched (p. 110). It quotes almost the whole of the poem by Antonio Telesio (Thylesius) of Cosenza[1] published along with his treatise *On Colours* in Paris in 1529 (p. 111). There are also a picture and record by John White, the artist and governor of Virginia, of a species of *Pyrophorus* observed by him there and in Hispaniola (p. 112); reminiscences of Thomas Candisius (Cavendish) 'Orbis totius mensor' (he had sailed round the world in 1586–8) and Sir Robert Dudley, son of the Earl of Leicester, who had seen Fireflies in Trinidad and thought that they were natives with torches;[2] and a record from Brewer of how on a hot night he was wiping his face with a towel when suddenly it flamed into a phosphorescent glow, and how in it he discovered a Centipede (probably *Geophilus electricus*), the source of the 'shining vapour' (p. 113). Chapter 16 has a number of pictures of locusts and grasshoppers of no special interest; and then several of the Mantis (*M. religiosa*) 'so called because they hold up their front feet like hands in prayer as if they were prophets who with this gesture pour out their supplications to God' (p. 118). Penny states that he very often saw this creature at Montpellier, but that its picture was sent to him by Antonius Saracenus, a distinguished doctor at Geneva. Of the Cicada we are told that Penny had one brought from Guinea by Ludovic Atmar the surgeon, and one from Virginia by John White the painter (p. 128); and of the Cricket that James Garret had produced their chirruping sound by rubbing together their torn-off wings (p. 134). Of the Cockroach (a name apparently unknown: it is here called 'Blatta') we are told the story of a huge specimen caught on the top of Peterborough Cathedral, and that the species is found in London in wine-cellars and more freely in bakeries (p. 139); it is presumably *Stilopyga orientalis*.

The chapter on Cantharides Mouffet declares that Penny and Gesner entirely omitted and he has therefore supplied. He states that they fall into two families; describes three species; but devotes almost all his space

1 By a misprint he is called Bonsentinus.

2 Cf. *Robert Dudley's Voyage to the West Indies narrated by Captain Wyatt* (Hakluyt Soc. II, vol. III), p. 25. This was in February 1595. It must therefore be one of the additions to the *Theatrum* made by Mouffet in 1603—though even so it is not clear how he got the story; for Wyatt's narrative was not published till 1899.

to the 'caustic power' of *Lytta vesicatoria* and to its uses for love-potions and poisons, citing one case which came to his notice when he was in Basel in 1579 (p. 146), and advertising to the nobility 'who seem more studious of sex than the commonalty' his own possession of a safe aphrodisiac.

Penny resumes again with the Scarabs, beginning with the Stag-beetle (*Lucanus cervus*) and discussing the size of the sexes; then come Longicorns 'nasicorus' including Camerarius's large Rhinoceros beetle (? *Golofa* sp.) with the verses which declare that like the Phœnix it dies yearly and is reborn out of its own putrefaction (p. 152); and then, after various other beetles briefly described, the 'Scarabaeus' (*S. sacer*). On this Mouffet produces nearly five pages of tradition and legend, prescription and fancy, concluding with the words, 'I seem to have made a giant out of the scarab...but I am surprised at the brevity and poverty of Penny at this point when so many memorable wonders about the ball-rolling scarab are to be found in Lucian, Pliny, Homer, Aristophanes, Theocritus, Alexandrinus, Erasmus and countless others' (p. 158)—which illustrates effectively the incompetence of Mouffet as Penny's executor. Then the record returns to descriptions of various other large beetles, the only note of interest being the story that on 24 February 1574 so many 'Dorrs' (*Melolontha vulgaris*) fell into the Severn that they blocked the wheels of the water-mills (p. 160). The small beetles have a chapter of two paragraphs and thirty-nine pictures (p. 161). Then follow Oil-beetles and Water-beetles, the former (apparently *Meloe proscarabaeus*) seen very commonly in May near Heidelberg.

A short but valuable chapter on the 'Fen cricket or Evechurre' (*Gryllotalpa vulgaris*), and a long and literary one on the Firefly; another, also mainly derivative, on the 'Tipula or Water-spider' (*Gerris* sp.); the next informative but brief on the Earwig (*Forficula auricularia*) in which Knyvet's proof to Penny that it carries wings under its 'elytra' is fully described (p. 172); the next on the Scorpion and the Ant, disappointing; and finally on winged Hemiptera—so the first book ends.

The second with a separate prefatory note and scheme of classification deals with the wingless insects, and first with caterpillars ('which, if hairy, Northcountrymen call Oubutts and Southerners Palmerworms', p. 179), the Silk-worm, 'viridium nobilissima' (*Sphinx ligustri*), 'vinula', 'porcellus' (*Choerocampa elpenor*), and other smooth ones; then the Pityocampes or Pine-caterpillar, a record derived from Gesner; and the 'ambulones', 'palmerworms' or hairy bombycids; then their origin, Penny rejecting Pliny's statement that they spring from dew and Mouffet blaming him for so doing, and their 'uses' in which is a note of how 'William Turner theologian and very learned doctor, the fortunate father of our friend Peter, administered an emetic to an English lady who had swallowed a hairy caterpillar' (p. 192). It is in this section of his work (p. 182) that he states that the head of the caterpillar becomes the tail of the butterfly.[1]

The 'Sphondyle' of Chapter VI includes all the large maggots that feed in rotten wood or the soil—the first described and pictured is that of the Goat-moth (*Cossus ligniperda*); the second is the Cockchafer (*Melolontha vulgaris*). Next come beetle-larvae that resemble imagines—Cockroaches, the Devil's Coach-horse (*Staphylinus olens*) and Knyvet's moth-larva (*Stauropus fagi*). The Centipedes and Woodlice have nothing of much interest; nor do the Scorpions, though there are a vast collection of lore and many pages of prescriptions against their bites. Brewer's note on the shining centipede; Penny's use of woodlice in wine as a cure for asthma; and a reference to Wolf of Zürich, Gesner's heir, and to John Arden (Ardenne) 'in his day [the fourteenth century] the most skilful of British surgeons' (p. 216) are alone worth mention. There follow fifteen pages about Spiders which show us the book as Mouffet would have written it. They discuss the 'phalangia' or poisonous and 'domestica' or harmless types, the former in extracts from traditional medicine, the latter in a fanciful eulogy which sees the household Spider as a type of physical and constructive perfection. Penny is only mentioned once, when his disgust at the eating of Spiders 'although he knew that Phaer[2] often did it without harm' is condemned (p. 227): Brewer and Peter Turner are mentioned in connection with threads and webs (p. 232): otherwise these chapters are purely pre-scientific. They are based upon the compilations of Wotton and Gesner, worked up into literary essays by Mouffet. A brief chapter on different species of Spiders with a reference to a discovery by Penny at Colchester follows. Then come twelve more pages of Mouffet, mainly a eulogy of the Ant; and another short chapter on the wingless female Glow-worm, and various beetle larvae, one of them found by Penny in a cornfield at Colchester. Then worms are discussed; and first Mouffet accepts Penny's statement that they are found in crumbling stone, although this contradicts ancient belief. Penny's brother is quoted for the discovery of worms in worn-out mill-stones; so is Felix Platter the President of the medical faculty at Basel for the existence of a Toad in a stone which he had cut open, and William Cave of Leicester[3] as confirming this (p. 248). 'Timber-worms', including the 'teredo' which attacked the hull of Francis Drake's *Pelican* (p. 250), but mostly beetle-larvae, and one a carnivorous carabid; 'Meale-wormes' (*Tenebrio molitor*) and the 'Bowde or Weevil' (*Calandra granaria*) and many others come next; and for getting rid of Weevils it is advised that an Ants' nest be put in a

1 Mr G. Taylor tells me that this error was repeated as late as March 1832 in Loudon's *Magazine of Natural History*.

2 Presumably Thomas Faire, author of *The Regiment of Life*.

3 Presumably of Pickwell, cf. *Visitation of Leicestershire*, 1619, p. 128.

bag and placed in the granary for ten days during which the Ants will kill the beetles, or that a brood of chickens be let loose in it (p. 259).

Chapter XXII dealing with human Lice, XXIII with the Lice on animals and plants, XXIV with 'Whealewormes' who were held responsible for ringworm and eczema and the 'dermestes' that live on furs (probably *D. lardarius*) and XXV with the Bed-bug contain little of interest except Penny's discovery that the Bug, which Aristotle had supposed to originate like the Louse from human sweat, bred its young from eggs: for when studying plant-life at Orléans with Noël Caperon he had to cut from its sheath a knife which had rusted, and found the sheath full of colonies of the insects at all stages of growth (p. 269). 'In 1583 while Penny was writing these records he was summoned to Mortlake, a village on the Thames, to two ladies terrified by traces of bugs and afraid of infection: he discovered the insects and cured the ladies by his laughter' (p. 270). Ticks, Moths (or rather the larvae of clothes-moths and other *Tineinae*) and Fleas occupy the next three chapters, the story of Mark, an Englishman who harnessed a Flea,[1] being the only notable record (p. 275). Earthworms ('Duggs beloved by fishermen' and 'yellow-tailes' being specially named); intestinal worms (Mouffet relates the story of an internal stone removed by Thomas Lorkin, the Regius Professor of Physic at Cambridge, p. 283; and adds various notes by Penny, by Randolph, a London doctor, and by Timothy Bright,[2] and in particular a story of a worm ten ells long ejected by one John a book-binder of Basel while he, Mouffet, was studying there under Theodor Zwinger and Felix Platter in 1579, p. 297); and long discussions of their uses and treatment fill pp. 278–316. Eulae or Gentles; Lendes or Nits; Aureliae or Chrysalids; Squillae or aquatic larvae, mostly of Dragonflies; various other water-insects; and finally Horseleeches and Water-worms—with a story of the cure of Richard Cavendish's[3] gout by blood-letting, p. 324—conclude the volume, save for four extra pages of woodcuts.

The *Theatrum Insectorum* deserves full treatment not only as the outstanding, indeed the only, important contribution of Englishmen to zoological studies until the latter half of the seventeenth century, but on account of the problem of its sources and authorship.

It is as we have seen plainly a composite work—in part a pandect collecting the references to the life of the 'smaller animals' in Classical authors in the fashion of Wotton and Gesner. It is also a medical treatise incorporating a number of cases like those noted in our survey or those of the impotent man at Basel in 1579 (p. 146) and of the elderly wife of

1 For 'Flea chains', see below, p. 306.

2 Of whom Mouffet says that he is famous for an epitome of Church History, presumably his *Abridgment* of Foxe, 1589.

3 Uncle of Thomas Cavendish the 'Cosmonauta'.

the knight Penruddock[1] (probably Anne, second wife of Sir George Penruddock of Ivychurch) who, having drunk too much goat's milk for fear of phthisis, bred 'wheal worms' and became incurable (pp. 266–7); and a mass of prescriptions against various bites and poisons. These elements are combined with much material that is purely descriptive—a great collection of pictures and short notes of shape, colour and sometimes habits, most of which enable the species to be identified, while several draw attention to points of real interest.

To disentangle the source of these different elements is a task that does not call for any high degree of skill in the higher criticism. We have Wotton's book, and can easily see that in certain details, e.g. in the statements that Wasps called 'ichneumons' kill Spiders and rear their young on them (Wotton, p. 184, *Theat.* p. 45) and as to the nests and food of Hornets (p. 184*b*, *Theat.* p. 51), or in the quotation of three lines of Latin verse (beginning 'Estque caput minimum') on the Spider (Wotton, p. 186, *Theat.* p. 217), Mouffet has copied it exactly. For dependence on Gesner proof is more difficult since with the exception of his work on Scorpions which was edited by Wolf and printed in 1587 all Gesner's original papers on the subject have perished. But it is clear that there are pages in the account of Bees (*Theat.* pp. 1–21)[2] and many other details both traditional and more recent such as the records of the Pityocampes (*Theat.* p. 185) and of Gesner's 'stinking' caterpillar, dated 1550 (*Theat.* p. 190), and the long lists of synonyms in many languages which are drawn from the scripts taken by Penny from Zürich.

Mouffet's own work is easily separated from that of his sources. He has a touch of fancy, a flair for a telling phrase, a literary gift which is wholly unlike the dry scientific precision of Penny's style. He loves a sentence like that with which he finishes the description of the Death's-head Hawk moth: 'As great tyrants prey upon and exhaust the nobles of lesser races, so these night-fliers beat with their wings and slay the day-flying butterflies sheltering under the foliage' (p. 89). Moreover he is incurably medical, and loves to accumulate masses of lore and learning scarcely relevant to his theme, and reminding us of his books on diagnosis and dietetics.

Penny alone has claim to merit as a scientist; and if we had got the whole body of his work—those note-books to which Mouffet so freely alludes, from which he claims to have cut out 'more than a thousand tautologies and trivialities', and whose contents he arranged on a system

1 Cf. R. C. Hoare, *Modern Wilts.* III, 'Dunworth', p. 81. Sir George was the father of Sir Edward, M.P. for Wilton (which Mouffet afterwards represented) in 1586, and builder of Compton Chamberlain.

2 Much of this is also in Topsell, *History of Serpents*, pp. 70–83, but this is almost certainly borrowed from the *Theatrum* rather than from Gesner's notes.

of his own—it seems clear that he would rank very high. Certainly he grasped the importance of beginning with identification, with the proper discrimination and description of the different species and types, and with observation of them rather than traditions about them. In this he was something of a pioneer—certainly the first Englishman to anticipate the Baconian insistence upon the collection of data as the essential preliminary to theorising. Evidently he delighted to give the dates and places of his captures—no doubt these are among the trivialities with which Mouffet was so ruthless: evidently he had searched diligently and been careful not only to get his specimens accurately drawn but to link up description and picture so that misplacement was on the whole avoided.

What his method of classification would have been if it had been left to us; how much of his descriptions were rejected as superfluous; to what extent his dry objective records have been embellished with his editor's flights of imagination, can only be conjectured. It is not likely that he avoided the two main difficulties which made a scientific entomology almost impossible—the persistent belief in spontaneous generation, that bees were bred by dead lions or putrefying calves, that lice were the product of sweat, that plants produced galls as they bore fruit; and the consequent failure to form any adequate concept of the life-cycle and its metamorphoses. But, so far as we can judge, if Mouffet had been content to edit instead of rewriting we should have had a very much fuller and more accurate catalogue; and subsequent students would have been able to build upon a foundation 'well and truly laid'.

As it is the book has still much value; and if it is scientifically disappointing, Mouffet's literary touch has charm and interest. His quotations from a lost writing by Franciscus Stancarus[1] which dealt with the seven kinds of Locusts (p. 122) or a lost poem by Edward Monins[2] (pp. 152, 180, 231)[3] and from recent or contemporary doctors, Thomas Linacre (p. 315), Johann Lange (pp. 133, 266), the veteran of Heidelberg, and Laurent Joubert of Montpellier (pp. 260, 267); his allusions to the famous surgeon Ambroise Paré of Paris (p. 204) and the pharmacist Valerand Dourez of Lyons (p. 271); his history of Grey, bishop of York, and his meanness to the poor and the weevils that destroyed his granaries—a variant on the story of Walter de Grey, Archbishop 1215–55, told by

1 Francesco Stancari of Mantua (A.D. 1501–74) whose 'heresies' fluttered the churches, especially in Poland.

2 Presumably Edward Monins of Waldershare in Kent whose eldest daughter Elizabeth married Sir Henry Crisp of Quex Park in Thanet, commemorated by Gerard and Johnson (cf. below, p. 209). He matriculated at Queens' College, Cambridge in 1565, was knighted in 1595, and died in 1602 (Venn, *Alumni Cantab.* III, p. 199 and Hasted, *History of Kent*, IV, pp. 188–9n.). The poem quoted was apparently called 'Berefish' (*Theat.* p. 180) and was in Latin hexameters.

3 And probably also p. 191 (from 'our poet').

Roger of Wendover and Matthew Paris (p. 257); and his *obiter dicta*—that ladies interested in gardening caught earwigs by putting pots filled with straw on the top of sticks (p. 172) and that others interested in clothes saved them from moths by wrapping up in them the skin of a kingfisher (p. 274) or that 'Ireland had a bad name for lice; the whole island is said to be crawling with them' (p. 262); it is in its wealth of such details that Mouffet's contribution justifies itself.

If we compare it with the only contemporary work on the same theme, Aldrovandi's seven books *De Insectis*, our admiration of the *Theatrum* is manifestly increased. The Italian naturalist writes eloquently and no doubt truthfully of the vast labours that he has undertaken, of his sojourn outside the city, of the draftsman and secretaries who accompanied him, and of the rewards offered to country folk for specimens and information. But when we look for the results apart from a number of not very recognisable pictures and brief descriptions,[1] there is hardly a single fact or observation which is of value;[2] and what there is of first-hand knowledge is swamped in vast accumulations of quotations, epigrams and proverbs, of moralisings, fables and ancient traditions. Of its author's erudition and industry, here as in the rest of his pandects, there can be no question: occasionally, as in his long list of locust-invasions in Europe during the fifteenth and sixteenth centuries, he is useful and informative: for those who wish to see how strange were the ancient beliefs and how fascinating the symbolisms the book is always of interest. But it is the work of a humanist of the Renaissance, not of a naturalist and observer, a monument to the past rather than a searchlight into the future. Penny's work, despite the sad accident of its delayed publication, is the true foundation of entomology.

1 The best are the three plates of Diptera (and others), several of which can be plausibly identified: but even here the information is very scanty (cf. pp. 137–8, ed. Frankfort, 1623).

2 An exception is the account of the Swallowtail (*Papilio machaon*) bred by him from a larva found in July 1592: but even here he says that he took it on Tamarisk or Myrtle!

C. THE POPULARISERS

CHAPTER XI. HARRISON, BATMAN AND LYTE

The vitality and many-sided activities of the Elizabethan age made it certain that the work of explorers in natural history as in other fields of effort would not go unrecognised. There were, as we have seen, among the druggists, perfumers and gardeners of the day a number of men of more than local influence. James Garret, the friend of De l'Ecluse, inheriting from his father James Garret the elder[1] his profession as an apothecary, developing an interest in the manufacture of scents, and acting as distributing agent for the curiosities and treasures which Drake and Raleigh were bringing back from their voyages, had in his brother Peter at Antwerp a link with the great Dutch savants of Leyden and the great Dutch bulb-growers of Haarlem. The Garrets lived in Lime Street, which runs from Fenchurch Street to Leadenhall Street, east of the markets, and had a garden near the Aldgate and beside London Wall. Jean-Henri Cherler, son-in-law of Jean Bauhin and joint author with him of the great *Historia Plantarum Universalis*, saw much of Garret on his visits to London in 1604, 1605 and 1609.[2] John Rich, Hugh Morgan and other professional druggists were extending the scope of their physic-gardens by exchange of seeds and plants with the continental experts in Paris, Louvain and Antwerp; and the world of fashion was beginning to encourage the raising of flowers for their borders and pleasances. Hampton Court, St James's Park and the corresponding purlieus of the great houses which the nobility were creating out of the demesnes of the monasteries gave large encouragement to horticulture and fruit-growing; and a demand for novelties came as much from the lovers of flowers as from the doctors. If Britain never succumbed to the tulipomania which infected Holland, competition was nevertheless keen and development was rapid. Colour variations as in the Cornflowers, the Pinks and a few other herbaceous species, double specimens in the Buttercup, the Columbine, the Cuckoo-flower, freaks, like the green campanulate Primrose, competed with the varieties of Daffodils and Tulips which the bulb-growers were beginning to distribute. Information was eagerly collected and spread abroad.

1 Susanna Garret, widow, presumably James the elder's wife, is mentioned as one of his tenants in Lime Street in the will of Alderman Hugh Offley dated 2 October 1594 (Stow, *Survey of London*, II, p. 404).

2 Cf. L. Legré, *Les Deux Bauhins*, p. 31. From the fact that he took no part in the foundation of the Society of Apothecaries it looks as if Garret had died before 1617.

If interest in medicine and horticulture gave botanical studies a special attraction, the interest in them was indeed only a type and symptom of what was taking place over the whole range of man's earthly environment. Antiquaries like Leland and Camden, chroniclers like Holinshed and Stow, Latinists like Elyot and Cooper, poets like Spenser and Drayton were beginning to display a concern for nature, an interest in the world around them, a delight in noting and naming its flowers and birds, that were wholly new. No doubt the Englishman had always cared for his byre and stack-yard and garden, his hunting and fishing. But he had taken it for granted, and shown little of the appreciation and curiosity which now began to be in evidence and have ever since so powerfully affected our national character and achievements.

When William Camden turns aside from his archaeology and heraldry to record his version of John Caius's story of the miraculous crop of peas appearing after the famine in 1555 on the beach at Aldborough (*Britannia*, ed. 1607, p. 339) he was only following a precedent which goes back as we have seen to Giraldus and Matthew Paris—though he discounted the miracle first by the suggestion that the peas were thrown ashore from a wreck, and secondly by the statement that he had already recorded similar peas growing at Dungeness in Kent. But when William Harrison, preparing his *Description of England*, penned his account not only of its physical features, its rivers and gardens, but of its beasts and birds—and evidently took pains and found pleasure in so doing—the new attitude stands out clear. That his work, appended to Holinshed's *Chronicle* and published in its two editions in 1577 and 1587, is in the second version much amplified in its references to nature, is proof of the growing enthusiasm for the subject. His chapters, especially those in Book III on fishes and birds, contain not only a very adequate survey of the subject, but several particular points of real interest. Thus he devotes some space in his chapter on 'Wilde and Tame Fowles' to the Barnacle 'whose place of generation we have sought oft times so farre as the Orchardes [Orkneys]....If I shoulde saye howe either these or some such other Fowle not muche unlyke unto them doe breede yearely in the Thames mouth, I doe not thinke that many will beleve me, yet such a thing is there to be seene, where a kinde of Fowle hath hys beginning upon a short tender shrubbe standing uppon the shore from whence when theyr time commeth they fall doun either into the salt water and live or upon the dry land and perish as Pena the French Herbarien hath also noted in the very ende of hys Herball'[1] (in Holinshed, *Chronicles*, I, p. 110, ed. 1577). To counterbalance this is another and much less fantastic record: 'I have seene Crowes so cunning also of theyr owne selves that they have used

1 Cf. *Adversaria* (ed. of 1605), p. 456, where special reference is made to Barnacles on a ship in the Thames.

to soare over great rivers (as the Thames for example) and sodenly comming downe have caught a small fishe in their feete[1] and gone away withall without wetting of their wings. And even at this present the aforesayde ryver is not without some of them, a thing in my opinion not a litle to be wondred at' (l.c. p. 111). And more exciting still is his account on the same page of a Viper: 'I did see an Adder once my selfe that laie (as I thought) sleeping on a moulehill, out of whose mouth came eleven yoong adders of twelve or thirteene inches in length a peece, which plaied to and fro in the grasse one wyth another, tyll some of them espyed me. So soone therefore as they sawe me they ran againe into the mouth of theyr damme whome I kylled, and then founde eache of them shrowded in a distinct celle or pannicle in hyr belly much like unto a soft white jellie.' These two last notes at least raise matters which are not mere legend or tradition. Harrison had seen the incidents even if he has misinterpreted and exaggerated them.

In the later edition there is a notable addition, the chapter on Gardens and Orchards, Book II, ch. 19, in which he not only pays a generous tribute to 'Carolus Clusius the noble herbarist' (Holinshed, p. 210, ed. 1587), who had evidently been known to him during his visit six years previously, but speaks with interest and knowledge of the simples and their uses, the vegetables old and new, and the crops and fruits of the country. His story of the rose with 180 petals grown at Antwerp in 1585 of which 'I know who might have had a slip or stallon if he would have ventured ten pounds upon the growth of the same'; and of his own garden 'little above 300 feet of ground, and yet, such hath beene my good lucke in purchase of the varietie of simples, that notwithstanding of my small abilitie there are verie neere three hundred of one sort and other contained therein, no one of them being common or usuallie to bee had'; these are evidence of a real zest for nature and its ways.

And yet, as if to remind us how far he is from any scientific outlook, he has inserted into the chapter on 'Ravenous Fowles' a passage on the Osprey in which after saying that they 'breed with us in parks and woods' and describing how the keepers take the young, tether them on the ground, and then collect the fish which the old birds bring, he goes on: 'it hath not beene my hap hitherto to see anie of these foules and partlie through mine owne negligence: but I heare that it hath one foot like an hawke to catch hold withall, and another resembling a goose wherewith to swim; but whether it be so or not so, I refer the further search and triall thereof unto some other' (l.c. p. 227). This ancient belief which Gesner declared

1 Mr B. W. Tucker informs me that 'it is a not uncommon trick of Crows to pick up dead fish or other floating food from the surface of water, but this is usually done with the bill'. He quotes evidence (*Science*, II, p. 265 and *Zoologist*, 1883, p. 470) that Raven and Crow do on occasion carry off food in their feet.

that he had heard from certain Englishmen (*Hist. Anim.* III, p. 179), but which Turner ignored, and Caius quoting it from Giraldus Cambrensis denied, ought surely to have been repudiated or at least omitted by any serious historian.[1] But with Harrison, as with most of the Elizabethans, tales of wonders from the Tropics

> And of the Cannibals that each other eat
> The Anthropophagi, and men whose heads
> Do grow beneath their shoulders

were so rife and so often in fact true that the borders of credulity would have legitimately been widely stretched, even if there had been, as there were not, any concepts of nature and of probability which set a limit to fancy. In the medieval world the supernatural had been so stressed and the abnormal so magnified that there was literally no general agreement, such as we cannot escape, as to what might properly be believed or expected. It was the task of the Elizabethans to survey the world of nature and so to prepare for the formulation of a concept of natural law. Harrison and his colleagues, in chronicle and drama, held a mirror up to nature, even if the resultant image was neither plainly seen nor interpreted in terms free from fantasy and superstition.

Despite the interest of these chapters of his work, and his own powers of observation and description, he has little claim to be a naturalist, still less a scientist. Though he refers once to Pena as we have seen, and at least once to William Turner whom he describes in his chapter on Baths as 'the father of English physicke and an excellent divine', he gives no evidence of familiarity with their writings, or indeed of any serious study of the subject. Rather he was obviously gifted with a real love of plants and animals; and under the new conditions of life under Elizabeth found this love worthy to be expressed and shared. The fact that without special knowledge of the literature he can write so fully and happily of it is in itself no small testimony to the awakening that was taking place. The New World of the Americas was not the only land to be freshly discovered and investigated. The home-country, the fields and woods, the rivers and coasts of England, were being explored with a similar sense of excitement and even of wonder. Men and women who under the old culture and the old religion had been taught to regard a love of nature as brutish and profane, and who as we have seen had consequently paid no attention to plants or birds objectively and in themselves, now began to discover in them not only aesthetic satisfaction but keen intellectual delight. To name the familiar objects of their gardens and estates, to treasure those that had special brilliance of colour or strangeness of form, to collect the rarer

1 In fact it survived for another century, appearing in William Lawson's comments on *The Secrets of Angling*, London, 1653.

products of remote counties or countries, this became an enthusiasm; and in it was found a rich and rewarding experience. There was not yet any reasoned justification of the observation of nature: Francis Bacon and the 'new philosophy' still lay ahead and unforeseen: but whereas, as J. et J. Tharaud puts it in the letter to Jacques Delamain, 'for Catholicism nature always remains more or less the enemy',[1] the Reformation had enabled Protestantism to recover that delight in 'the works of the Lord', that sympathy for beast and bird and flower which was always characteristic alike of the Hebrew and of the Greek, and which no student of the Gospels can ignore.

That there is a profound cultural significance in this new attitude to nature no one who realises that it is a main source of the whole scientific movement is likely to dispute. In Italy it had first appeared, a child of the Renaissance; and for a century or more it flourished there, though indeed it can be argued that the love of nature was never largely dispersed among the people. But when the Counter-reformation had succeeded in crushing intellectual and spiritual adventuring Italy dropped behind. Then as we have seen France and the Low Countries took the lead; and by the mid-sixteenth century were producing a widespread movement which was to give rise to the great Dutch achievements of the next age. With Britain the impulse came late. Turner alone in the sixteenth century was fit to be mentioned in the same breath with the savants of Western Europe; and even he is an inconspicuous figure beside Gesner or Aldrovandi, Rondelet or De l'Ecluse. But with us, if there was as yet no outstanding merit, there was something in our inheritance and circumstances which created a general response to the new adventure. We began late: we had much lost ground to make up: but in spite of this the interest became rapidly both intense in character and wide in extent. And a century after Turner it produced a galaxy of brilliant workers in all the fields of nature-study. For that result some at least of our gratitude must go to the men who, without any claim to greatness or originality, yet so presented the charm of plants and animals as to arouse popular enthusiasm and diffuse a knowledge which if not highly scientific was at least sincere and aspiring.

As symptomatic of what was happening it is not out of place to refer to one almost casual piece of evidence. In October 1674 Anthony à Wood, inveterately inquisitive, took occasion to consult the trivial diary of a forgotten Yorkshireman Richard Shanne of Woodrowe. An entry made in 1588 attracted him: 'There was taken at Crowley in Lincolnshire in the winter time 5 strange fowles of divers colours, having about their necks as it were great monstrous ruffs and had underneath those ruffs certaine quills to beare up the same in such a manner as our gallant dames

1 *Why Birds Sing*, p. xvii, a very striking discussion by a Catholic of the acknowledged indifference to nature in Catholic countries.

have now of wier to beare up their ruffs (which they call supporters). About their heads they had feathers so curiously set togeather and frisled, altogeather like unto our nice gentlewomen who do curle and frisle their haire about their heads. Three of these strang fowles was brought unto Sir Henrie Leese...Mr Richard Shann of Wodrow in Medley [Methley] Yorks., drew a picture of one of them which he placed in his herball. Two men that had set lime twigs to catch birds withall did find them taken therein.' Such a description of the Ruff, whether or no it is the first in English as Gunther suggests,[1] takes us back to Matthew Paris's comment upon the Crossbills and reveals the native interest which at this time was beginning to demand and receive recognition and books to assist it.

An indication of this increase of interest and of knowledge is shown by the attention paid to the correct identification of plants in Thomas Cooper's important dictionary, *Thesaurus Linguae Romanae et Britannicae*, published in London in 1563 and re-issued many times in the next twenty years. Cooper, the son of a poor man at Oxford, had done brilliantly at school and university and was preparing for ordination when Queen Mary's accession checked his plans. He then turned to medicine and obviously did some work on Ruel's *De Natura Stirpium* and perhaps on Turner's *Names*. The result shows itself in a series of interesting notes on the more obscure Latin plant-names; we shall see examples of them when we consider the work of Henry Lyte. He also shows a real interest in birds, and some knowledge in translating their names. His work being based on that of Elyot, it is not surprising that in places he transcribes Elyot's notes as on Onocrotalus 'a large Swan-like bird' and on Platalea. But in this latter case he adds, 'We call him a Shovelar'; and this seems derived from Turner who had recorded Pliny's account of its preying upon other fish-catching sea-birds under the heading 'Platalea or Shovelar'. So, too, there are plain resemblances to Turner in his identification of Boscha with Pochard, Haliaëtus with Osprey, Rubicilla with Bullfinch, and Attagen as 'most like to that we call a Godwitte'. The plainest of them is under Upupa, 'a birde no bigger than a thrush and hath a creste from his bill to the uttermost parte of his heade which he strouteth up or holdeth downe accordynge to his affection. Wherefore it can not be our lapwynge as it hath been taken for. It is rather to be called a Houpe.' In view of this, even though unlike Turner he calls Fulica a Coot, it seems probable that he knew Turner's book at first hand—though it is possible that he got his acquaintance with it from Gesner. The *Thesaurus* is said to have impressed Elizabeth so deeply that she determined to prefer its author as quickly as possible. In consequence Cooper became Bishop of Winchester and a religious controversialist; and his scientific interests were not developed.

A further indication of popular concern is to be found in the sumptuous

1 Cf. *Early British Botanists*, p. 265, whence this record is drawn.

volume *Batman uppon Bartholome His Booke De Proprietatibus Rerum, newly corrected enlarged and amended with such Additions as are requisite.... Taken foorth of the most approved Authors, the like heretofore not translated in English. Profitable for all Estates, as well for the benefite of the Mind as the Bodie*, published by Thomas East in London in 1582. That Bartholomew had been the text-book for three hundred years and was very widely used by Shakespeare and his contemporaries has been generally recognised since Douce noted it in his *Illustrations of Shakespeare*, I, p. 9. An attempt to bring it up to date was natural enough; and Stephen Batman (Bateman) was a suitable editor. Born at Bruton in Somerset, educated there and at Cambridge, he had been domestic chaplain to Archbishop Parker and largely responsible for the collection of his famous library. He had written a number of books including the *Travayled Pilgreme* in verse; and an encyclopaedia of 'leaden goddes' dedicated like his *Bartholome* to his patron Henry Carey Lord Hunsdon.

Unfortunately, his qualifications for a revision of the *De Proprietatibus* did not extend to any close knowledge of nature. He confined himself to cutting down some of the longer and less interesting chapters, and to adding in brackets notes or further particulars, generally of a disappointingly trivial nature. In the sections with which we are concerned he draws mainly upon Cooper's *Thesaurus* as on pp. 185*b*, 280*b*, 310*b*, 345*b*, upon Gesner very frequently, Dodoens in Book XVII, Ludovico di Varthema for details of a few Eastern species as on pp. 276*b*, 299, 363*b*, 368, and upon Sir Thomas Elyot for Olives on p. 308. After a note on Artichokes (p. 285) he adds 'Reade D. Turner'; so too, on 'Zea, Spelt', he quotes 'Fol. 131 Turner' (p. 293*b*); on Holyhock (p. 306); on Spikenard, 'Reade more of this at large in the second booke of D. Turner fo. 62*b*' (p. 306*b*) and after Hellebore (p. 290*b*), 'Read Fuchsius, Mathiolus, Turner or Dodoneus'. On occasion he speaks personally, often in the merest moralising as in his note on hawking, wherein as he says gentlemen spend twenty marks to catch a prey worth sixpence (p. 185), or his rhapsody on the English Rose (p. 315*b*), or his comparison of the Basilisk with the spoilers of the clergy—'I speak by experience' (p. 351). Sometimes his notes are more valuable—'the Goodwike [Godwit] commonly taken in the Ile of Eley' (p. 187); 'Oat bread is not agreeable for mankinde' (p. 280); 'With the strawe of Wheat they thatch houses, and with Rie strawe they commonly make strawen hats' (p. 283*b*); 'With the said Segs [Sedge] is made Hambroughs [Burghams, collars] for the necks of horses instead of Lether harnesse' (p. 284*b*); 'With Bran and the dragges of Ale is made the famous potage in Devonsheere called Drouson'[1]

1 Cf. Gervase Markham, *Farewell to Husbandry*, p. 133: 'Boyling Oatemeale with barme of the dregges and hinder ends of youre Beere barrells...of great use in all the parts of the West Countrie, called drousson pottage.'

(p. 294); 'Of Lin commeth the Linseed wherof is made Painters oyle' (p. 302*b*); 'The paper that is now common is made of olde lynnen rags, wrought in a mill, and brought to a perfection, wheron is written the help of memorie, the bewrayer of colution and the treasure of truth' (p. 310*b* after a note on Papyrus and before a long and pointless tale); 'Yew is altogether venemous...the birdes that eate the redde berryes eyther dye or cast theyr feathers' (p. 322*b*); 'Of late yeares there hath bene brought into England the cases or skinnes of such Crocodiles to be seene and much money given for the sight thereof' (p. 359*b*); 'There is no meat more wholsome than Rabets' (p. 372); 'The Mole hath eyes but they are very small' (p. 382). In one place (p. 361) on Dragons he refers to his own book *The Chronicle of the Doome*; and in another (p. 364) on Elephants he translates a long account of 'the Empire of the Abyssines or of Presbiter John' by 'Ortelius', Abraham Ortel, the geographer.

Probably when the work was being done Batman was already an old man. He had apparently taken his LL.B. degree at Cambridge in 1534 after six years' residence. He died in 1584 two years after *Bartholome* was published. There is no reason to suppose that he had been interested in natural history or that this task was anything more than a piece of hack-work, an attempt to bring up to date a book which had become a classic. His work bears a strong resemblance to several of the Victorian editions of White's *Selborne*. But there are two notes which reveal the change that despite Batman had made the *De Proprietatibus* an anachronism. On the Onocentaur he writes, 'As auncient men spent their time in writing of follyes to make the common people wonder...so in the last discovered Indies the barbarous people seeing a far of the Spaniards on horseback...supposed they had bene monstrous devourers' (p. 375*b*); and on the Sirena or Mermaiden, 'The nature of divers fishes is to pray upon man as the Conger the Mackrell and the Crab: myselfe in the yeare 63 sawe the experience; and as for the Mermaide, that is the sea-fish, shapes appere after diverse formes that some grose head imagine to be lyke a maide, as the Munke fish' (p. 380*b*). His statements are neither logical nor lucid: but they make it plain that the age of mythical beasts, and the naïve credulity which could accept them without question, were passing away. Batman was no great naturalist, but he could not swallow the natural history of the Middle Ages.

Far more important than these attempts to furbish up the old are the books which introduce the Englishman to the new knowledge as declared on the Continent. The first was by Henry Lyte (1529–1607), who translated De l'Ecluse's French version of the *Cruydeboeck* of Rembert Dodoens and published it as *A Niewe Herball or Historie of Plants* in London in 1578. Of an old family,[1] educated at Oxford and by travel, Lyte wished,

1 Cf. *Visitations of Somerset*, p. 44.

as he states in the dedication to Queen Elizabeth and in the prefatory pages of the *Niewe Herball*, to contribute something to the welfare and renown of his country as well as to the health of its people. Dodoens's book, which is partly a record of plants with special reference to those growing in the Low Countries or in the gardens of its 'herboristes' but partly an elaborate treatise on their medicinal uses, had been published in Flemish in Antwerp in 1554 and in a French translation by De l'Ecluse four years later. Lyte produced a careful version of it, comparing at certain points the 'last Douch copy' with 'my French copy the which is in divers places newly corrected and amended by the Author him selfe' (*Niewe Herball*, p. 345). This he 'augmented' (p. 311) by a number of additional notes, mainly distinguished by being printed in Roman type instead of Black letter. Lyte's original copy of De l'Ecluse's version with his notes written in red or black ink and in a fine script is preserved in the British Museum, and fully bears out his claim to have revised and amplified it.

The book itself, especially in De l'Ecluse's version, was well deserving of translation. Dodoens, more than his two younger Flemish contemporaries, was a physician rather than a botanist. Though he had had some experience of travel this seems to have been confined to Universities; and field-work is not very evident: one allusion, 'In Province about Marselles whereas I have seene great store' (p. 543), is almost the only sign of his visits to Montpellier, North Italy and Germany. But he knew the gardens and herbaria and literature of the subject; and he could produce a clear, condensed and objective record, arranged in regular paragraphs. Each species or group of closely related species has a chapter to itself and this is split into headings, description, place, time, names, nature and medical uses. Occasionally he interpolates a full and well-told legend from the Classics giving the origin of Crocus or Lily or Narcissus, of Daphne or Rose: occasionally he tells of the first discovery of a strange plant as of the 'Musa or Mose Tree', the Plantain or Banana (*M. paradisiaca*) (p. 705), which 'was found by a certain Fryer named Andro Thevet in the country of Syria by the great towne Aleph [Aleppo] so called of the first letter of the Hebrue Alphabet, where as is great resort and traffique of marchants, as well of Indians, Persians and Venitians as of divers other strange nations':[1] occasionally he adds a note like that on p. 39, 'the people of the countrey delight much to set it ["Orpyne", *Sedum telephium*] in pots and shells on Midsomer Even...' or on p. 108 of Alysson (*Alyssum saxatile*), 'the same hanged in the house, or at the gate, or entry, keepeth both man and beast from enchantments and witching': at the end of Book IV after a short notice of the 'Ficus Indica'

1 Cf. F. André Thevet, *Cosmographie de Levant* (Lyons, 1554), p. 182. He adds that its fruit resembles a Cucumber and is thought to be that eaten by Adam.

he inserts pictures and a brief account of the 'worme called Buprestis', which judging by the woodcuts is certainly the Mole Cricket (*Gryllotalpa vulgaris*) which he says is 'founde in certayn places of Holland, and lykewise sometimes in Brabant and Flaunders, where the Kyen sometimes are bitten of them' (p. 544): but in the main he gives us a sober and orderly catalogue.

This must not be taken to mean that he shows any approach to a scientific outlook, either in his taxonomy or in his general ideas. The arrangement of the several books and of the species within them seems almost fortuitous. There are signs of an originally alphabetical arrangement: but this is not consistent even if allowance is made for the habit of grouping similar species together. There are also signs of a desire to classify like with like: but here too there is no sort of continuity. Neither colour nor leaf-form nor habitat nor medicinal use supplies any clue to the sequence.

Nor does he show any appreciation of structural resemblance even when he groups species under a single heading. In Chapter 51 of the first book 'Of Lysimachion' four 'kindes' are pictured together—and they belong in fact to four different families—*Lysimachia vulgaris*, *Epilobium hirsutum*, *Lythrum salicaria*, and *Veronica spicata*: on p. 326 *Centaurea scabiosa* and *Erythraea centaurium*, and on p. 334 *Gentiana cruciata* and *Saponaria officinalis* are linked; and on pp. 412–13 *Drosera rotundifolia*, *Lycopodium clavatum* and *Zostera marina* with several mosses and liverworts appear in the same chapter.

Moreover his outlook is revealed by the fact that his introductory preface consists of a series of paragraphs reproducing Pliny and dealing with the deterioration of one species into another, with the spontaneous generation of midges, caterpillars and worms, and with remedies and methods which are sheer superstition. He is in fact a member of the generation which produced the pandects; and his merit is just that he described and identified a large number of recognisable plants.

The additions made by Henry Lyte are not always easy to detect. It is clear that in the matter of localities the words 'in this country' or 'in the gardens of our herborists' usually refer to Flanders and the gardens of Coudenberg and his fellows. But there are a few such in which England seems to be meant, and there is one specific addition 'in Maister Riches garden' in Roman type on p. 495. Of such additions the note on p. 66, 'The Authors of Stirp. Advers. nova do affirme that Androsemum groweth by Bristow in England in S. Vincentes Rockes and woody Cleves beyond the water. But if Androsemon be Tutsan or Parke leaves, it groweth plentifully in woodes and parkes, in the west partes of England' and another on p. 174, 'They use about Coventrie in England where as great store of these plantes [*Campanula trachelium*] do grow, to eate their rootes in Salads, as Pena writeth in his booke intituled *Stirpium adversaria nova*. Fol. 138',[1] although not in Roman but in Black letter, both plainly derive from Lyte. Similarly the many quotations in the later part of the work from Turner's *Herbal* and Cooper's *Dictionarie* are sometimes in one type sometimes in the other. So are the notes of English localities.

To Turner the first allusion by name is on p. 235, though the locality of Pennywort 'plenteously in Sommersetshyre and about Welles' on p. 38 perhaps derives from him. Thereafter he is freely named—pp. 268, 279, 290, 298 (of *Trinia glauca* 'Dr Turner saith he founde a roote of it at S. Vincentes rocke by Bristowe', cf. Turner, *Herbal*, II, p. 83*b*), 367 (of *Colchicum autumnale*, 'Turner nameth it Mede Saffron and wild Saffron', *Herbal*, I, p. 155), 433, 438, 455, 465, 474, 476, 499 ('Of Horned Claver or Medic Fother wherof both Turner and this Author do write'), 611, 658, 664 ('Turner, lib. 2, fol. 72 calleth it Orobanche'), 680, 727 ('I referr you to the second part of Maister Turner's herbal, fol. 143'), 729, 746. And several of the localities, e.g. for 'Sea Purcelayne' (*Atriplex portulacoides*) 'bysides the Ile of Purbeck' (p. 575, cf. *Herbal*, I, p. 123); 'Kneeholm' (*Ruscus aculeatus*) 'as in Essex, Kent, Barkeshire and Hamshire' (p. 674, cf. II, p. 122, where Hants is omitted); Linden (*Tilia cordata*) 'by Colchester in the parke of one maister Bogges' (p. 753, cf. II, p. 153*b*), are obvious quotations though Turner's name is not given. Curiously enough Lyte seems to take no interest in Turner's northern English plants: the 'Trol' (*Trollius europaeus*) and 'Herbe Paris' (*Paris quadrifolia*) (pp. 419, 425) are given Swiss and Belgian localities respectively, but without any suggestion that they occur in Britain; and none of the plants that are in Turner but not in Dodoens is mentioned at all.

The allusions to Cooper are much fewer and naturally less significant: on p. 235, 'Turner calleth Clinopodium Horse tyme and so doth Cooper' and the two are similarly coupled on p. 611; p. 527 (of *Acanthus spinosus*), 'Cooper in his Dictionarie calleth it Branke Ursine, Beare Briche'; p. 535 (of *Carduus crispus*), 'Cooper calleth this wild Artichoke and Cowthistel'; p. 612, 'Cooper calleth it Bastarde Parsley and sayth that it is an herbe like Fenill with a white flower and commeth of noughtie Parsly seede'; p. 729, 'Cooper in his Dictionary sayth that the fruite of Celtis or Lotus is called in Latine Faba Graeca' and above all p. 530,[2] 'Cooper saith that Leucacantha is a kinde of Thistel with white prickle leaves, called in English Saint Marie Thistel [*Carduus Marianus*]. Wherein he followeth Matthiolus...', and Lyte adds a note referring to the fact that Mattioli's Carline Thistle is not the same as the plant here illustrated by Dodoens. Cooper's own note went on to say 'Ruellius is deceyved that taketh it for Spinam albam contrary to the opinion of sundrie learned men'.

1 'Pena' is also quoted on p. 163.

2 This note is all in Roman type.

The most interesting of Lyte's notes are naturally those which show some originality. Perhaps the only one of these that deals with the locality of a wild plant is that on the Hyacinth (*Scilla nutans*) 'especially about Wincaunton, Storton [Stourton] and Mier [Mere] in ye West partes of Englande' (p. 206), which refers to places not far from 'my poore house at Lytescarie'[1] from which he dated his dedication to the Queen—Lytes Cary being some two miles due east of Somerton and seven or eight west of Wincanton. Two others, less easily explicable, are on p. 113 of *Verbascum nigrum*, 'The wilde Mulleyn is not common in this countrey but we have seen it in the pleasant garden of James Champaigne the deere friende and lover of Plantes' and on p. 226 of *Orchis maculata*, 'The roote of Royall Satyrion...cureth the old fever Quartayne...as Nicholas Nycols writeth Sermone secundo.' Identification is very seldom attempted, but on p. 468 under 'Bockwheate' (*Fagopyrum esculentum*) Lyte adds in Roman type, 'I think this to be the grayne called in some places of Englande Bolimonge', and on p. 604 under 'Skirwurtes' (Skirrets, *Sium sisarum*) is this comment on the statement in Dodoens that the 'seede is somewhat broade', '(as I reade in my copie) but the Skirworte that groweth in my garden which agreeth in al things els with the description of this Skirwort, hath a litle long crooked seede of a browne colour, the which being rubbed smelleth pleasantly, somewhat lyke the seede of Gith, or Nigella Romana,[2] or lyke the savour of Cypres wood'. One on the 'Savoie Colewurte' (p. 552) which 'cannot abide the colde...neverthelesse the winter being caulme as it was in the yeare of our Redeemer MDLX after winter it bringeth forth...' refers to a date later than De l'Ecluse's version and is probably by Lyte. There are a few other notes giving cross-references, e.g. p. 332, or noting omissions, e.g. p. 517: but that is all.

From this we can only infer that Lyte in his translation used Cooper's *Thesaurus* to help him with the English names; used Turner for this purpose and probably from his own interest in the plants of the West Country; and had looked into De l'Obel's *Adversaria* for the same reason. He had done as much as a country gentleman, who had no knowledge of any plants outside those of his own neighbourhood and garden and no large acquaintance with the literature of the subject, could reasonably be expected to do. He had produced a book less loaded with argument and therefore much more valuable to the uninstructed than Turner, a book in the main concise, clear, and packed with information as to medicinal and other uses, a book which deserved and received wide popularity. By having it printed by 'Henry Loë' in Antwerp (a fact notified on the last page though the title gives London) he had secured the excellent illustra-

1 For which cf. W. George, *Lyte's Cary Manor House*.

2 Cf. Turner, *Names*, p. D iii*b*, 'Git or Nigella Romana'.

tions which the house of Van der Loe had obtained from Fuchs's *Herbal* and which Plantin the great printer of Antwerp took over after Loe's death in 1581. The pictures are thus very often similar to those used by Turner. One or two of them seem to be misplaced; one, that of 'Zea, the Corne called Spelt', p. 455, appears again three pages later as 'Zeopyron', and another, that of *Melampyrum arvense*, is twice printed on pp. 164 and 471. But in the main the book is a very praiseworthy piece of work.

There is little evidence to justify us in regarding its author as a naturalist. A man of some education, of culture and taste, with hobbies and interests outside the merely bucolic and parochial, he produced first his *Herball*, then a record purporting to trace the early history of Britain, and finally a table of the descent of his own family from Leitus the Trojan, who had accompanied Brutus, son of Priam, to Western England at the dawn of our history. He is the forerunner and type of a class who without much claim to scientific status have greatly promoted the interest in nature among us.

CHAPTER XII. JOHN GERARD

Lyte was an honest amateur who produced a translation of some merit and thus made one of the masters of European botany familiar to English readers. John Gerard, whose much more famous *Herball* was published in London by John Norton in 1597, was less of an amateur and less honest. Indeed, it is hard to acquit him of almost all the sins of which a man of letters or of science can be guilty.

Gerard, whose life is best told by B. D. Jackson in his edition of the Garden Catalogue of 1596 (pp. xi–xvi), was born at Nantwich in Cheshire in 1545: he went to school at Wistaston, two miles away, and then, perhaps, travelled in 'Denmark, Swevia, Poland, Livonia or Russia' (so *Herball*, p. 1223, cf. pp. 1175, 1181). In 1562 he was apprenticed to Alexander Mason, a barber-surgeon in London, Warden of the Company 1556 and 1561 and Master 1567; in 1569 he was admitted to the freedom of the Company; in 1607 he became its Master;[1] in February 1612 he died and was buried in St Andrew's, Holborn. He was married, and his wife helped him in his business; but we know no more of her and very little else of him. He seems hardly to have been out of London except on occasional visits to Margate and Rye, to Harwich and Cambridge. His garden in Holborn, probably on the south side and at the corner of

1 He had been Warden in 1597 (F. Weston, *Some Account of the Barbers' Company*, p. 38).

Fetter Lane,[1] and the gardens of Lord Burghley in the Strand and at Theobalds[2] in Hertfordshire near to Waltham Cross of which he had charge occupied much of his attention. Of his own garden he issued a list of plants in 1596 dedicated to Burghley and with a commendation from De l'Obel, and this was expanded into a substantial Catalogue in 1599: these have both been reprinted exactly in Jackson's book. His only other adventure into literature was the *Herball* and how much of this is his own is still a matter in dispute.

As an illustration of successful piracy the story is worth repeating. Here are the facts of it. Gerard published the *Herball* as his own with no word of acknowledgment for special obligation. In his letter 'to the Readers' he pays tribute to 'that excellent work of Master Doctor Turner'; says that after him 'Master Lyte, a worshipful gentleman, translated Dodonaeus out of French into English; and since that Doctor Priest, one of our London Colledge hath (as I heard) translated the last edition of Dodonaeus which meant to publish the same; but being prevented by death, his translation likewise perished'; and adds, 'lastly myselfe, one of the least among many, have presumed to set foorth unto the view of the world the first fruits of these mine own labours'...'faults I confesse have escaped some by the printer's oversight, some through defects in myselfe...and some by means of the greatnesse of the labour....' It is all very naive and convincing—a busy man putting out with proper diffidence a monumental work—and the same note is sustained as for example on p. 315 where he writes: 'Other distinctions and differences... I leave unto the learned phisitions of our London Colledge as a thing far above my reach, being no graduate but a countrie scholler, as the whole framing of this historie doth well declare: but I hope my good meaning will be well taken, considering I do my best, not doubting but some of greater learning will perfect that which I have begun according to my small skill, especially the ice being broken unto him, and the woode rough hewed to his handes.' Such ingenuousness, especially when followed by the artless and charming story of how John Bennet 'a chirurgeon of Maidstone in Kent, a man as slenderly learned as my selfe' cured a butcher's boy of an ague by dosing him with the rhubarb that Gerard had supplied, cannot but create a desire to believe in its author's honesty.

Yet even in the *Herball* itself there was evidence of a less unsophisticated authorship. The friend to whom Gerard freely refers, 'Master Stephen

1 So H. B. Wheatley in Furnivall, *Shakespeare's England*, I, p. ci. 'All the plot of ground to the back of Fleet Street between Shoe Lane and Chancery Lane in the reign of Elizabeth consisted of gardens with a cottage here and there being intersected by these two lanes and Fetter Lane as at present', J. T. Smith, *Streets of London* (1861), p. 266.

2 Exchanged for Hatfield with the King in 1607.

Bredwell practitioner in physic'[1] and finder of Horse Radish 'wilde at a small village neere London called Hogsdon' (p. 187), had written in his letter of commendation: 'D. Priest for his translation of so much as Dodonaeus, hath hereby left a tombe for his honorable sepulture: Master Gerard comming last, but not the least, hath many waies accommodated the whole worke unto our English nation'—a sentence hardly compatible with Gerard's claim to unaided authorship. Moreover, though the arrangement differed from it, whole sections of the text were recognisable as mere translations of the *Pemptades*: indeed, when the English localities, the gardening and other personal notes were excepted and certain citations from De l'Ecluse and De l'Obel removed, little remained except Dodonaeus; and that little like the personal notes was often wrong. With pictures from Bergzabern[2] (whom Gerard insists on calling Taber montanus) and Latin names from De l'Obel, the amount of ice broken by Gerard melts into very little.

Then De l'Obel took a hand, a discreet and at first generous hand, in the game. He was, as his friend and publisher testifies (*Correspondance*, IV, p. 193), a man of sterling honesty. He was also a friend of Gerard—had they not together discovered *Papaver argemone* at Southfleet in Kent (*Herball*, p. 301) and had he not written an attestation of the Garden Catalogue printed by Gerard in 1596—even if in the copy in the British Museum he afterwards inserted and signed the words *haec esse falsissima*?[3] His story is revealing—he tells it in *Stirpium Illustrationes*, pp. 3, 4. Norton the publisher was warned by James Garret that Gerard's work was full of blunders. He therefore asked De l'Obel to check and correct the placing of the pictures and to undertake other editorial revision. He claims to have corrected more than a thousand mistakes until Gerard got irritated and made rude remarks about English idiom and De l'Obel's ignorance of the language. In telling the story De l'Obel speaks of Gerard as 'rapsodus', that is, one who stitches together patches of other people's work. This charge he elaborates: for in writing his notes on Rondelet's *Pharmacopoeia* and on p. 59 blaming Dodoens for his use of *Seta* instead of *Sericum* he added that Dr Priest when engaged by Norton to translate the *Pemptades* had been misled into rendering it as bristle instead of silk—a mistake which is to be found in Gerard's *Herball*, p. 1160. De l'Obel did not actually underline the inference which was plainly to be drawn from his words: nor in the testimonial which

1 Parkinson, *Theat. Bot.* p. 449, who quotes 'a receipt of the Country Empericke' given to him, calls him 'a Chirurgion of London who practised physicke'. He was physician of Christ's Hospital and a man of standing in the profession.

2 G. Bauhin, *Pinax*, Nomina Auctorum, says that the pictures came mainly from Bergzabern with some from De l'Obel and about 16 original—about 2190 in all.

3 So Gunther, *E.B.B.* p. 247.

he wrote for Gerard's book did he make any allusion to it. But his charge was evident for all who had eyes to see; and in his later writings he roundly accused him of plagiarism.[1]

Finally, Thomas Johnson, of whom more anon, when he produced his emended edition of the *Herball* in 1633, set out the facts in plain terms. 'For the author Mr John Gerard I can say little....His chiefe commendation is that he out of a propense good will to the publique advancement of this knowledge, endeavoured to performe therein more than he could well accomplish; which was partly through want of sufficient learning as may be gathered by the translating of divers places out of the *Adversaria* [there follow instances to show that when Gerard rendered Latin without Dr Priest's script he made howlers]....He also was very little conversant in the writings of the ancients....But let none blame him for these defects, seeing he was neither wanting in pains nor good will....Now let me acquaint you how this worke was made up. Dodonaeus his Pemptades comming forth Anno 1583 were shortly after translated into English by Dr Priest a Physition of London, who died either immediately before or after the finishing of this translation. This I had first by the relation of one who knew Dr Priest and Mr Gerard: and it is apparent by the worke itself which you shall finde to containe the Pemptades of Dodonaeus translated so that divers chapters have scarce a word more or lesse than what is in him. But I cannot commend my Author for endeavouring to hide this thing from us, cavilling (though commonly unjustly) with Dodonaeus wheresoever he names him,[2] making it a thing of heare-say that Dr Priest translated Dodonaeus....' Johnson then describes Bredwell and De l'Obel's remarks and continues: 'Now this translation became the ground work whereupon Mr Gerard built up this worke: but that it might not appear a translation he changes the generall method of Dodonaeus into that of L'Obel, and therein follows almost all over his Icones both in method and names as you may plainly see in the Grasses and Orchides.' Finally he narrates how Norton got from Frankfort Tabernaemontanus's pictures and Gerard did his best, a bad best, to arrange them after l'Obel's order and Dodoens's script; he 'oft times by this means so confounded all that none could possibly have set them right, unlesse they knew this occasion of these errors'.

Gerard was a rogue: of that there can be no doubt. But like many such he was a pleasant fellow, and had a number of good friends. Some of these—an unusually large selection—wrote commendatory letters or verses in his praise; and the names of these are of interest. Lancelot

1 Cf. *Stirp. Illustr.* (ed. How, 1655), p. 95, 'Insulsus Adversariorum nostrorum et observationum raptor': cf. p. 13.

2 Johnson speaks still more severely of this in discussing Gerard's comment on Dodonaeus' account of Beans, *Gerard*, p. 1210.

Browne of St John's College and Pembroke Hall, Cambridge, the first of them, was a physician to Queen Elizabeth and to James I, and held high office in the College: he was also William Harvey's father-in-law. Guillaume Delaune[1] (G. Launaeus) who contributed Latin verses was a French protestant who came to this country in 1582 and was licensed to practise physic: his son Gideon became apothecary to Queen Anne wife of James I, and a benefactor of the Society: he died in 1659 reputed to be 97 years old. Anthony Hunton and Francis Herring were both graduates of Christ's College, the former licensed to practise in 1589, the latter an M.D. in 1597 and a doctor of some distinction, definitely interested in botany. William Westerman of Gloucester Hall and Oriel College, Oxford, was at this time Vicar of Sandridge in Hertfordshire: he afterwards became Chaplain to Archbishop Abbot. Thomas Newton 'of Cheshire' and Trinity College, Oxford, was a man who made a regular habit of writing commendatory verses, but he was interested in Gerard not only as from the same county but owing to his own book *An Herball for the Bible* which had appeared in 1587. Last and in some ways most important is George Baker 'one of her Majesties Chiefe Chirurgions in ordinarie' and at this time Master of the Barber-Surgeons of London. Baker was a man of some fame in his profession and was known to the great French surgeon Ambroise Paré, some of whose work he translated into English.[2] Similarly he had done part of the revision and correction of Bartholomew Traheron's translation of the works of another great surgeon, Giovanni de Vigo, and had written an introductory letter for it when it was published in London in 1586. He had himself been born at Tenterden and had married Anne the daughter of William Swaine, whom Gerard describes as a worshipful gentleman with a house at 'Howcke green' (Hook Green) or Southfleet, where he found *Neottia nidus-avis* 'in a wood belonging to one Master John Sidley' (p. 176) and *Anagallis femina* 'in a chalkie corne fielde' (p. 494). It is possibly this connection that made north Kent so favourite an area for Gerard and his friends.

Gerard was a rogue. Moreover, botanically speaking he was, as has been indicated, a comparatively ignorant rogue. In spite of the testimony of George Baker as to his skill when matched with Ambroise Paré's French herbarist, his experience of searching for plants seems to have been almost exclusively in the home counties: in Kent where he knew the neighbourhood of Gravesend (p. 333) and Southfleet; Whitstable (p. 295); the Isle of Thanet, Margate and 'Queakes house [Quex Park]' the seat

1 Isaac de Laune, 'a learned physition', presumably a brother, sent Gerard plants of *Gentiana lutea* from Burgundy (*Herball*, p. 352).

2 So Johnson in preface to his translation of Paré, published in 1634: see below, p. 286.

of Sir Henry Crisp, and 'Byrchenton' (pp. 192, 219); Rye and 'Winchelsey Castle' (p. 946); in Surrey 'the parke of Sir Fraunces Carewe neare Croidon'[1] (p. 1066) praised by Camden,[2] and Groutes the seat of Richard Garth in the same neighbourhood (p. 1388); in Essex 'Landamer landing' near Thorpe le Soken and 'Langtree Point' opposite Harwich (p. 1000) which Camden calls Langerstone;[3] Clare and 'Henningham' (p. 1034); Much Dunmow (p. 977); 'a village called Graies upon the brink of the river Thames' (p. 1088); and a multitude of fascinating localities in London and its surrounding villages, localities which with their plants are so significant as to be worth quoting at length. *Cotyledon umbilicus* 'groweth upon Westminster abbay over the doore that leadeth from Chaucer his tombe to the olde palace' (p. 424); *Coronopus ruellii* 'in Touthill fielde neere unto Westminster' (p. 347); *Saxifraga tridactylites* 'upon the bricke wall in Chauncerie lane, belonging to the Earle of Southampton' (p. 500); *Salvia verbenaca* 'in the fields of Holburne neere unto Graies Inne' (p. 628); *Potentilla reptans* 'upon the bricke wall in Liver-lane' (p. 839); *Ophioglossum vulgatum* 'neere the preaching Spittle adjoyning to London' (p. 327); *Tilia europaea* 'in my Lord Treasurer's garden at the Strand; and in sundry other places, as at Barnelmes [the home of the late Sir Francis Walsingham, cf. p. 501], and in a garden at Saint Katherines neere London' (p. 1299); *Lysimachia vulgaris* 'along the medowes from Lambeth to Battersey' (p. 388); *Lythrum salicaria* 'under the Bishops house wall at Lambeth neere the water of Thames' (p. 388); *Solanum dulcamara fl. albo* 'in a ditch side against the garden wall of the right honourable the Earle of Sussex his house in Bermonsey Streete by London' (p. 279);[4] *Dianthus armeria*, still called the Deptford Pink, 'in the great field next to Detford by the path as you go from Redriffe to Greenwich' (p. 476); *Saxifraga granulata* 'from the place of execution called Saint Thomas Waterings unto Dedford [Deptford] by London,...also in the great fielde by Islington called the Mantels' (p. 693); *Mentha pulegium* 'in the common neere London called Milesende' (p. 546); *Malva moschata* 'on the left hand of the place of execution called Tyborne...among the bushes and hedges as you go from London to a bathing place called the Old Foorde...in the bushes as you go to Hackney' (p. 786); *Populus alba* 'in a lowe medow turning up a lane at the further end of a village called Blackwall' (p. 1302); *Sanguisorba officinalis* 'upon a causey betweene

1 At Beddington where Plat, *Garden of Eden*, p. 165, records that Carew entertained Elizabeth by leading her to a Cherry Tree 'whose fruit he had of purpose kept back from ripening...by straining a cover of canvas over the whole tree and wetting the same'.

2 *Britannia*, p. 216.

3 L.c. p. 338. It is called Langor Point in *Briefe Description*, p. 3 (Camden Misc. XVI, 1936).

4 Parkinson, *Theat. Bot.* p. 350, reporting this adds 'now is not there to be found'.

Paddington and Lysson greene [Lising Green on Morden's map of Middlesex 1696; and Lisham Green in 1755]' (p. 889); *Rosa spinosissima* 'in a pasture as you go from a village hard by London called Knightsbridge unto Fulham a village thereby' (p. 1088); *Linaria elatine* 'in the next fielde unto the churchyarde at Cheswicke' (p. 501); *Sambucus ebulus* 'in the lane at Kilburn Abbey by London' (p. 1238); *Epipactis purpurea* 'in a woode five miles from London neere unto a bridge called Lockbridge' (p. 358); *Anthyllis vulneraria* 'upon Hampstead Heath neare London right against the Beacon neere unto a gravell pit' (p. 1061); *Solidago virgaurea* 'in Hampsteed Wood, neere unto the gate that leadeth out of the wood unto a village called Kentish towne not far from London' (p. 349); *Scutellaria minor* 'upon the bog or marrish ground at Hampsteed heath neere unto the head of the springs that were digged for water to be conveied to London 1590 attempted by that carefull citizen Sir John Hart Knight, Lord Mayor of the Citie of London,[1] at which time my selfe was in his Lordships company and viewing for my pleasure the same goodly springs I found the said plant not heretofore remembred' (p. 466). One could forgive many misdeeds in one who can so bedeck the metropolis.

Of these London localities there is a reasonable prospect that Gerard spoke with accuracy. It is when he gets further afield that his evidence becomes unreliable if not demonstrably false. This seems due, in most cases, to mere ignorance. Thus he makes a notable series of errors over the Gentians, claiming that *Gentiana cruciata* (never found in Britain) 'groweth in a pasture at the west ende of little Rayne in Essex on the north side of the waie leading from Braintrie to Much Dunmow' (p. 352); that *Gentiana verna* 'groweth plentifully in Waterdowne forest in Sussex in the way that leadeth from Charlewoodes lodge unto a house of the Lord of Abergavenie called Eridge house'—which may be a mistake for *G. pneumonanthe* (p. 354); and that *Gentiana pneumonanthe* 'groweth upon Longfielde downes in Kent...upon the chalkie cliffes neere Greene-Hythe and Cobham'—which is no doubt *G. amarella* (p. 355). To these he adds the famous 'Gentiana concava' found 'in a small grove called the Spinnie neere unto a small village in Northamptonshire called Lichbarrow' (p. 353)—a plant which perplexed botanists for the next hundred years: cf. Ray in Camden, *Britannia*, ed. Gibson, c. 442: De l'Obel apparently saw his specimens and wrote to Jean Bauhin that it was not in his opinion a Gentian but a Saponaria, as G. Bauhin then named it (cf. J. B. *Hist. Plant. Univ.* III, p. 522). He is still more mistaken when he declares of *Saxifraga cotyledon* (or perhaps *S. lingulata*), 'I founde the same growing upon Bieston Castell in Cheshire' (p. 424), where he also claimed to have found *Bupleurum falcatum* (p. 485); that *Soldanella*

1 Son of Ralph Hart of Sproston Court, Yorks, grocer; Lord Mayor of London 1589–90, represented City in Parliament 1593–1601.

alpina 'groweth upon the mountaines of Wales not farre from Coumers Meare in Northwales' (p. 690); and that *Paeonia officinalis* 'groweth wilde upon a conie berrie [rabbit warren] in Betsome [Betsham], being in the parish of Southfleete, two miles from Gravesend, and in the grounde sometimes belonging to a Farmer there called John Bradley' (p. 831).

But his botanical misdeeds are insignificant compared with the final crime by which, as Mrs Arber rightly but sternly insists, his credit is hopelessly shattered—his preposterous tale about the Barnacle tree. He begins with a flare of trumpets: 'Having travelled from the Grasses growing in the bottome of the fenny waters, the woods, and mountaines, even unto Libanus it selfe; and also the sea, and bowels of the same: we are arrived at the end of our Historie, thinking it not impertinent to the conclusion of the same, to end with one of the marvels of this land (we may say of the world)' (p. 1391). Then he tells of the 'Orchades and the trees whereon do grow certain shell fishes...which shels in time of maturitie doe open, and out of them grow those living things which falling into the water do become foules whom we call Barnakles, in the north of England Brant Geese and in Lancashire Tree Geese...thus much by the writings of others'. This is hearsay: he continues, 'But what our eies have seene and hands have touched, we shall declare'—and the Biblical language adds gravity to his intention. 'There is a small island in Lancashire called the Pile of Foulders[1] wherein are found the broken pieces of old and brused ships...wheron is found a certaine spume or froth, that in time breedeth unto certain shels,...wherein is conteined a thing in forme like a lace of silke...one ende whereof is fastned unto the inside of the shell, even as the fish of Oisters and Muskles are; the other ende is made fast unto the belly of a rude masse or lumpe which in time commeth to the shape and forme of a Bird: when it is perfectly formed, the shel gapeth open, and the first thing that appeereth is the foresaid lace...next come the legs of the Birde hanging out; and as it groweth greater it openeth the shell by degrees, till at length it is all come foorth, and hangeth only by the bill; in short space after it commeth to full maturitie, and falleth into the sea where it gathereth feathers and groweth to a foule, bigger than a Mallard, and lesser than a Goose.'[2]

He follows up this tale with another, again 'of mine owne knowledge', that 'betweene Dover and Rumney' he found similar shells on an old

1 Harrison (Holinshed, 1587, p. 38) described it as lying south-east of Walney and as called Fouldra, having on it a pile or castle. It is now called Piel Island.

2 It is presumably from this account that Drayton, *Poly-olbion*, XXVII, ll. 301–11, derives his story of the 'Tree-geese called Barnacles by us' growing on the trees in 'Fournesse'.

rotten tree, took the shells to London, opened them, and found in them living things in all stages from shapeless lumps to 'birds covered with soft downe, the shell half open and the birde ready to fall out' (p. 1392).

It is difficult to treat as a serious witness one who could so solemnly state what was in fact false; and the value of his records must be written down accordingly. Yet he has preserved the evidence of some otherwise unknown students which is worth notice. The most interesting of these is Thomas Hesketh 'a worshipful and learned gentleman of Lancashire, a diligent searcher of simples and fervent lover of plants' who found *Narthecium ossifragum* 'neere unto the towne of Lancaster and unto Maudsley [Mawdesley] and Martom [Martholme where Hesketh was born in 1561] two villages not far from thence' (p. 89); *Tragopogon porrifolius* 'upon the bankes of the river Chalder [Calder] neere unto my ladie Hesketh hir house' (p. 596); *Viola lutea* 'by a village in Lancashire called Latham [Lytham] four miles from Kyrckam [Kirkham]' (p. 704); *Andromeda polifolia* 'in a fielde called Little Reede neere Maudsley' (p. 1110); *Rubus chaemaemorus*, here called Cloud Berries and recorded on a different page from the Knotberries of Penny and De l'Ecluse, 'upon the tops of two high mountaines one in Yorkshire called Ingleborough, the other in Lancashire called Pendle' (p. 1368). He also introduced into London gardens the double wild Crowfoot (*Ranunculus auricomus fl. pl.*) which he found wild 'in Hesketh not far from Latham' (p. 811); gave to Gerard seed of *Silene maritima* which 'groweth by the sea side at a place called Lythan five miles from Wygan' and reported 'of the same kinde some with red flowers which are very rare to be seene' (p. 385—the geography has got mixed—Lathom six miles from Wigan being inland, and Lytham on the coast near Kirkham some twenty miles off); and found a double Mayweed (*Matricaria inodora fl. pl.*) 'in the garden of his Inne at Barnet (if my memory faile me not) at the signe of the red Lion, or else neere unto it in a poore woman's garden, as he was riding into Lancashire' (p. 617).

Two of his friends were at Cambridge, Nicholas Belson a fellow of King's College in 1557–9[1] and then apparently a schoolmaster in Suffolk, who proved that 'distilled water of Daffodils doth cure the palsie' (p. 116), found a fern 'in a gravelly lane in the way leading to Oxey park near Watford' (p. 977) and told Gerard the strange tale of Master Mahewe [Mayhew] of Boston who poisoned himself with an unknown root[2] found in the fens (p. 821); and 'a learned doctor in physicke called Master John

1 Cf. Venn, *Alumni Cantab.* He is presumably the subject of the curious story about Yarrow in *Herball*, p. 915.

2 Gerard regarded this as *Aconitum napellus*: Johnson, *Gerard*, p. 970, suggests *Ranunculus lingua*.

Mershe'[1] who reported *Botrychium lunaria* 'in the bishops fielde at Yorke neere unto Wakefielde in the close where Sir George Savell [Savile, afterwards 1st Baronet, d. 1622] his house standeth, called the Heath Hall' (p. 329). Evidently he had visited Cambridge, for he reports *Onobrychis viciaefolia* 'upon the grassie balkes betweene the landes of corne two miles from Cambridge, neere unto a water mill towardes London' (p. 1064) and gives a charming account of *Anemone pulsatilla*, 'those with purple flowers do growe very plentifully in the pasture or close belonging to the personage house of a small village sixe miles from Cambridge, called Hildersham; the parsons name that lived at the impression heereof was Master Fuller [George Fuller, Fellow of Christ's 1554–61, Rector of Hildersham 1561–91], a very kinde and loving man, and willing to shewe unto any man the saide close, who desired the same' (p. 315). Judging by his references to other places in this area Gerard went to Clare (p. 347) and Sudbury and so to 'Henningham' [Hedingham] and Pedmarsh (e.g. p. 1034), noting especially the great abundance of *Melilotus arvensis* 'overgrowing the earable pasture'; and perhaps on by way of Much Dunmow where he found a variety of *Scolopendrium vulgare* 'in the garden of Mr Cranwich a chirurgeon, who gave me a plant for my garden' (p. 977). Probably at this time he visited Colchester, learnt of the eye-salve prepared by 'a learned Phisition called Master Duke and the like by an excellent apothecarie of the same towne called Master Buckstone'[2] (p. 347), and brought for his garden plants of *Potentilla comarum* 'growing in a marrish ground adjoining to the land called Bourne pondes' (p. 839). He found *Cynoglossum montanum* 'as you ride Colchester high way from Londonward betweene Esterford [Kelvedon] and Wittam' (p. 659) and *Solidago virgaurea* 'in a wood by Rayleigh harde by a gentle man's house called Master Leonard dwelling upon Dawes heath' (p. 349).

At Oxford there was 'a learned Gentleman of S. Johns in the towne of Oxenford, a diligent φιλοβοτανὸν my very good friend called Master Richard Slater' who found *Teucrium scordium* 'in a medowe by Abington called Nietford' (p. 534).

In London there were many—some like James Garret (p. 145), Hugh Morgan (p. 1308) and Richard Garth (p. 757) already known to us as 'expert in the knowledge of plants'; others like William Martin[3] 'barber and chirurgion' (p. 187), John Bennet of Maidstone (p. 315), Thomas Edwards of 'Excester' (pp. 89, 143), or William Gooderus[4] 'the Queenes

1 Cf. Cooper, *Ath. Cantab.* II, p. 242.

2 To whom he presumably gave one of his plants of 'Scammonie' (p. 718).

3 Warden of Barber-Surgeons Company 1599–1601, Master 1605, a close friend of Gerard (*Annals*, pp. 193–4).

4 Cf. above, p. 177, where he is mentioned by Mouffet.

chief chirurgion' (pp. 341, 719), members of his own profession; others like John de Franqueville[1] (p. 307) or Nicholas Leate,[2] in whose company 'walking in the fields next unto the Theater by London' Gerard found a double Buttercup (p. 804), 'worshipful merchants' and his very good friends; others, Richard or Vincent Pointer of Twickenham (pp. 1269 and 1313) whose real name was Corbet and whose epitaph was written by Ben Jonson;[3] Henry Banbury of Touthill Street (p. 1269); Master Gray 'under London Wall' (p. 1308);[4] Master Warner (probably the Thomas Warner whose suburban garden is mentioned in *Stirpium Illustrationes*, ed. W. How, p. 122, and after whom Warner's Rose catalogued by young Tradescant (*Museum*, p. 163) was named) 'neere Horsey Downe (but beware the Bag and Bottle[5])' (p. 1269); Master Fowle, Keeper of the Queen's house at Saint James where melons were ripened (p. 772); and Master Huggens, Keeper of Hampton Court (p. 977), who had gardens for flowers, herbs or fruit-trees. Two references speak of his servant William Marshall 'whom I sent into the Mediterranean sea as chirurgion unto the Hercules of London' and who saw 'divers Plane trees in Lepantae' (p. 1304) and the 'Indian fig' (*Opuntia ficus-indica*) 'in an island called Zante about a day and nights sailing with a meane winde from Petrasse' (p. 1330). He alludes also to Jean Mouton of Tournai (p. 443) known to us through De l'Ecluse, and to Jean Robin of Paris, botanist to the King of France, from whom he got seed of 'Nasturtium indicum' (p. 196).[6] These and many others have their memorial in his pages.

And how charming these pages are! Rogue their author may have been: but when we have ceased to respect him as a botanist or esteem him as a man of honour we cannot fail to enjoy him. How delightful the interludes with which from time to time he brightens what else might become a

1 Cf. De l'Obel, *Adv.* p. 513: he had a garden in London: cf. below, pp. 241–2.

2 He imported many plants from Syria and Turkey, became a leading merchant and citizen, and was responsible for the draining and laying out of Moorfields, cf. Stow's *Annales*, p. 945 (quoted in Furnivall, *Shakespeare's England*, p. 27). He is presumably the 'Captain Nicholas Leet of the Turky Company' whose ship, the *Vineyard*, sailing from Constantinople in 1623 was seized in Sardinia; cf. Howell, *Familiar Letters*, II, pp. 47–8, 52, 62, etc.

3 Of Pointer Plat, *Garden of Eden*, p. 141, records that he 'keepeth conies in his orchard onely to keepe downe the grasse low'. His real name was Vincent Corbet and he was the father of Richard Corbet, Bishop of Oxford, 1628, and Norwich, 1632, the poet and wit. He died in 1619 and Ben Jonson wrote for him the charming epitaph printed in his *Underwoods*, x. His garden and nurseries were noted for their trees.

4 Cf. De l'Obel, *Adv.* p. 413.

5 I wish I could explain this interjection. The only hint is the story on p. 942 of one Cornwale, 'dwelling by the Bagge and Bottle neere London', who swindled the Apothecaries by selling them common Ami as true Wormwood.

6 He describes Robin as 'dwelling in Paris at the signe of the blacke head in the streete called Du bout du Monde' (p. 389).

bare catalogue—the story of the three boys poisoned by Deadly Nightshade at 'Wisbich in the Isle of Ely' (p. 270); the 'ridiculous tales whether of olde wives or some runnagate surgeons or physickmongers, I know not a title bad enough for them' about the Mandrake (pp. 280–1); the delicious description of 'Clounes Alhealle' (Marsh Woundwort, *Stachys palustris*) how 'a very poore man, in mowing of Peason, did cut his leg with the Sieth wherin he made a wound to the bones and withal very large and wide...and crept unto this herbe which he brused in his hands and tied to the wound with a peece of his shirt...and presently went to his daies worke againe', and moreover how, taught by this, Gerard had himself 'cured many greevous wounds and some mortall with the same herbe, one for example done upon a gentleman of Graies Inne in Holburne, Master Edmund Cartwright[1] who was thrust into the lungs...' (p. 852); or again how 'that woorthie Prince of famous memorie Henrie the eight King of England was woont to drinke the distilled water of Broome floures against surfets and diseases thereof arising' (p. 1133), or in different mode the record of the 'firre trees in Cheshire, Staffordshire and Lancashire where they grew in great plentie as is reported before Noah's floud, but then being overturned and overwhelmed have lien since in the mosses and waterie moorish grounds very fresh and sound untill this day' (p. 1181). The rogue like so many of his contemporaries could write.

But we are concerned not with the charm of his writings but with their value as natural history; and beyond the defects already noted there are others.

The man who could appropriate Priest's work as his own was obviously not scrupulous about acknowledgments to others. Of Turner he always speaks with respect; and in referring to his account of the Mandrake (p. 282) and of Herb Paris at Cottingwood (p. 329), to his identification of 'Pliny's Cantabrica' as Caryophyllus (p. 473) and his discovery of *Trinia glauca* (p. 896) he pays tribute of warm praise. So, too, with Penny, in noticing the *Swertia* (p. 352), the *Hypericum* (p. 1105), and the *Cornus suecica* (p. 1113) there is a proper eulogy attached. But just as we have already seen that he included material from Turner unacknowledged, so the account and picture of Butterwort (pp. 644–5) and of 'Knotberries' (pp. 1090–1) are lifted without any hint of obligation. To De l'Obel he refers freely in his earlier pages, but later he seems to go out of his way to give Pena, De l'Obel's unknown colleague in *Adversaria*, the credit for all the discoveries, including, for example, the Bristol plants that Pena certainly never knew (cf. p. 896); and this fact bears out the story of the coldness arising between Gerard and his corrector.

1 Admitted, but without details, 1578 (*Gray's Inn Admission Register*, c. 53).

His quarrel with De l'Obel is corroborated by the fact that mistakes are on the whole more frequent and glaring in the third than in the earlier portions of the book. But this is not to say that any of it is blameless. Many of the mistakes arise from the failure to link up the pictures with the descriptions: this, considering that the pictures came from Frankfort, having appeared in Jacob Dietrich of Bergzabern's *Eicones*, while the descriptions were from the *Pemptades* published by Plantin in Antwerp, was almost inevitable: but it does not justify such blunders as inserting a picture of a Club-moss, *Lycopodium alpinum*, for 'Lavender Cotton' (*Santolina chamaecyparissus*) on p. 948.[1] Moreover, the distinction between different species is wholly arbitrary: five colour-varieties of *Polygala vulgaris* (pp. 449–50) and seven of *Centaurea cyanus* (pp. 592–4) are listed and pictured as separate; and double-flowered specimens are similarly segregated.

Of his mistakes as of his shortcomings in general his 'emaculator' Thomas Johnson has given so generous a sample that we need not illustrate them further. Rather while admitting his faults it may be better to say a word on his behalf. If he copied from others, so did everyone of his contemporaries from Shakespeare downwards; and few of them were as conscientious as De l'Ecluse in acknowledging their borrowings. If he re-arranged Priest's translation on De l'Obel's system, this may have been from genuine desire to improve the arrangement according to his friend's ideas. If he claimed to have found plants unknown to our flora, such ignorance is not necessarily criminal. If he thought he saw young goslings in barnacles, it is familiar to all of us that even scientists see what they expect to see. A man whose zeal outran his abilities, who had never learnt to appreciate the need for accuracy or a high standard, who was provided by his publisher with the raw materials for a book, a manuscript and illustrations, and whose enthusiasm for gardening and confidence in his own powers spurred him on to a task beyond his strength—that is, I think, his defence; and considering how much his work has contributed to our national love of flowers we need not be too hard upon him. It is no small testimony to his achievement that within a few years of his death Michael Drayton could finish the long passage descriptive of herbs and their effects (*Poly-olbion*, XIII, ll. 195–230) with the lines:

> Of these most helpful herbs yet tell we but a few,
> To those unnumbred sorts of simples here that grew.
> Which justly to set down, even Dodon short doth fall;
> Nor skilful Gerard yet shall ever find them all.

1 This blunder is noted by Parkinson, *Theatrum Bot.* p. 97. J. Bauhin, *Hist. Plant. Univ.* II, p. 159, notes a similar error, the picture of a convolvulus under the title of 'small yellow horned Poppie': cf. Johnson, *Gerard*, p. 368, and Gerard, *Herball*, p. 294.

Nor was Drayton the only poet to be influenced. John Milton in the famous passage, *Paradise Lost*, IX, ll. 1101–12, in which he describes the tree from which Adam and Eve got their fig-leaves, borrows his account of it from Gerard's chapter (pp. 1330–1) on the 'arched Indian Fig-tree' the Banyan (*Ficus bengalensis*), several lines being almost verbally reproduced from his prose.

CHAPTER XIII. EDWARD TOPSELL

Few illustrations reveal the difference in progress between botany and zoology at this period more clearly than the comparison between Lyte's or Gerard's Herbals and the 'Histories' of Edward Topsell, the earlier, *Of Foure-footed Beastes*, published in London by William Jaggard in 1607 and the second, *Of Serpents*, a year later. Medical and horticultural interests had compelled the popularisers of plants to keep their fancy in check, to describe real species and to study their habit of growth and curative properties. In the description of animals and reptiles there was no such need for accuracy. Aristotle was indeed more reliable about them than Pliny and Dioscorides were about plants: but since Aristotle a mass of fable and legend had been fastened upon them; and travellers' tales added contemporary but hardly less extravagant details. The beginning that Wotton and Caius in their widely different contributions had made for the study of animal life had not been followed; and, except for Caius's book on Dogs, neither of them had exerted much influence upon English readers.[1]

Nevertheless, Topsell's work was certainly encouraged and perhaps even suggested by the famous translation of Pliny by Philemon Holland. The son of a Marian exile, educated at Chelmsford and Trinity College, Cambridge, where he matriculated three years before Mouffet and was a fellow until his marriage in 1579, Holland did not begin his great series of translations until after he had settled in Coventry in 1595 and taken his M.D. at Cambridge in 1597.[2] In 1600 he published his *Romane Historie*, a version of Livy with some additional material; and in 1601 *The Historie of the World commonly called the Naturall Historie of C. Plinius Secundus*, in two volumes, dedicated to Sir Robert Cecil and the most successful of all his works, a new edition of it appearing in 1634. His gifts as a translator, his medical training and his temperament combine to give

1 Wotton was largely quoted by Aldrovandi; and Caius, as we have seen, had been publicised by Gesner: but this represents a very indirect influence upon England.

2 So *Trinity College Admissions*, II, p. 65.

quality to the work. It took the place which until then had been filled by Bartholomaeus Anglicus, and gave Englishmen a readable account of animals and plants in their own tongue.

On the whole the book is well done. Based apparently upon Jacques D'Aléchamps's edition published at Lyons in 1587 it reproduces though without acknowledgment some of the notes on plants in that volume,[1] and its references to Dodoens and Mattioli.[2] Holland once refers (in II, p. 62) to Jean Ruel, *De Natura Stirpium*, p. 703: his note seems to be original but to be based on a misunderstanding. He was evidently more interested in botany than in zoology, no doubt as a result of his medical training: but there is no sign of expert knowledge.

That the great 'Omni-gatherum', as Turner had called it, no longer received the reverence which the men of the Renaissance had shown to it, is indicated by the letter from 'a grave and learned preacher' 'H.F.' attached to Holland's own preface, and defending him against the charge of popularising the work of a heathen. Pliny is not quite the authority to Holland that he had been to Wotton. Nevertheless, there was still need to appeal from the fables of the Middle Ages to the relative accuracy and objectivity of the Classics. The *Naturall Historie* would at least reveal the absurdity of the *Bestiaries* and create a demand for more exact records. So Holland prepared the way for Topsell; and Pliny led on to Gesner.

To say so is not to imply that the later work is less fabulous or imaginative than the older: indeed by comparison Holland is temperate and truthful. Rather it is to suggest that a real interest in nature was being rapidly developed; that this encouraged the production of books; and that these, however quaint their contents, were meant to be serious histories derived from the most learned and reliable of recent scholars. There was a public able to enjoy and ready to pay for the sort of books which Topsell set himself to produce. His work has been made accessible in the well-selected and beautifully printed volume called *The Elizabethan Zoo* (Cambridge, 1926); but students of his sources will know that he must not be blamed too severely for credulity. He was in fact a man of very little originality, and reproduced what his authorities gave to him.

Of his life Miss M. St Clare Byrne, in the introduction to the volume already mentioned, has gathered a few more details not found in previous accounts. Speaking of 'Sevenoke' in the dedication of his first book, *The Reward of Religion*, London, 1597, to the 'Lady Margaret Baoronnesse Dacres of the South', he says, 'I had my beeing where your honour

1 For example, D'Aléchamps's peculiar statement on p. 452 that he thinks the Calathian Violet of Pliny (usually identified as *Gentiana pneumonanthe*) to be the yellow Foxglove: cf. Holland, II, p. 85.

2 D'Aléchamps, p. 454; Holland, II, p. 90.

hath your dwelling'—the lady, who was sister of Gregory Fiennes Lord Dacre, inherited from him on his death in 1594 and was declared Baroness Dacre in 1604, had married Sampson Lennard of Chevening, Kent. Topsell was born in 1572, educated locally, and entered at Christ's College, Cambridge, in 1587 where he graduated four years later. He was ordained and held in succession a number of livings, Framfield and East Hoathly in Sussex, Datchworth in Herts, Syresham in Northants, Hartfield and East Grinstead in Sussex and finally Little Bytham in Lincolnshire. From 1604 he was perpetual curate of St Botolph's, Aldersgate, where he was buried in 1625.

Such a career suggests that he was not a man of high distinction, intellectual or practical. But much of his time seems to have been given to authorship. *The Reward of Religion delivered in sundry lectures upon the Booke of Ruth* was followed in 1599 by two similar volumes, *A Commentary upon the Lamentations of Jeremy* and *Times Lamentation or an Exposition on the prophet Joel*, the latter dedicated to Sir Charles Blunt (Blount), Lord Mountjoy, who is described as 'the meane of my preferment'. These were followed by a volume begun by Henry Holland, Preacher at St Bride's, *The Historie of Adam*, edited and completed on Holland's death by Topsell with a dedication to Richard Neile, Dean of Westminster. In 1610 he published *The Householder or Perfect Man* and in 1612 a preface to Charles Richardson's *The Repentance of Peter and Judas*. All these are similar volumes, expository comments upon Scripture; and their intention is defined in an interesting preface to the *Commentary*. This, after stating briefly the various modes of preaching, explains the method by which the text is read and interpreted and then its meaning and lessons are expounded and illustrated. Topsell is a careful and on the whole sober exegete: his lessons, though usually superficial and sometimes far-fetched, are free from the fantastic allegorisings and speculations too often in vogue, and show a mind well-stored with anecdote and quotation, if lacking in imagination or distinction or any deep insight. He is mainly concerned to condemn 'atheisme, paganisme and papisme'—particularly the last-named—and to inculcate a solid and pedestrian virtue.

Such a man was not ill-fitted for the task by which alone he is remembered. A phrase in the Epistle Dedicatory (addressed once more to Neile of Westminster) sums up his purpose: 'Surely when Salomon saith to the sluggard go to the Pismire, he willeth him to learne the nature of the Pismire.' His concern is to teach the nature of animals and serpents as understood and described by the ancients and now made the subject of his vast *Historia Animalium* by Gesner.

This does not of course involve any such interest in the animals themselves or any such objective study of their structure and habits as a modern biologist would take for granted. To Topsell, as to his contemporaries, the relatively scientific approach of Aristotle, who was genuinely curious as to the processes of nature and a remarkable observer and recorder of fact, had been distorted both by the credulity and anthropomorphism of the later Graeco-Roman writers and by the prevalent conviction of the Middle Ages that the animal world existed only in order to supply rewards and punishments, examples and warnings to mankind. The large and growing collection of Aesop's Fables, the edifying anecdotes of the Bestiaries, the traditional tales carved on so many miserere seats and canopies, had made lion and fox, stork and goose and many other creatures symbolic and wholly familiar. As agents of God or embodiments of the devil they had played a large part in the preaching, the art, the literature of medieval Europe; and if, as Topsell's work shows, some remnants of genuine knowledge as to their anatomy and physiology still survived and other details were rapidly being gathered, the old attitude towards them still persisted, and even the newly acquired facts of observation were liable to be twisted into conformity with legendary ideas as to the particular beast's temperament or usage. The lion and the eagle are always brave and benevolent—royal in their behaviour to their subjects; the elephant is wise and chaste; the goat passionate and lecherous; the fox, the wolf, the tiger, the weasel, each of them is credited with a typical quality, and all records of them are interpreted in terms appropriate thereto. The same applies to birds, where from the days of the auspices a lore of symbolism had been built up, and characters attached to the commoner or more noticeable species. Even insects had their place, and the industry of the ant, the loyalty of the bee, the rakishness of the wasp are still proverbial. Topsell, preacher and commentator, could find in the literature of the animal kingdom exactly what he had found in Scripture and the Classics, a mass of material well-suited for the amusement and edification of his readers.

It is important to grasp the character of the man—conscientious, unimaginative, commonplace, fond of his tags from the Classics, his moral anecdotes, his pleasant turns of phrase. Otherwise we shall mistake and misrepresent the character of his book. It is almost impossible for us to see his quaint pictures of 'satyre' and gorgon and lamia, of 'boas' and flying dragon, without assuming that he like ourselves knew these creatures to be fictitious or at least included them merely to delight his readers with their strangeness. No doubt men were beginning to be sceptical about the centaurs and gryphons and sphinxes of mythology: but it is absurd and unfair to suppose that Topsell was not serious in his selection of material or that he felt any greater hesitation about accepting the Eale or the Mantichora or the Unicorn than the Antelope or the Hyaena or the Su. The contrast which we so easily observe between the pictures of fabulous creatures and those like Dürer's Hare or even Rhinoceros[1] or Caius's Ounce was not yet recognisable. The greater contrast between the plain prose of the chapters on Cats and Horses and the highly coloured fantasias on the 'Camelopardal', the Elephant and the Tiger was to him merely a matter of geography. He is not, indeed, devoid of critical capacity: but like his authorities he has no fixed idea of natural law or probability, and little aptitude for weighing evidence.

Once this is understood the book and its sources can be considered more closely.

Topsell himself modestly ascribes to Gesner the whole credit for the collecting and marshalling of the material; and it is therefore easy for the editor of *The Elizabethan Zoo* to treat his book as a mere translation. A few minutes spent in comparing it with the *Historia Animalium* will show how inexact is such a description. Not only is there a fair amount of material derived from other sources—not, be it noted, those so impressively listed by Topsell as his authorities, all but three of which are copied *verbatim* from Gesner and were quite unknown to our country parson—but the actual text of Gesner is treated with freedom, paraphrased, re-arranged, condensed, expanded in a way only possible in an age when plagiarism was universal and copyright unknown.

Topsell in fact set himself to produce books about beasts. He had probably never seen anything more romantic than the Lions in the Tower of London or the Fallow Deer in the park at Knole. He was not an observer or a naturalist, except at second hand and in books. But in Gesner there was a vast store of ancient wisdom, of tradition and fable, description and anecdote, the science of Aristotle and the legends of the Middle Ages. Gesner had drawn upon all the available sources of information, old and new, and was himself a very considerable student and naturalist: Topsell printed his list of authorities and, apparently, added two or three otherwise unknown names to it. But he used other works: John Caius on Dogs, the script of which was written for Gesner but sent to him too late and which Topsell prints as 'translated by A.F.' (Abraham Fleming[2]) (*Historie*, pp. 164–81); Thomas Blundeville on *The Art of Riding and Breakinge great Horses*, published in London in the 1570's and Gervase Markham whose *Discourse on Horsemanshippe* had appeared in 1593—and from these two Topsell drew most of his long section

1 This woodcut, W. Kurth, *Complete Woodcuts of A. Dürer*, Plate 299, was designed not from life but from a sketch sent to him in 1513 by a friend. The Rhinoceros, the first seen in Europe, was sent from India to the King of Portugal, and drawn near Lisbon. It was afterwards drowned on a voyage to the Pope: cf. W. M. Conway, *Literary Remains of A. Dürer*, p. 144. Another apparently reached Lisbon in 1535: so Aldrovandi, *De Quadrup.* p. 191.

2 His version was published in 1576 by Richard Johns.

(pp. 340–431) on diseases of the horse; Thomas Bonham, the doctor,[1] who had apparently written something on Worms and whose prescriptions were published posthumously in 1630 by his assistant Edward Poeton as *The Chirurgeon's Closet*; Thomas Tusser, the Hesiod of Elizabethan days, whose poem *Five hundred Pointes of good Husbandrie*[2] gave him quotations on Sheep (pp. 624–5; cf. Tusser, pp. 48*b*, 49, 15, 45) and on Swine (pp. 681–2; cf. Tusser, pp. 18, 22, 23); and 'Gulielmus Camdenus Clarentius', the only one besides Caius and Bonham to appear on the list of authorities, whose *Britannia* he quotes for its evidence on the English wool-trade (p. 625).

It is in fact only in relation to sheep that he shows any sign of breaking away from his authorities. In the paragraph on p. 624 which begins, 'Now concerning the manner of our English nation and the customs observed by us about this businesse...I cannot containe myself from relating the same, considering that we differ from other nations', he gives a pleasant and so far as can be judged an independent account of the washing and shearing of the sheep, and of the quantity of the wool—'in the least a pound, in the middle sort of sheepe two pounds or three pounds, as in the vulgar in Buckingham, Northampton and Leicester shieres, but the greatest of all in some of those places and also in Rumney marsh in Kent foure or five pounds...and I have credibly heard of a Sheepe in Buckinghamshiere in the flocke of the L.P. that had shorne from it at one time one and twentie pound of wooll'. So he runs on to Tusser and Camden.

Otherwise, except for one sentence about Lions having bred in captivity in England (p. 459) and another about a She-Wolf mated to a Mastiff in the Tower of London in 1605 (p. 735), Topsell adds hardly anything of his own. Even on matters dealing with Scotland he not only draws solely from Gesner but draws ignorantly. Thus in his account of the Norwegian monsters reported by the ambassadors of James IV the Scottish king, he speaks of 'James Ogill that famous scholer of the University of Abberdon' (p. 16) which is not a very literate version of 'Jacobus Ogilvius Aberdoniensis' (*H.A.* 1, p. 866).[3] In the paragraph devoted to the 'White Scotian Bison'—our old friend the Chillingham Bull (*Bos primigenius*)—

1 Mentioned also as a source by Parkinson, *Theat. Bot.* (see below, p. 255). A remarkable man: St John's, Cambridge, B.A. 1584–5; M.A. 1588; and apparently M.D.: incorporated M.D. Oxford, 1611. In 1604 he championed the Barber-Surgeons in their petition to be allowed to give 'inward remedies' to their patients. Soon after he was examined by the College of Physicians, and eventually after long proceedings on refusing to obey them he was imprisoned. In 1609 he brought an action for false imprisonment and was successful. A full record of the case, with Coke's verdict and other documents, is in C. Goodall, *Royal College of Physicians and Historical Account of Proceedings against Empiricks* (London, 1684), pp. 359–63 and 164–220.

2 Refs. are to ed. of 1580 published by Henry Denham.

3 Apparently James Ogilvie first 'Civilist' of the University: cf. P. J. Anderson, *Officers and Graduates of Aberdeen*, p. 31.

his record not only adds nothing to Gesner's but in its version of the place names 'in the woods called Callendar or Calder which reacheth from Monteth and Erunall unto Atholia and Loquhabria there are bred white oxen...this wood was once full of them but they are now all slaine except in that part which is called Cummirnald [Cumbernauld]' is less intelligible; for Erunall is in Gesner, *H.A.* p. 130, Ernevallem or Strathearn, and Loquhabria is Loquabria or Lochaber—Monteth and Atholia are of course Menteith and Atholl. Gesner specifies what Topsell omits, that this account is from Hector Boece's *Description of the Kingdom of Scotland*. It is found in Holinshed's edition on p. 13.

As with the text so with the pictures. Topsell's version is a poor substitute for Gesner. The cuts are redrawn and in general much depraved, being coarsely outlined and crudely shaded. In a few of the simpler pictures there is little loss: in others as in the Marmoset an attractive likeness has become grotesque. In at least one case Topsell has blundered badly: for he inserts into his section on the Otter on p. 573, along with a tolerable copy of Gesner's picture, a copy of the fantastic cut of the Scythian Wolf which Gesner had taken from Olaus Magnus (*H.A.* p. 683): although 'Lupus Scythicus' comes next to 'Lutra' in the *Historia Animalium* it is difficult to believe that anyone who had the slightest appreciation of animals could have thought this creature an Otter.

Topsell's second book, *The Historie of Serpents*, published in 1608 and reprinted along with the former volume and with a translation of Mouffet's *Theatrum Insectorum* by John Rowland in 1658, raises several questions much more difficult to answer than those suggested by its predecessor. Much of it is borrowed as before from Gesner; and these portions follow the method already discussed. But in addition there are a number of sections dealing with insects—Bees, Wasps, Hornets, Caterpillars—and those on Spiders and Earthworms, which are not in the *Historia Animalium*: and these show very evident relationship partly with Wotton's *De Differentiis* and more frequently with the *Theatrum*. With Wotton several of the resemblances (e.g. the Latin lines on Spiders, *De Diff.* p. 186) are found also in the *Theatrum*; and Topsell may have borrowed them from either. But at least one passage, the three lines from Ovid in *De Diff.* p. 193, though copied exactly in *Historie of Serpents*, is not in Mouffet. This indicates, even if it does not prove, that Topsell had looked at Wotton's book. But the similarities between the *Historie* and the *Theatrum* are not only far more numerous but are demonstrably due to deliberate copying not of our printed version but of the earlier form of the script.[1] Not only do they include long passages, indeed whole sections in which, as for example in the record of the Fox and the Wasps or of the Stinking Caterpillar, the Classics and Gesner are the immediate

1 Cf. above, pp. 180–1, for proof that Topsell copied the first version of Mouffet.

sources, but they reproduce incidents personal to Penny or Mouffet in terms so similar as to make plagiarism certain.

Thus, for example, of Wasps Topsell writes: 'These kinde I did once see in a Wood in Essex where going unwarily to gather simples with another physitian and offending one of this fumish generation, the whole swarme of them presently rushed forth about mine eares, and surely had I not had in my hand some sprigs or branches of broome for my defence I had undoubtedly payde deerely for my unadvisednes' (*Hist. of Serpents*, p. 87), which is an exact rendering of the story told of Penny and Mouffet in *Theatrum*, p. 45.[1] Still more obvious is the passage in *Serp.* p. 94, 'I my selfe beeing at Duckworth in Huntingtonshire, my native soyle, I saw on a time a great Waspe or Hornet making after and fiercely pursuing a Sparrow in the open street of the Towne who at length beeing wounded with her sting, was presently cast to the ground, the Hornet satisfieng herselfe with the sucked bloud of her quelled prey, to the exceeding admiration of al the beholders and considerers of this seldome seene combate': this is an exact rendering of the incident in *Theatrum*, p. 51, which Penny saw in Peterborough.[2]

Such copying is not only from Penny's observations. Between his chapters on Bees, Wasps and Hornets and on Caterpillars Topsell has four and a half pages (pp. 96–102) on Cantharides. These contain all the material in Chapter xx of the *Theatrum* (pp. 144–7) except the case which Mouffet reported from Basel in 1579, and for this a longer but somewhat similar story is quoted (pp. 98–9) from Johann Lange, 'in his first booke *Epistola Medicinal.* forty eyght'; and this material Mouffet expressly states to have been omitted by Gesner and Penny and supplied by himself. This seems to involve the conclusion that between the compilation of the *Theatrum* and the publication of the *Historie* and probably after Mouffet's death in 1604 Topsell obtained access to the manuscript in its earlier form or to the version edited and perhaps printed by Scholz at Frankfort in 1598, and filled out his version of Gesner's second volume by inserting these chapters. That he claimed Penny's records as his own even to the extent of altering Peterborough into 'my native soyl, Duckworth' (Duckworth being non-existent unless a misprint for Buckworth, and his own soil being Sevenoaks) does not increase respect for his veracity.

So, too, in the chapter on Spiders (*Historie of Serpents*, pp. 246–76) the whole is a paraphrase of Mouffet, copied from the actual text of the *Theatrum* though omitting all references to Penny or Mouffet by name. He omits the story from Brewer with which the account of Spiders in *Theatrum* concludes (p. 237) and replaces it by a reference to 'one Henry Lilgrave living not many yeeres since, beeing Clarke of the Kitchen to the right noble Ambrose Dudley Earle of Warwicke, who would search every corner for

1 Cf. above, pp. 176–7.

2 Cf. above, p. 167.

Spyders, and if a man had brought him thirtie or fortie at one time he would have eaten them all up very greedily'. But he includes and translates the long passage from 'Edouardus Monimius' quoting it as from *Heptam. Lib.* VII (pp. 268–9), and in the long fable of the Lady Podagra quotes twenty-three lines from Chaucer's description of the Franklin (p. 263).

The final chapter 'Of Earthworms' has in the margin a reference to 'Doctor Bonhams discourse of Wormes', but is nevertheless similarly derived from *Theatrum*, pp. 278–82, though much expanded by quotations from Lucretius and Homer and a considerable section from Aristotle. It also contains a paragraph that seems to be original. 'At Harlestone [four miles N.W. of Northampton] a myle from Holdenbie in Northamptonshiere there was a quarry of free stone found out, of which they digged for the building of Sir Christopher Hattons house, where there was taken up one beeing a yard and a halfe square...and being cloven asunder there was found in the very midst of it a great Toade alive, but within a very short space after, comming to the open ayre, it dyed....This stone was cut asunder by one whose name is Lole, an old man yet living at this day, it was seene of five hundred persons Gentlemen and others, the most part of them living at this howre' (pp. 314–15).

This record, with the one about Lilgrave, stands almost alone in the *Historie*. In these sections only two other passages seem to be due to Topsell. The first of these is in the chapter on Wasps (p. 90): 'I will now set before your eyes and ears one late and memorable example of the danger that is in Wasps, of one Allens wife, dwelling not many years since at Lowick in Northamptonshire [two miles N.W. of Thrapston], which poor woman resorting...to Drayton the Lord Mordants house,[1] being extreamly thirsty, finding by chance a black Jack or Tankard she set it to her mouth never suspecting what might be in it, and suddenly a Wasp in her greedinesse passed down with the drink and stinging her....' The second is of Caterpillars (p. 111) after reference to 'a caterpillared hook' in fishing, 'Which kind of fishing fraude, if you would better be instructed in, I must refer you to a little booke dedicated to Robert Dudley, late Earle of Leicester, written by Ma: Samuell Vicar, of Godmanchester in Huntingtonshiere'.

In the rest of the book there are but few similar passages. One on p. 191 deals with a Toad: 'I have heard this credible History related from the mouth of a true honourable man, and one of the most charitable Peeres of England, namely, the good Earle of Bedford,[2] and I was requested

1 Presumably Lewis third Lord Mordaunt who succeeded to Drayton Hall in 1571 and died there in 1601.

2 Presumably Edward Russell third Earl who succeeded in 1585 and lived until 1627, but who neglected his Woburn property: cf. G. Scott Thomson, *Life in a noble Household*, pp. 19–21.

to set it downe for truth, for it may be justified by manie now alive which saw the same. It fortuned as the said Earle travailed in Bedfordshire, neere unto a market-towne called Owbourne [? Woburn] some of his company espyed a Toade fighting with a Spyder...and the Toade divers times went backe from the Spyder and did eate a peece of an herbe which to his judgement was like a Plantine....The Earle having seen the Toade doe it often, commaunded one of his men with his dagger to cutte off that herbe....Presently after the Toade returned to seeke it, and not finding it according to her expectation swelled and broke in peeces: for having received poyson from the Spyder in the combat, nature taught her the vertue of that herbe, to expell and drive it out.' On p. 213 of the Newt: 'Some say that if in Fraunce a Hogge doe eate one of these he dyeth thereof....But in England it is otherwise, for I have seene a Hogge without all harme carry in his mouth a Newte and afterward eate it.'

For the rest the book is drawn almost entirely from the sources already familiar. The chapter on the Scorpion, for which Gesner's posthumous volume v was available, has followed it in preference to the *Theatrum*: there is some close resemblance and overlapping; but this is due to common dependence upon Gesner, not to copying of Mouffet. It is worth noting that though both Topsell and Mouffet mention Wolf of Zürich, Topsell tells of how he kept a Scorpion in a box at Montpellier and fed it on Flies and how when put in the sun in a glass it died (*Serp.* p. 226), while Mouffet relates on Penny's authority that he shut up a Scorpion with a Viper and each killed the other[1] (*Theat.* p. 206). But in the main the book is uninteresting. It is much less rich in memorable anecdotes and phrases than its predecessor; and its dulness though mainly due to the character of the subject-matter is partly caused by a lack of freshness in the author. He is now a man doing a bit of hack-work, and not doing it very well.

Nevertheless, the incidents recorded on his own responsibility give this second volume a special value. That they are almost all located in Northants or Huntingdon is not surprising, for from 1602 till 1608—that is during the whole time when this book was being written—he was living at Syresham on the main road between Brackley and Towcester; his son Abel was born there; and it was the longest settled period of his working life. That he dated both volumes from St Botolph's, Aldersgate was no doubt proof that he divided his time between town and country, and due to his dedication of them to Neile of Westminster. But the trivial tales which he gathered from local gossip indicate that he was not wholly without a living as opposed to a literary interest in the subject. And if he was himself no naturalist, his books had a share in fostering the love of nature in others.

1 The conflict of Viper and Scorpion is traditional: Mouffet quotes Aelian on it; and C. Singer, *Journal of Hellenic Studies*, XVII, Pl. iii and p. 36, gives two pictures of it from MSS.

D. THE EXPLORERS

CHAPTER XIV. THE DAWN OF THE 'NEW PHILOSOPHY'

The reign of Elizabeth had promoted (though it had not achieved) a change of outlook in Britain as momentous as the change in the status of the country. The medieval tradition with its magnificent superstructure of Christian feudalism resting upon fundamental beliefs in angels and devils, elements and principles, legendary saints and fabulous beasts was being undermined as the world of theological and ecclesiastical dogma, of alchemy and the bestiaries was dispossessed first by the knowledge of Biblical and Classical literature and then by the observation and study of nature. Except in the specific field of Church-reform the assaults upon the old way of life were almost unconscious. Men did not set themselves deliberately to demolish the mythology of Bartholomew or the pharmacology of the *Ortus Sanitatis*: they merely replaced them by more accurate knowledge of plants and animals and of the treatment of disease. Little by little nonsense was recognised, fables were exploded, superstitions were unmasked, and the world-outlook built up out of these elements fell to pieces. The seemingly irrelevant labours of men like Turner or Penny to identify and name and describe bore fruit in a refusal to accept tradition on authority and in an insistence that statements must be based upon observation and capable of verification.

So for a time there grew up an unrecognised conflict between the old and the new, a conflict which found expression in outbursts of seemingly uncaused violence before it was given precision by the writings of Francis Bacon and fought out in the life and teaching of the schools. Men still professed to accept the Weltanschauung of the Middle Ages even if the Pope had been replaced by the King or Rome by Geneva. But the fundamental assumptions and tenets of those ages had lost their reality and relevance: they no longer satisfied the intelligence or explained the problems of mankind: and there was in consequence a deep if subconscious resentment against them, a resentment which might blaze up into ferocity at any moment.

This conflict between the old and the new is an important element in the horrible and sustained outbreaks against witchcraft during the first half of the seventeenth century; and though it would be a mistake to over-estimate its importance or to ignore other causes and motives, it deserves and has not yet received due recognition.[1] The sufferings of the ill-fated victims who were sought out, tortured, and done to death by John Darrel or Matthew Hopkins were not due solely to the Christian zeal to obey the text forbidding a witch to live (Exodus xxii, 18) nor to the recognition that witchcraft was a survival of paganism. Rather they were the price paid by mankind for its error in maintaining a cosmology which though traditional was now plainly erroneous, a supernaturalism which explained whatever it did not understand by reference to magic, and a demonology which had been preached and exploited for nearly a thousand years. For the Elizabethans the climate of medievalism was not yet uncongenial; apart from a few outbreaks in East Anglia magic was accepted without revulsion: indeed, in the contemporary landscape Prospero and his Ariel or their living counterparts Dr John Dee and his Anael were not out of place; witches were not confined to blasted heaths in the wilds of Scotland; and as we have seen Satyrs and Gorgons might well be found along with Caliban in the Zoo. But their successors knew enough to recognise that all this wizardry was out of keeping with the sanities of life and with the reign of law which they were beginning to disclose: they did not yet know enough to explain the phenomena associated with magic, and could not therefore dismiss the traditional explanation with flat denial and purgative laughter. They did not believe, but they could not refute: the only course was to destroy. It was not until the study of nature supplied a concept of the sequence of cause and effect, and a framework of reference verified by experiment, that the old irrational fears and folk-lores, the old underworld of pre-Christian and sub-Christian devil-worship, were brought to light and cleansed. It is by a strange irony that the insistence on the reality of witchcraft often came from those whose piety and intelligence should surely have ridiculed any such belief.[2] They clung to this distorted remnant of the old supernaturalism in fear of losing their hold upon Scriptural authority and medieval tradition. But Henry More, Joseph Glanvill and Sir Thomas Browne could not long delay a whole-hearted acceptance of the New Philosophy. For John Ray and the younger members of the Royal Society magic was an unnecessary hypothesis: with them the new age of science had been fully inaugurated.

That the change of outlook from the old to the new in regard to zoology came so slowly is largely due to the influence of the sports of hunting and falconry. As we have seen, medieval interest in nature, where it was not purely artificial and mythological, was confined to the art and technique

1 Thus, e.g., W. Notestein, *A History of Witchcraft in England*, represents witchcraft itself as if it were a new and sudden development, and speaks of the superstition gaining ground when his evidence deals only with the growth of opposition to it.

2 This is not to ignore the fact that Reginald Scot, Samuel Harsnett and other critics of the reality of witchcraft down to Francis Hutchinson were deeply religious.

of hawk and hound and horse. Here was created a cultus with its proper ritual of action and language, into which every gentleman was duly initiated and in which the whole countryside took a more than religious interest. Around the hart and the heron arose a wealth of lore and legend, handed on from generation to generation and constituting a tradition as rigid and as exacting as the liturgy. The successful pursuit of sport demanded a nice knowledge not only of the habits and uses of the various dogs and birds employed in it, of their care and doctoring, their food and tempers, but of the weather and soil, the lie of the land and the signs of game, the behaviour of the hunted and the place and manner of the kill. It would seem that this must involve an interest in wild life and an outlook favourable to the growth of science. But in fact the whole business was so overlaid with tradition and so conservative in its insistence upon rules that often bore little resemblance to the principles from which they had been developed, that instead of promoting natural history and the scientific outlook it was as much a hindrance as a help. In fostering a love of the open air, in encouraging a real if strictly limited interest in the birds and beasts peculiar to the chase, and in making the breeding and management of horses and dogs a matter of high social and financial importance, it was of value. But in maintaining the traditional superstitions, in restricting interest in nature to the technique of killing, and in absorbing in relatively worthless pursuits a vast amount of human energy and skill, the loss was perhaps as great as the gain.

The Elizabethans, as Shakespeare's language and their own daily activities testify, maintained the traditional sports, hunting the deer and coursing the hare, 'astringing' in the woodland and 'flying at the brook', in all their medieval splendour. A book like D. H. Madden's *The Diary of Master William Silence*,[1] or the contemporary literature, John Caius on dogs, Gervase Markham and Thomas Blundeville on horses and horsemanship, which we have already named, George Turberville, *The Booke of Faulconrie*, 1575, Symon Latham, *Latham's Faulconry*, 1615, and Edmund Bert, *An Approved Treatise of Hawkes and Hawking*, 1619, shows how deep and general was their devotion. With the traditional usages and vocabulary of sport, those nice customs which are second nature to those brought up in their observation and in which the parvenu so speedily betrays himself, the old attitude towards nature, the folklore and fable of the Middle Ages, was inextricably entangled. There could be little scientific progress while this interest dominated society.[2]

1 Longmans, 1897. Admirable in its knowledge of Elizabethan sport and literature, less so in its natural history—as witness the statement about Goshawks in Ireland l.c. p. 158)—although he could have quoted *Fynes Moryson's Itinerary*, IV, p. 194: 'The hawkes of Ireland called Gosshawkes.'

2 It is notable that in order to sell the English edition of Willughby's *Ornithology* in 1678 John Ray was urged to include in it a treatise on Falconry. The result, abridged from Turberville and Latham, well illustrates the jargon and traditions of the sport.

Hart and heron, hunted with all the panoply of tradition, supplied sport only for the wealthy; and by the end of the century jests were being directed at the contrast between the cost of a falcon's upkeep and the worthlessness of its prey.[1] But with the breakdown of feudalism the smaller gentry and yeomen in the country and the merchants and apprentices in the towns began to develop sporting interests of their own. The spearing of the Otter, the drawing of the Badger, the coursing of the Hare and above all the hunting of the Fox began to acquire a ritual and a following. Wild-fowling with guns and trapping with nets and bird-lime ('The mistle thrush excretes its own death' said the proverb quoted by William Turner[2]) began to take the place of hawking.[3] A new era of game-laws and a new squirearchy came into existence.

Fishing maintained its importance and *The Booke of Fishing with Hooke and Line made by L.M.*, published in London in 1590 and no doubt written by Leonard Mascall of Plumstead,[4] clerk of the kitchen to Archbishop Matthew Parker, preserves the lore of the *Treatyse*. It is in the main a free rendering of the earlier—and more attractive—work: but it adds the claim about the Carp that 'the first bringer of them into England (as I have beene credibly informed) was Maister Mascoll of Plumsted in Sussex [presumably an ancestor of the author] who also brought first the planting of the Pippin: but now many places are replenished with Carpes, both in poundes and rivers'.[5] As he goes on to repeat the *Treatyse's* statement of ignorance as to baits, and as in other cases his accuracy is conspicuously unreliable, there seems little need to take the claim seriously.[6] He also adds to the list of fishes the Loach (*Cobitis barbatula*) and the Cull or Miller's thumb (*Cottus gobio*); then certain sections on the catching of Herons, Otters, Water-rats (which he regards as destructive of fish and spawn, and would poison with 'powder of Orsonike') and Sea-pies (presumably *Haematopus ostralegus*—to be taken

1 Cf. the quotations from J. Harrington, *Metamorphosis of Ajax*, 1596, and from *Pasquil's Jestes*, 1609, cited by J. Haslewood, Introd. to *Boke of St Albans*, p. 31.

2 *Herbal*, II, p. 165.

3 The Le Strange accounts (*Archaeologia*, XXV) illustrate the beginning of the change. At Christmas 1526 we have '8 malards, a bustard and 1 hernsewe [heron] kylled wt the crosbowe' (p. 487); in 1533 'a watter hen, a crane, a wydgyn kylled wt the gonne' (p. 530) and 'twyn for the ptriche net' (p. 542)—these alongside of 'larks kylled wt the hobbye' and 'fesands kylled wt the hawke' (pp. 527, 529). For pictures of these nets in use and a condensed account of wild-fowling and the game-laws, cf. Ray, *Ornithology*, pp. 29–53.

4 The subject of a pleasant book by W. A. Woodward, *The Countryman's Jewel*, London, 1934.

5 Harrison in Holinshed, 1587, p. 46, in the passage in which he describes the abundance of 'fat and sweet salmons' in the Thames says that it has few carp since 'that kinde of fish was brought over into England and but of late into this stream'.

6 Cf. J. W. Hills, *History of Fly-fishing*, p. 39, 'Mascall in such parts as he pirated is so careless that often he does not trouble to see that what he writes makes sense.'

with limed twigs fastened to a bait); then some dozen sections dealing with the breeding and preserving of fish which T. Satchell in the preface to his edition (London, 1884) identifies as taken from C. Estienne, *L'Agriculture et Maison rustique*, Paris, 1570, but which are well adapted to English readers; then a longer section 'to breede Millars-thumbes and Loches', and one on 'the breeding of Crevis' (*Astacus fluviatilis*); and finally more notes on destructive birds, the Kingfisher, Cormorant, 'Dobchicke', 'More coote or bauld coote' and Osprey. Of the Kingfisher he narrates that 'being dead if he be hanged up by the bill with a thread in your house where no winde bloweth, his breste will alway hang against the winde', and of the Osprey, 'they say he hath one foote like a Ducke and the other like a Hawke, and as he flies nie over the water the fish will come up unto him'.

In 1613 the craft was made the subject of a long and attractive poem, the *Secrets of Angling* by J.D.—this being the posthumous work of John Dennys. It is not perhaps great poetry, though it has not only real charm but much valuable information to impart. Whether or no it derives directly from the *Treatyse* is disputed: R. B. Marston, *Walton and Earlier Writers*, p. 69, regards it as 'not evident that John Dennys knew Dame Juliana's work': but there is very little indeed that is obviously independent. The notes appended by William Lawson to the second edition in 1620 show more first-hand knowledge—Lawson was a north-countryman and evidently a skilful angler—but neither Dennys nor Gervase Markham, who produced in 1614 one of his usual pieces of hack-work on the subject, adds much that is new or of interest except to the student of piscatorial archaeology. In point of fact there is little or no sign of development in knowledge of fish between the fifteenth and the eighteenth centuries. The books deal with a pursuit which is not only sporting but practical and profitable: they are not in any serious sense contributions to science.

In the towns 'sport' took on a more 'gladiatorial' form. Conflicts between dogs and captive beasts had been occasionally staged at the Tower of London. Caius[1] had described how the Hunting Leopard had disposed of a Mastiff which had been let into her cage, and Stow[2] records the tests of the valour of the Lions in the Tower in 1604 and again in 1609 by King James I who was sadly disappointed in his desire to show his courtiers how the king of beasts dealt with refractory subjects. Bear-baiting which had been practised ever since early Plantagenet times became immensely popular: Strype's record of the Sunday's baiting on Bankside in 1554 when a 'blind bear brake loose and caught a serving-man by the calf of the leg and bit off a great piece of it';[3] Robert Laneham's description of

1 *De Rariorum*, p. 4.
2 *Survey of London* (ed. 1764), I, p. 123.
3 *Memorials*, III, I, p. 327.

the baiting at 'Killingwoorth' (Kenilworth) before Queen Elizabeth in 1575; Thomas Dekker's account of the Paris Gardens in Southwark in 1609 in his *Worke for Armourers*;[1] and Slender's boast that he had 'seen Sackerson loose twenty times and have taken him by the chain' in the *Merry Wives of Windsor*[2] prove the cruelty and popularity of the sport. And bull-baiting was as cruel and almost as popular. The mob came to its witch-hunts already blooded.

Nevertheless, during the sixteenth century there had been a change. Not only were the professional classes and the townsmen increasing in importance as against the landed gentry, but even among the aristocracy the devotion to 'venery' was beginning to pall. It is at least symptomatic that other forms of pageantry, the masque and the drama, the dance and the horse-race, were beginning to take their place alongside the chase as an occasion for royal display, and that two of the most striking figures of the latter part of the period, Philip Sidney, the paragon of chivalry, and Francis Bacon, the chief founder of the new philosophy, openly expressed their contempt for the traditional pursuits of their order; and this was echoed later by Samuel Pepys (*Diary*, 14 August 1666) and John Evelyn (*Diary*, 16 June 1670). Their condemnation prepared the way for a truer appreciation and a more intelligent knowledge of birds and beasts.

That the study of plants was already beginning to be recognised as important and indeed as part of a gentleman's education is plain from the *Autobiography* of Edward Lord Herbert of Cherbury.[3] Discussing the content of a sound education, under the year 1599, he writes: 'I conceive it is a fine study and worthy a gentleman to be a good botanic, that so he may know the nature of all herbs and plants, being our fellow-creatures and made for the use of man; for which purpose it will be fit for him to cull out of some good herbal all the icones together, with the descriptions of them and to lay by themselves all such as grow in England, and afterwards to select again such as usually grow by the highway-side, in meadows, by rivers, or in marshes, or in cornfields, or in dry and mountainous places, or on rocks, walls, or in shady places, such as grow by the seaside; for this being done, and the said icones being ordinarily carried by themselves or by their servants one may presently find out every herb he meets withal, especially if the said flowers be truly coloured. Afterwards it will not be amiss to distinguish by themselves such herbs as are in gardens, and are exotics, and are transplanted hither....' He adds: 'I must no less commend the study of anatomy which whosoever

1 Quoted by J. A. de Rothschild, *Shakespeare and his Day*, p. 149.

2 Act I, Sc. I end.

3 For similar evidence coming from the same time and neighbourhood cf. Gunther's account of Sir John Salusbury's interest in botany, *Early British Botanists*, pp. 238–45.

considers I believe will never be an atheist; the frame of man's body and coherence of his parts being so strange and paradoxal that I hold it to be the greatest miracle in nature' (ed. S. L. Lee, London, 1886, pp. 57–9). Lord Herbert's *Autobiography* is never wholly ingenuous, and this passage by itself does not represent his whole mind. We may set alongside it the Tutor's remarks on divination by herbs and the use of herbs for magic in the *Dialogue*, which is almost certainly his work. Here on pp. 172–4 of the edition of 1768 he makes it plain that his interest in herbs and herbals arises from regard for their 'admirable faculties and virtues', not from a concern with what we should call botany.[1] Like all his contemporaries he stands between the old and the new: but unlike many of them he has moved a long way towards the point at which sound knowledge compels a rejection of fable and superstition. To this the love of gardens which he shared with his mother[2] and with his stepfather Sir John Danvers[3] must have largely contributed. The gardener like the fisherman must constantly check his ideas about nature by reference to actual living things.

For the medieval attitude to nature, kept alive by the tradition and ritual of sport, was not less perpetuated by the doctrines and practices of alchemists, astrologers and certain at least of the herbalists. We have already noted how deep-seated was the belief not only that all nature was created with immediate reference to mankind but that its particular use or virtue, its 'correspondence' or meaning, was discernible, at least to the initiate. Among animals it was less easy to discover 'uses', but among plants ingenuity was more fully rewarded. It had become almost an axiom that every plant had some special value or threat to mankind, that this was what gave the plant its character, and that this was stamped upon it by a 'signature' or identifying feature. It is only necessary to glance at the older names, whether in Latin or English, to see how very large a proportion of them are related either physiologically or emblematically or astrologically to the plants' supposed 'vertues'. Lungwort or Feverfew, Houndstongue or Marigold, Mercury or Moonwort are examples which could be multiplied almost indefinitely. Many of these embody very ancient and widespread beliefs; and the principle was so congruous to the whole outlook that it persisted even when its factual basis (if ever there were any) had been disproved.

The lengths to which such symbolisms can be carried will not be strange to anyone who considers how large, varied and fantastic is the

1 His brother, George Herbert, shows a similar interest in the medicinal value of herbs, believing that 'home-bred medicines are familiar for all mens bodyes', *A Priest to the Temple* (*Works*, ed. Hutchinson, p. 261). George through his marriage to Jane Danvers had a second link with the Danvers family.

2 Cf. George Herbert, *Memoriae*, v (*Works*, p. 425).

3 Cf. below, p. 271 for his acquaintance with Parkinson.

similar literature that to-day gathers round the great pyramid. Nor shall we be shocked when De l'Obel states in his last, unfinished work[1] that Adderstongue is 'sought by chymists and carried round by herb-women as an amulet against the evil eye'. For in the Elizabethan age men were hardly beginning to question the principles on which such speculations were based, and it must often seem as we look back upon it that sense and nonsense, fantasy and fact were so inextricably confused that only by a wholesale scepticism, a total rejection of all received ideas, could the beginning of a truer interpretation be made possible. Fortunately the age was too vital to be daunted by the difficulty of its task.

But for a time—indeed until well after the age of science had begun—there were many men who carried on the old superstitions and elaborated them in their writings. Some of them like Nicholas Culpeper,[2] whose *Physicall Directory* was published in London in 1649, were glib and largely uneducated charlatans;[3] others like William Coles,[4] whose *Adam in Eden* followed in 1657, were men of real learning obsessed by an absurd but at the time excusable idea. Robert Turner's dictum 'God hath imprinted upon the Plants, Herbs and Flowers as it were in Hieroglyphics the very signature of their Vertues'[5] is after all only a particular application of Sir Thomas Browne's 'The finger of God hath left an inscription upon all his works, not graphical or composed of letters, but of their several forms, constitutions, parts and operations, which, aptly joined together, do make one word that doth express their natures.'[6] Even when, as with Culpeper or indeed with Robert Lovell[7] and Robert Turner, astrology played a large part, the belief in it was so widespread that it is not easy

1 MS. *St. Ill.* III, p. 192.

2 Died in 1654 a victim, according to his wife, 'of the destructive Tobaco he too excessively took'.

3 As Coles put it in *Adam in Eden*, p. 281: 'Mr Culpepper seldome hit the naile on the head as to the matter of plants.'

4 Wood, *Ath. Oxon.*, Foster, *Alumni Oxon.*, the *D.N.B.*, and others call him Cole: his own books in their title-pages and commendatory letters and verses invariably call him Coles.

5 Quoted by Mrs Arber, *Herbals*, p. 254, from *Botanologia, the British Physician*, London, 1664, 'To the Reader', p. A 5*b*. Turner was a Hampshire man and knew the plants about Holshot (cf. e.g. l.c. pp. 274, 296, 357); graduated Christ's College, Cambridge, 1639–40; based his work on Petrus Morellus, *Methodus praescribendi*, Basel, 1630, in its English translation, on W. Coles's two books, and on Culpeper from whom, however, he often differs in astrological matters. He quotes Gerard and Parkinson freely as well as Mattioli and a few others.

6 *Religio Medici* (*Works*, ed. G. L. Keynes, I, p. 75), quoted by Miss R. Freeman, *English Emblem Books*.

7 Of Christ Church, Oxford, author of *Pambotanologia*, an alphabetical but elaborate compilation of all recorded 'uses' of herbs, Oxford, 1659, and of its sequels *Panzoologia* and *Pammineralogia*, 1661. In the introductions to each volume the species are listed under the different planets: cf. above, pp. 46–7.

to condemn it. When men like Elias Ashmole could devote themselves to what we can only regard as extravagant absurdities, and when (as is the fact) every herbal and every pharmacology is loaded with the same rubbish, the remarkable thing is not that the medieval outlook persisted, but that so many of the early botanists escaped almost entirely from it except in their prescriptions. In the animal kingdom, as we have seen from Topsell, progress was much less evident: the attitude of Aesop persisted until at least John Johnstone's *Historia Naturalis de Quadrupedibus* published in 1652: but in the vegetable a good start had been made in William Turner's generation, and the first half of the seventeenth century saw the foundations of botanical science well and truly laid.

Meanwhile though zoology remained until Ray's time entangled with what he describes as 'Hieroglyphics, Emblems, Morals, Fables, Presages, or ought else appertaining to Divinity, Ethics, Grammar or any sort of Humane Learning' (Preface to *Ornithology*)—entanglements from which Sir Thomas Browne in his *Pseudodoxia Epidemica* sought to set it free—medicine and horticulture gave an impetus to botany which soon led to a marked advance among English students.

The man to whom chief credit for this should probably be given is Mathias de l'Obel who, as we have seen,[1] had come back to England after the death of William the Silent in 1584, and a year or more in Antwerp 1584–5, and had helped to remove the mistakes from Gerard's *Herball*. Of his life here during its last long phase we know very little. But it seems probable that soon after his arrival he became attached to the household of Edward Baron Zouche, the ward of William Cecil, educated at Trinity College, Cambridge, under John Whitgift. Zouche, who is known at this time from his correspondence with Henry Wotton,[2] was in England until 1587; then abroad till 1593; then at Hackney and working hard at his garden there;[3] and then sent by Elizabeth to Denmark in 1598. On this journey De l'Obel accompanied him,[4] and did a considerable amount of botanising, publishing the results in the Appendix to his *Adversaria* in 1605, while others appear in *Stirpium Illustrationes*. He refers frequently to 'Hafnia' (Copenhagen), Elsinore and the royal palace of Cronenburg (e.g. *St. Ill.* pp. 89, 112) and to the 'portus Flecheri' (Flekkefjord) (pp. 99, 102, 158) in Norway, both in the part of his work published by How and in the manuscript in Magdalen College (cf. MS. II, pp. 65*b*,

1 Cf. above, p. 167.

2 Cf. L. Pearsall Smith, *Sir H. Wotton: Life and Letters*, passim.

3 Plat, *Garden of Eden* (ed. 1660), pp. 143–4, records that in the winter of 1597 Zouche transplanted 'divers apple-trees damson trees etc. being of thirty or fortie years growth at Hackney...and all put forth leaves at Michaelmas 1598'.

4 So *Theat. Bot.* p. 615, though the date is given as 1592. According to *Theatrum*, p. 1183, they also visited Norway.

108, III, pp. 186, 194). It is therefore established that the attachment had been made either before 1587 (which in view of Zouche's devotion to horticulture and consequent impoverishment is not unlikely) or after his return in 1593. If the former, it may well have been through Gerard who was superintendent of the Cecils' gardens at Theobald's and in the Strand. If the latter, it may be that Wotton's friendship for Charles de l'Ecluse, whom he met in Frankfort in 1590 and with whom he kept up a correspondence, led to the linking up of Lord Zouche with De l'Obel.

After his return from Denmark Zouche was made President of Wales in 1602 and an appeal against him in 1604 failed to dispossess him.[1] As such De l'Obel dedicated to him the second edition of his *Adversaria* in 1605. Through his patron's influence he received the title of Botanicus Regius in 1607 from James I. He was now settled and prosperous. The famous physic garden on Zouche's estate at Hackney had become his headquarters. As superintendent of it he obtained a position of importance among the doctors, apothecaries and gardeners of London; and his influence both as an experienced worker in the field and as one of the earliest systematists was of great value in maintaining contact between English students and the more advanced savants of the continent.[2] His second wife helped him in his work.[3] His son Paul was an apothecary in business in London. One of his daughters married James Cole, a London silk-merchant living at Highgate and with a town house in Lime Street, who seems to have shared the interest in plants (so *St. Ill.* p. 119, where his name is spelled by De l'Obel Coel, and Gerard, *Herball*, p. 168,[4] where he is described 'as exceedingly well experienced in the knowledge of simples') and who is described as 'lately deceased'[5] in Parkinson, *Paradisus*, p. 401. Another daughter married a notable London apothecary, Lewis Myres (cf. MS. *St. Ill.* II, p. 64), who did a considerable trade in drugs and plants with Syria. Moreover, his second wife, who had been married before, had a son Abraham Hoguebart (or Hugobert—so C. R. B. Barrett, *History of Apothecaries*, p. 10), who seems to have drawn from life in the Welsh mountains the picture of a Sundew (cf. MS. *St. Ill.* III, p. 194). So the family all took an interest in botany.

These and many other botanists and correspondents of De l'Obel are mentioned in the notes which formed the large work *Stirpium Illustrationes* that he was compiling during the last years of his life. This ap-

1 There are very critical comments upon his pomp and oppression in Henry Machyn's *Diary*.

2 G. Bauhin (e.g.) in his *Theat. Botan.* c. 112 records plants sent to him by De l'Obel from London.

3 Cf. E. Morren, *M. de l'Obel*, p. 8.

4 Cf. p. 342, where Cole's discovery of 'Rose Ribwoorte' (*Plantago lanceolata* var.) 'neere London by a village called Hoggesdon [Hoxton]' is recorded.

5 His will is dated 31 December 1627 and has a codicil of May 1628: cf. Gunther, *E.B.B.* pp. 247–8.

parently passed by purchase into the hands of John Parkinson who already had a garden in London and who used De l'Obel's manuscript in his *Theatrum Botanicum*. After Parkinson's death William How, a young doctor interested in botany, published in 1655 a selected part of this manuscript and put forward a violently phrased charge of plagiarism against Parkinson. We shall discuss this later. But whatever the justice of the charge the notes and the manuscript now preserved in the Goodyer collection in the Library of Magdalen College, Oxford, contain important evidence of the spread of botanical knowledge and of De l'Obel's influence. This is fully borne out by the diploma or certificate given to him by the President and Council of the Royal College of Physicians when Thomas Moundeford was President and presumably in the year 1614. This is printed by How and speaks warmly of the great value of De l'Obel's services to botany and medicine. The verses prefixed by Alexander Reed, the doctor who found the green Primrose of *Stirpium Illustrationes*, p. 119, at Nantwich, tell the same story.

The first of these notes supplies a link with the most illustrious of botanists before Ray, the brothers Jean and Gaspard Bauhin. De l'Obel had gained acquaintance with Jean the elder of them when he was working at Lyons (*Adv.* p. 439), and had received seeds from Gaspard which he grew at Hackney (MS. *St. Ill.* II, p. 133). Now he was visited by the distinguished Scottish doctor 'very learned in this and all parts of medicine', James Cargill of Aberdeen, who had been attending Gaspard's lectures at Basel[1] and who showed De l'Obel a grass new to him and growing in London (*St. Ill.* pp. 30–1). Cargill had already sent to De l'Obel from Scotland a plant like the yellow Norwegian asphodel (*Adv.* Appendix, p. 485)—no doubt *Narthecium ossifragum*; and in 1603 he sent the first recorded British specimen of *Trientalis europaea*, at least four other plants and several sea-weeds to Bauhin who reported them in 1620 in his *Prodromus*.[2] De l'Obel became friendly also with an even more famous Scottish doctor, James Nasmyth, 'chiefe Chyrurgeon in his time to King James', to whom he gave seeds of a Plantain from Montpellier (Parkinson, *Theatrum Bot.* p. 495).[3] Nasmyth was interested in plants, had a garden in London, and in April 1605 was one of the first to flower the 'Black Fritillary'. Somewhat similar is his contact with Matthew Lister, afterwards the famous doctor, who, on his return from his studies at Basel and Montpellier, gave him a plant and re-established

1 He contributed complimentary verses to Bauhin's edition of Mattioli's *Works* (Frankfort, 1598), and is mentioned as a helper in the Preface.

2 Pp. 100; 25, 107, 113, 119; 154–5; cf. also his later *Theatrum Botanicum*, cc. 273, 552, and Johnson, *Gerard*, p. 37.

3 Parkinson, no doubt copying De l'Obel's MS., calls him John Nesmit. In *Adv.* pp. 487, 489, 496, 506, etc., he is often J. Nasmyt.

his links with the school there (*Adv.* App. p. 501). He maintained close touch with Jean Robin, botanist to the French King and in charge of the physic garden in Paris of which he printed an alphabetical plant list in 1601 (l.c. p. 487), who sent him a Cyclamen from the Pyrenees in 1602 (MS. *St. Ill.* II, p. 37*b*), and with his younger son Vespasian (*St. Ill.* pp. 72, 95); with Pieter Pauw, the Professor of Anatomy at Leyden, who sent him seeds (*St. Ill.* p. 104) and published a catalogue of the Leyden garden in 1601; with Jan van Brancion of Mechlin and Jean Boissot of Brussels, the close friends of De l'Ecluse (*Adv.* App. pp. 494–5); and with Cornelius Coorne, Jean Somer and Thomas de la Fosse, cousin of Jean Mouton of Tournai, at Middelburg (*Adv.* pp. 492, 488; *St. Ill.* p. 102), whom he visited in 1603 (p. 503). A Flemish friend of his, Bartholomew Seelen, had settled in London and had a garden there (*St. Ill.* p. 125, etc.). The young Frenchman, N. C. Fabri de Peiresc, an admirer of De l'Ecluse, visited him at Francis Street in 1606 (*Lettres de Peiresc*, VI, p. 675 and Rand, *Life*, p. 100).

The rest of the book from which How printed this material, together with his and De l'Obel's other scripts, passed into the hands of John Goodyer of Mapledurham near Petersfield and thence to the library of Magdalen College, Oxford. A brief account of it is given by R. T. Gunther in *Early British Botanists* (Oxford, 1922), pp. 252–3, but the script itself hardly bears out either How's charges or the claim that it is a completed work. It consists first of a large number of brief descriptions of Grasses, the 200 or so from which How made his selection of 98, and then of slips and pages from the *Adversaria* intercalated with leaves from the *Observationes* (1576) and illustrations from the *Icones* (1591), to which are added two or three pen and ink drawings and a large number of marginal notes giving new and alternative information, written in a neat and beautiful script but often ill-arranged and hard to attach. Much of the material consists of brief and casual jottings, the preliminary to a book but wholly incomplete and largely unimportant. Hardly any of it chronicles original observations and the only finished paragraphs are condensed versions of the work of D'Aléchamps and De l'Ecluse. If the rest of the plants had been written up as the Grasses have been, the whole might have made a valuable Herbal: but in what remains to us there is no evidence even that the necessary material had been gathered. Evidently he was adding slips and fresh notes right up to the end of his life; for there is a paragraph on 'Cyperus angulosus descriptus Adversariorum pag. 37' (*Cyperus longus*) in which after referring to the quantity of it which he saw at Martinelli's in Venice he continues: 'I saw it this year 1615 in London at Parkinson's the skilful apothecary' (MS. *St. Ill.* I, p. 79); and a similar reference to seeds not yet ripe in August 1615 is in *St. Ill.* p. 131.

In addition to his work in London and by maintaining contacts with other botanists and gardeners he carried on some field-study himself. He visited Colchester in 1596 and sent to De l'Ecluse a report of a *Limonium* seen there (*R.P.H.* p. lxxxii). About this time, too, he was certainly at Richard Garth's house at Drayton[1] near Portsmouth and was the first to find *Frankenia laevis*[2] in that neighbourhood (*Theat. Bot.* p. 1485). He also seems to have revisited the West country where he found *Euphorbia pilosa* 'in a wood belonging to Mr John Coltes nigh unto Bath' (l.c. p. 189),[3] and, according to Jean Bauhin (*Historia Plant. Univ.* III, p. 729), *Asplenium marinum* 'on sea rocks in Cornwall near the house of Mr Muyle [Moyle]', that is at Bake near St Germans (cf. *Theat. Bot.* pp. 1044–5). It is tempting to assume that he was responsible for a series of observations made in Dorsetshire around Portland and also in the neighbourhood of Chepstow and Tidenham which we shall record when we deal with Parkinson:[4] but there is no evidence for this in the *Stirpium Illustrationes*. He clearly journeyed, perhaps in company with Lord Zouche, into Wales; for there are in *Theat. Bot.* pp. 1190, 1194, records of plants, probably *Sesleria caerulea* and *Juncus squarrosus*, found 'on a high hill called Berwin [Berwyn, south-west of Llangollen]', the second 'found by Dr Lobel in his lifetime' there. With the same tour may be associated the references to *Meconopsis cambrica* on p. 370 which he discovered at many places in Wales,[5] and the record of a many-berried Mistletoe (*Viscum album*) which he noted four miles from Chester 'at Lee Hall, the seat of Mr Calvelle'[6] (MS. *St. Ill.* II, p. 68). It was perhaps at this time that he collected the Welsh ('Cambro-British') names for plants which he inserts in his MS. *Stirpium Illustrationes*.[7]

But though he did great service both by linking up English studies with the experts of Europe and by promoting local investigation, his own work is not always free from blunders. Parkinson, who notes in his *Theatrum Botanicum* a large number of new discoveries as made by him, includes among them some which have never been seen since and probably were wrongly named; he recorded as British, for example, *Verbascum sinuatum* at Bath (p. 62), *Sedum arboreum* 'in the Iland of Holmes by Bristow' (p. 732), *Antirrhinum asarina* at Clewer 'towards Mendiep'

1 See below, p. 243 and De l'Obel, MS. *St. Ill.* III, p. 184, where is recorded his finding of *Asperula cynanchica* on the chalk there.

2 The identification of this find was discussed by Ray, *Corr.* p. 140.

3 In this same wood 'three miles from Bath neere the house of one Mr John Colt' he found a Burdock: so Johnson, *Gerard*, p. 810.

4 For these and the confusion of their place-names, see below, pp. 258, 260.

5 See below, p. 264.

6 Presumably Lea, N.N.W. of Chester, then seat of Hugh Calveley, Sheriff of Cheshire.

7 E.g. III, pp. 194, 228.

(MS. *St. Ill.* II, p. 35),[1] and an unidentifiable fern 'at Bristow' (p. 1047); and he is possibly responsible for some of the non-existent localities attached to some of Parkinson's records. This bears out Ray's comment upon him as 'in his descriptions inexact, and in localities often mistaken through trusting to his memory' (*Historia Plantarum*, I, *Auct. Nom.*).

The portrait taken of him by François Dellarame in 1615 shows him in his seventy-seventh year[2] and hardly supports the reputation for arrogance which attaches to him in certain quarters. This is not the place to discuss the value of the classification which he introduced into his plant-books or to attempt an estimate of his status in the history of science. But at a period when English botany was not strong, his residence here and his wide knowledge of the relevant books and men made a great contribution to its development.

Unfortunately his last years were clouded by the unpleasant connection of his son Paul[3] an apothecary, who lived near James Garret in Lime Street,[4] with the notorious case of Sir Thomas Overbury's death in the Tower in 1613. Paul had married the sister of Theodore de Mayerne the King's physician, and when Mayerne was appointed to look after Overbury's health he instructed his brother-in-law to supply whatever medicines were required—no doubt, as was stated at the trial, because Lime Street was near the Tower. In 1615, when the suspicion of poison was accepted, it was stated that Paul's servant had been bribed to administer poison instead of the drugs ordered by the doctor. One deposition, that of Edward Rider, describes an interview with 'Dr L'Obell' (our botanist) and his wife, then a very elderly couple living in a 'walled house', and a further meeting when Rider declared that an apothecary's boy in Lime Street was guilty, 'speaking as if I knew not that it was his son's boy', and when, according to Rider, Mrs De l'Obel gave a very suspicious exclamation in French.[5] That the suggestion of De l'Obel's complicity was never seriously made, and that there is not a shadow of evidence to support it, is obvious from the conduct of the whole of the lawyers involved, Sir Edward Coke, Sir Francis Bacon and the rest. Sir Philip

1 'In a lane by which one goes to Clower, seat of Thomas Horner, towards Mendiep where they dig lead.' For Horner, father of Sir John, knighted 1584, cf. *Visitation of Somerset* (Harleian Soc. XI), p. 57. The same find is also recorded in Parkinson, *Theat. Bot.* p. 267, as made by De l'Obel. It is recorded as the first discovery of *Asarum europaeum* by W. A. Clarke, *First Records*, p. 124, and by G. C. Druce, *Comital Flora*, p. 260, who both take it from How, *Phyt. Brit.* p. 12, where De l'Obel's Asarum is defined wrongly as Asarabacca.

2 Reproduced by Mrs Arber, *Herbals*, p. 90.

3 His name, rendered as Paul Lobello by Barrett, l.c. p. xxii, occurs in the list of apothecaries to whom the Royal Charter establishing their society was given in 1617.

4 Perhaps because Lord Zouche, his father's patron, had a 'messuage or tenement' there, Stow, *Survey*, I, p. 415; or through his brother-in-law, James Cole.

5 These depositions are quoted in A. Amos, *The Great Oyer of Poisoning*, pp. 168–70.

Gibbs, *The King's Favourite*, in reconstructing the case makes an unnecessary mystery out of De l'Obel's connection with it, and also confuses father and son. But the trial, although Lord Zouche was one of the judges, must have been an anxious end for the old man's life. He died a few months after its close, in 1616.

Contemporary with De l'Obel, though with no such claim to eminence, was Hugh Plat the inventor and gardener. In a study of naturalists Plat has little place except as illustrating the very large and rapid development of interest in such experiments as he recorded in his *Jewell House of Art and Nature*, first published in 1594,[1] and in agriculture and gardening. Third son of Richard Plat, a wealthy brewer and the founder of Aldenham School, he was educated at St John's College, Cambridge, became a member of Lincoln's Inn, but lived a man of leisure and wide interests with a garden at Bishops' Hall, Bethnal Green and another in St Martin's Lane. He was knighted for his books and inventions in 1605 and died in 1608. His writings are based upon a wide correspondence with men of all types and *Floraes Paradise* (London, 1608) contains a mass of allusions to and quotations from a number of helpers. Andrew Hill and 'Mistriss Hill' are the most prominent; Master Pointer of Twickenham,[2] Masters Colburne, Taverner, Melinus and Simson; the Lords Darcy and Zouche; 'Mr Stutfield that married my Lord North's brothers sister'; 'Master Jacob of the Glassehouse';[3] 'Mr Hunt the good Horseman'; Mr Nicholson, gardener; Tomkins the gardener; 'Garret the Apothecary and Pigot the gardener'; Sir Francis Carew of Beddington[4] and 'my cousin Matthews of Wales' make a goodly company.

In the second part appended by Charles Bellingham to the book, now called *The Garden of Eden*, in 1660 a further list of names occurs: 'my cousin Duncombe'; Sir Thomas Challenor (no doubt Chaloner of Steeple Claydon) and Sir Edward Denny of Ireland; 'Mr Jarret the Chyrurgeon in Holburn' and 'Mr Flower by Bednal Green'; and Mr Barnaby Googe's book on Husbandry[5]—these are laid under contribution. Plainly gardening was becoming fashionable and widespread when so many adherents of the craft could be cited.

One of the most prominent and valuable of the London botanists at this time was another refugee from overseas, John de Franqueville, the merchant whom we have already met as a friend of Gerard. His father had lived at Cambrai and had fled to England after the Edict of Nantes,

1 See below, p. 307.
2 I.e. Vincent Corbet, see above, p. 214.
3 Apparently in Portsoken Ward just east of Aldgate.
4 Cf. Gerard, *Herball*, p. 1066; Parkinson, *Theatrum*, p. 273, etc.
5 *The Whole Art and Trade of Husbandry*, a translation or expansion of Conrad Heresbach's book, in dialogue form, first printed in 1577, dedicated to Sir William Fitzwilliam and containing a list of eighteen Englishmen to whom the book is indebted.

bringing with him the double white Narcissus which his son afterwards distributed to English gardeners (so Parkinson, *Paradisus*, p. 86). De Franqueville figures largely in the botanical records of the first twenty years of the century; De l'Obel credits him with having shown him in 1600 the first root of sugar-cane, 'Acorus indicus' (*St. Ill.* p. 122), which he had obtained from Thomas Warner,[1] and the first plant of *Phytolacca decandra*, 'Solanum magnum Virginianum' (l.c. p. 88), in 1615, besides many other rarities. Goodyer obtained from him in 1617 'two small rootes' of the 'Heliotropium indicum' or Jerusalem Artichoke which brought him 'a peck of rootes wherewith I stored Hampsheire' (Johnson, *Gerard*, p. 754). A short list of plants in his garden is printed by Gunther, *E.B.B.* p. 326.

During the lean period that separates Penny from Johnson a link with the younger generation was maintained by Hugh Morgan, the veteran whose garden was known to Turner and who spent the last years of his life in Battersea. Parkinson (*Paradisus*, p. 437) reports on a Judas Tree (*Cercis siliquastrum*) growing in his garden there; and Strype records the inscriptions on the brass plates in Battersea Church (*Survey of London*, II, pp. 740–1): 'Hugh Morgan late of Battersey Esq sleepeth here in Peace: whom Men late did admire for worthful parts. To Queen Elizabeth he was chief pothecary till her death

> And in his Science as he did excell
> In her high Favour he did always dwell.
> To God religious, to all men kind,
> Frank to the poor, rich in content of mind.
> These were his vertues, in these dyed he,
> When he had liv'd an hundred year and three.'

A second plate states that he died on 13 September 1613 and that his brother's grandson, Robert Morgan, set up the memorial.

How rapid and widely known were the development of enthusiasm for gardening and the introduction and breeding of new and brilliant flowers may be seen from the production at Utrecht of an English edition of the *Hortus Floridus*, a picture-book of engravings by Crispin van de Passe senior, the artist, published at Arnhem in 1614. This was issued next year as *A Garden of Flowers wherein is contained a true and perfect description of all the flowers contained in these foure following bookes...faithfully translated out of the Netherlandish originall into Englishe for the common benefite of those that understand no other languages*;[2] and in front of each of the four seasons into which it is divided are detailed directions for the painting of each plate—as for the 'Jacinth' 'bluish purple and lack [lake] with ash-colour with a little white' for the flowers and similar instructions

1 For whom see above, p. 214.

2 For the various editions of this book, cf. D. Franken, *L'Œuvre Gravé des Van de Passes*, pp. 265–70 and Arber, *Herbals*, pp. 244–5.

for the leaves and stem. The Dutch were by this time developing the excitement which came to be known as 'tulipomania', but it is a surprise to find that at so early a date they thought it possible to sell such a volume in England.

Nor was the interest confined to plants appropriate to horticulture. Many of the most notable gardeners were also keen botanists. Richard Garth, the Principal Secretary to the Chancery, whom we have met as the friend and correspondent of De l'Ecluse,[1] had left his Surrey home, Groutes at Morden near Croydon, and gone to Drayton near Portsmouth and opposite the Isle of Wight (*St. Ill.* pp. 85, 150), where he had bought the manor from Robert Earl of Sussex in 1592 and laid out another garden. Here he found one of the Sea Oraches (probably *Atriplex littoralis*) on the saltings near his house. He had died in 1597 but his widow Jane, daughter of John Bushey, seems to have kept on the care of his garden, and was visited there by De l'Obel who found *Polypogon monspeliensis* and was given a brew of Metheglin (*Advers.* ed. 1605, pp. 469, 473, where he prints a recipe from her, another from Andrew Corbet of Shropshire and a third from 'Mr Throckmorton').

Several new men of great importance are beginning to appear: William Coys of Stubbers, North Okington, the eminent gardener (*Adv.* App. p. 487; *St. Ill.* pp. 117, 120), of whom De l'Obel speaks in the warmest terms;[2] John Goodyer of Mapledurham, of whose life and works R. W. T. Gunther has given so full an account in *Early British Botanists* and who was indebted to Coys for much of his enthusiasm; John Parkinson, the London apothecary, whose garden in Long Acre already contained plants of interest (*St. Ill.* p. 92) and who received as early as 1607, after the bitter winter of that year, his first plants from Wilhelm Boel the Frieslander, constantly mentioned by him in his *Paradisus*, who for many seasons supplied him and De l'Obel with seeds and roots from his trading trips to North Africa, Spain and Portugal and the South of France; and John Tradescant, said by Anthony à Wood to have been a Fleming but probably an East Anglian, who was living at Meopham in Kent in 1607 and was working for Robert Earl of Salisbury at his manor of Shorne some four miles away. The Earl, like his father Lord Burleigh, was a keen gardener and in 1605 had taken over from Gerard his share of a garden adjoining Somerset House. Tradescant had a large part in the work which Salisbury was doing at this time at Hatfield[3] and was in his service till his death in

1 Cf. De l'Ecluse's account of Brazilian fruits received from him in *Exotica* (ed. 1605), pp. 43–4.

2 There is an attractive account of Coys by Dr J. Ramsbottom in *Essex Naturalist*, XXVI, pp. 67–71.

3 Cf. the bills for seeds and trees paid to Tradescant in 1611, quoted by Alicia Cecil, *History of Gardening in England* (London, 1910), pp. 152–3.

1612, and then seems to have gone to Edward Lord Wotton of Boughton Malherbe eighteen miles from Meopham to the south-east. In 1618 he went with Sir Dudley Digges to Archangel, where he landed on 16 July, saw something of the Russian flora, started back on 5 August, and wrote an account of the voyage still preserved in the Bodleian.[1] In 1620 he joined Sir Samuel Argall's expedition against the Barbary corsairs; and this is referred to in *Paradisus*, pp. 190, 579. Then after a short period with George Villiers Duke of Buckingham,[2] 1625–8, during which he accompanied him on his attempt to relieve La Rochelle in July 1627, with supplies for the Huguenot leader, M. de Soubise,[3] brother of the Duc de Rohan, he set up his own garden and museum in South Lambeth, and during the next ten years amassed a great store of plants of which he printed a catalogue in 1634[4] and which his son, John the younger, returning from Virginia in 1638,[5] inherited and increased.[6] His death in that year seems to have prevented him from moving to Oxford as chief gardener for Lord Danby in the new Physic Garden of the University.[7]

Mention of the travels and collections of the Tradescants suggests that we bring our survey of the Elizabethan age to an appropriate close by a brief study of what was perhaps its most characteristic activity. It was the period of adventure and exploration, and Britain was mapped and studied and described with the same enthusiasm as Virginia or the East Indies. We have already seen the beginnings of this new interest in the land, its situation and climate, its soil and commodities: but Harrison's work was taken up not only by antiquarians and historians, but by cartographers and naturalists.

Christopher Saxton who first made maps of the English counties spent nine years in travel before he published his results in 1579. John Speed, thirty years later, not only amplified and improved the survey but in 1611

1 Cf. J. von Hamel, *Tradescant in Russland* (St Petersburg, 1847, translated in 1854 in *England and Russia* by J. S. Leigh), who quotes extensively from it, and gives a good account of its author; and Gunther, *E.B.B.* p. 330. Parkinson, *Paradisus*, p. 346, alludes to this journey, and the abundance of *Veratrum album* seen on it.

2 One of the chief authorities for his career is a brief paragraph in Parkinson, *Parad.* p. 152, dealing with the 'Ephemerum Virginianum Tradescanti' as Parkinson named it, the Spiderwort now called Tradescantia which he received from a friend in Virginia and introduced to British gardens.

3 Cf. Parkinson, *Theatrum*, p. 624; he brought *Matthiola sinuata* back from the island of Ree (Rhé): cf. also Johnson, *Gerard*, p. 1099.

4 A copy of this exists in Magdalen College and has been reprinted by Gunther, *E.B.B.* pp. 334–45. There is a paper on the garden by William Watson in *Phil. Trans.* for 1749, vol. XLVI, pp. 160–1, in which the few remaining trees and plants of interest are noted.

5 Cf. Parkinson, l.c. p. 1045 and J. S. Leigh, l.c. p. 293.

6 For the Tradescant epitaph in Lambeth churchyard, cf. A. Cecil, l.c. pp. 153–4: for the catalogue of 1656, cf. below, pp. 306–7.

7 Cf. Vines and Druce, *Morisonian Herbarium*, p. xv.

in his book *The Theatre of the Empire of Great Britaine*—the prelude to his *History*—added a condensed page of information on each county. Much of this is devoted to the antiquities, history, buildings and towns: but paragraphs are in most cases added on the geographical, geological and faunal peculiarities of the shire; and these, though often derived from others, and often relatively trivial, contain a good deal of new and interesting detail, and provided a basis for the subsequent work of Joshua Childrey and Christopher Merret. References to Speed's work will be given when we deal with the two later writers; very little of it bears directly upon natural history: but in his references to minerals and fossils and in a few other cases such as the account of the plague of mice near Southminster in Essex in 1581[1] which he borrows from John Stow, *Annales of England*,[2] his notes are of immediate interest to naturalists.

Much more significant for our purpose is one of his contemporaries and fellow-antiquaries in whom the spirit of Harrison and Lyte, indeed as some have suggested even of the age itself, found a worthy embodiment. He is a pioneer in the detailed exploration of the country,[3] the first in the great succession of county historians, Richard Carew of Antony, whose *Survey of Cornwall* was published by John Jaggard in 1602. Carew was born in 1555 and went up to Christ Church as a gentleman commoner at the age of eleven, being chosen in 1570 to debate extempore in Latin with Philip Sidney who was a year his senior. He married in 1577 and devoted himself to the management of his estates, the service of Cornwall, and the study of its antiquities and interests. Camden commended him warmly both in *Britannia*, p. 140, where he spoke of the fishpond at Antony open to the sea and stocked with sea-fish,[4] and in an epitaph written on Carew's death in 1620, and printed in his *Epistolae*, p. 106.

The book, dedicated to Sir Walter Raleigh and described by Fuller as 'a pleasant and faithful description', begins with a long and useful account of Cornish minerals and especially tin-mining (pp. 6*b*–18*b*). Turning to 'things of life' he has a paragraph on the herbs 'Seaholme and Sampire' whose 'roote preserved either in sirrup or by cauding [candying] is accepted for a great restorative'; and 'Rosa Solis' in the 'gaully grounds' (p. 19). Of 'breathing life' he tells a nice tale of Martin Trewynard and his snake; notes the abundance of Rats 'cumbersome through their crying and ratling while they daunce their gallop gallyards

1 *T.B.* p. 31.

2 Edit. 1601, p. 1166.

3 John Norden, whose *Speculum Britanniae* (first part *Middlesex*, 1593) is a series of county chorographies, included no account of natural history in his records; and Speed's notes are slight and seldom original.

4 Plat, *Jewell House*, p. 60, records a similar use of dykes in Sheppey for keeping sea-fish by Sir Edward Hobbie; and Googe, *Whole Art of Husbandry*, p. 163, gives a full account of how to construct one.

in the roofe at night' (p. 22); propounds a theory of the incidence of Lice; and mentions 'the Marternes Squirrels Foxes Badgers and Otters, the Hares, Conies and Deere' of the county. Of birds he comments principally upon the winter arrival of Woodcocks coming in on the north coast and drawing towards the south 'as the moyst places which supplie them food beginne to freeze up' (p. 25). He has a gibe for the time wasted over training the Sparrowhawk 'though shee serve to flie little above sixe weekes in the yeere and that onely at the Partridge where the Faulkner and Spanels must also now and then spare her extraordinary assistance' (p. 25). He notes the absence of Nightingales; has an account of 'a flocke of birds in bignesse not much exceeding a Sparrow which made a foule spoyle of the apples. Their bils were thwarted crosse-wise at the end, and with these they would cut an apple in two at one snap eating onely the kernels' (p. 25*b*)—which takes us back to Matthew Paris; and states that Swallows are found in winter 'sitting in old deepe Tynne-workes and holes of the sea cliffes'—a belief which Gilbert White would have joyfully endorsed.

Of fishes and fishing he has a fuller record. 'Shote[1]...in shape and colour he resembleth the Trowt' in the rivers (p. 26); 'Trowte and Peall[2] come from the sea and passe up into the fresh ryvers to shed their spawne' (p. 28); 'Sammons it hath been observed that they (as also the Trowt and Peall) haunt the same rivers where they first were bred' (p. 28*b*); and then a long list of sea-fishes—shell-fish, flat-fish and 'round fish'—a list well worth comparing with that compiled from an old fisherman at Penzance by John Ray in 1673. He gives a racy description of the fisheries, particularly of the Pilchards, who are 'persecuted by the Plusher [?Dogfish] and the Tonny' while 'certaine birds called Gannets soare over and stoup to prey upon them' (p. 34). Finally he has a paragraph on the Seale or Soyle 'not unlike a pigge ugly faced and footed like a Moldwarp, he delighteth in musike' (p. 34*b*), and others on the 'sea foule', 'Dip-chicke (so named of his diving and littlenesse), Coots, Sanderlings, Sea-larkes, Oxen and Kine,[3] Seapies, Puffins, Pewets,[4] Meawes, Murres,

1 Shote is a term widely used for the sedentary or 'Common' Trout (*Salmo fario*) as compared with the Sea Trout (*Salmo trutta*): so F. Day, *British Fishes*, II, p. 104.

2 That Peal are distinct from Sea Trout and from Salmon is a belief widely held—especially in the West country. To discuss the immensely difficult problems of the classification of the British Salmonidae is outside the scope of a note. But Carew's view has some modern support.

3 Presumably like Oxbird a name for the Dunlin (*Calidris alpina*); cf. J. E. Harting, Introd. p. xvii to E. H. Rodd, *Birds of Cornwall*. Sea-lark is *Charadrius hiaticula* and Sanderling, in view of Ray's very exact description of the Cornish bird, certainly *Crocethia alba*.

4 Not the Lapwing but the Black-headed Gull (*Larus ridibundus*): Meaw is *Larus canus* and Murre *Uria aalge*.

Creysers,[1] Curlewes, Teale, Wigeon, Burranets, Shags, Ducke and Mallard, Gull, Wild-goose, Heron, Crane and Barnacle' (p. 35), of which he notes that 'it is held that the Barnacle breedeth under water on such ships sides as have beene verie long at sea...but I cannot heare any man speake of having seene them ripe', that 'the Puffyn hatcheth in holes of the Cliffe' and that 'the Burranet [*Tadorna tadorna*] hath like breeding, and after her young ones are hatched shee leadeth them sometimes over-land, the space of a mile or better, into the haven'. He concludes with the 'Jacke-Daw peculiar to Cornwall and therethrough termed a Cornish Chough...when he *is kept tame*, ungratious in filching and hiding of money' (p. 36).

In the history of science Richard Carew may be a person of no importance: he made no great discoveries nor much contribution to knowledge. In the history of civilisation he is a portent, a social type, representing a way of life, a quality of culture, a relation to his environment that are altogether novel. It was this new ability not only to love the world of nature but to appreciate and understand and interpret it that supplied the creative impulse for the scientific movement which was thus brought to birth.

Finally, along with exploration of the general features of our land, there was begun at this time the systematic search for plants in England by those simpling expeditions which Thomas Johnson afterwards developed and John Ray carried to perfection. On page 38 of the *Stirpium Illustrationes*, De l'Obel writes: 'This elegant Grass [apparently only a very large specimen of *Phleum pratense*] was first found by us on the upland meadows between Islington and Highgate, on a public excursion and simpling-tour when we accompanied Dr Richard Forster sometime President of the famous London College of Physicians and Dr Francis Herring Fellow of the same College.' Forster, an M.D. of All Souls, Oxford, was President in 1601–4 and again in 1615; Herring was an M.D. of Christ's College, Cambridge, who had contributed commendatory verses to Gerard for his *Herball* and in 1604 produced *A modest defence of the Caveat given to the wearers of impoysoned amulets or preservatives from the Plague*: they were both outstanding men and that they encouraged these botanical explorations is a matter of some importance: it means not only that the doctors were no longer ready to leave their drugs to herb-women or untrained apothecaries, but that the flora of Britain would at last be scientifically investigated. This systematic exploration was made a regular part of its activities by the Society of Apothecaries, which became independent of

1 Rodd, l.c. p. 314, suggests Kryssat, Cornish for Kestrel, but this is not a sea-bird. Is it not rather a form of Scraye, the Common Tern (*Sterna hirundo*), or perhaps of Skrape, Scandinavian for the Shearwater (*Puffinus puffinus*), though this in Cornwall is less likely?

the Grocers Company in 1617 and received a new charter from the King. An annual Simpling Day became a regular institution, a date late in June being usually chosen,[1] and two Stewards being appointed to lead the party.[2] From this small beginning great results were to follow; and William Turner's dream of a full investigation of the native flora of Britain was to come true.

CHAPTER XV. JOHN PARKINSON

The two best known British botanists in the first half of the seventeenth century typify the two chief means by which knowledge of plants was advanced. They were both apothecaries and prominent in the activities of their profession. John Parkinson, of Ludgate Hill, had been an Assistant at the incorporation of the Society in 1617 and Warden in 1620: but he found the business too heavy, retired from office in 1622 and became increasingly a gardener.[3] De l'Obel as we have seen knew him in this capacity in Long Acre; and his first book with its punning title, *Paradisi in sole Paradisus terrestris*, 'The Park on earth of the Park in sun', published in London in 1629[4] is the first English gardening book. It has full introductory chapters on flower-gardens, kitchen-gardens and orchards, and contains nearly a thousand plants described and illustrated under these heads. On the strength of this publication he was given, by Charles I, the title of Botanicus Regius Primarius, and began to prepare his much larger work, the *Theatrum Botanicum*, 'An Universall and Compleate Herball'. This did not appear until 1640 when its author was already seventy-three. He died ten years later and was buried in St Martin-in-the-Fields. Thomas Johnson, a younger and abler man, was the first systematic explorer of England for plants. He had become a London apothecary in Snow Hill by 1628 and wrote some commendatory verses in Latin for Parkinson's *Paradisus*. A valuable study of him, *Thomas Johnson, Botanist and Royalist*, was published by H. Wallis Kew and H. E. Powell in 1932; and this gives a lively survey of his remarkable writings and career. To this we shall return later.

Parkinson's *Paradisus*, though not strictly speaking a work of im-

1 E.g. in 1627, 21 June, Barrett, *History of Apoths.* p. 26.
2 First appointed in 1628, Barrett, l.c. p. 27.
3 Cf. Barrett, l.c. pp. 3, 7, 15.
4 A 'second impression much corrected and enlarged' was issued in 1656. Apart from a few trifling transpositions and changes in spelling it is virtually a facsimile. The first edition was 'faithfully reprinted' by Methuen in 1904.

portance for natural history or the development of science, is interesting and valuable not only for the student of horticulture but as testifying to the development both of plant-breeding and of exploration. After a dedication to the Queen, an introductory commendation from Sir Theodore de Mayerne, the producer of the *Theatrum Insectorum*, and verses from Dr Ottwell Meverel (Othowell Meverall), another famous doctor and hero of a narrow escape from being buried alive while he was a B.A. in residence at Christ's College, Cambridge, William Atkins, William Brode (Broad),[1] and Thomas Johnson, and a letter to the Reader, he devotes nine chapters to 'the ordering of the Garden of pleasure', its site and design, its edging and plants, 'outlandish' or native, and their care and increase. It is an admirable survey and, like the *Boke of St Albans* on fishing, shows how large and sane an experience had been accumulated in this ancient craft. Many new plants, and some new methods and implements have been introduced since Parkinson's day: but it is easily arguable that of all human callings the gardener of his period would be most at home in the modern world.

Moreover, there is not to be found in any contemporary or previous writing so modern a note as that which is struck in the last chapter of this survey, wherein Parkinson rejects without compromise the belief that flowers can be made to grow double or with strange scent or new colour or different time of flowering by observation of the moon, or by doctoring the seeds with wine, or by injecting perfumes into the root, or by watering the plants with coloured water: 'in the contradiction of them I know I shall undergo many calumnies, yet notwithstanding I will endeavour to set down and declare so much as I hope may by reason perswade many in the truth' (l.c. p. 22). His theoretical arguments, it must be confessed, are not very convincing, but his robust commonsense ('I have been as inquisitive as any man might be, but I never could finde...I have made trial at many times, but I could never see') is the best vindication of the scientific method.

Then follows the treatment of the several plants beginning with the Crown Imperial (*Fritillaria imperialis*) and proceeding by way of the Lilies, 'Tulipas', Daffodils, 'Jacinths' (including *Scilla autumnalis* found wild 'at the hither end of Chelsey', p. 132), Garlics and Crocuses to the 'Flowerdeluces', 'Corn Flagges', Cyclamens, Anemones and 'Crowfoots'. He includes before Colchicum and Crocus a first account and picture[2] of the Virginian Spiderwort lately introduced by John Tradescant and now known by his name *Tradescantia*. There is a remarkable number

1 A friend of Johnson and a prominent apothecary, cf. below, p. 274.

2 *Parad.* p. 151 is an earlier picture than Johnson, *Gerard*, p. 49 (cf. Kew and Powell, *T. Johnson*, p. 61). By 1660 various colour-varieties had been raised from seed in England, cf. *The Garden Book of Sir Thomas Hanmer*, p. 84.

of Tulips,[1] not only species from Persia, Crete and Southern Europe, but garden varieties, early and late flowering, and of a wide range of colours and shapes. So, too, there is a large variety of Daffodils and of Anemones, in each case evidently the result of cross-breeding and careful selection. So, too, with Ranunculus, not only are species like the Globe flower (*Trollius europaeus*) 'which in the Northern countreys of England, where it groweth plentifully, is called Locker goulons' (p. 268) and the Cretan Cyprianthe (*Ranunculus asiaticus fl. albo*), which he apparently grew successfully,[2] described with accuracy, but here as elsewhere double-flowered and abnormally coloured varieties are well represented.

Then after a few Cranesbills, Saxifrages and the Soldanella which 'groweth on the Alpes...and will hardly abide transplanting' (p. 235) comes another of the great garden tribes, the Auriculas and Primroses. Of the former, although he states that none were known to the ancient writers, a very large number of colour varieties had already been raised; among them (included with apologies as 'the Blew Beares eares with Borage leaves') is *Ramondia pyrenaica* (p. 236). Of the latter the varieties are all 'sports' from the Common Primrose and Cowslip; and evidently the gardeners had searched carefully. Of Primroses there is the Single White, the Double Yellow, and the Single and Double Green—these being the large almost campanulate forms, still found, though rarely, in the wild: of Cowslips there are various double forms, the Primrose-cowslip hybrid, a form with the calix foliated called by him 'Jackanapes on horseback', and another, with calices narrow and divided and empty of flowers, called Green Rose Cowslips: to these he adds the Red and White Birdseye (*Primula farinosa*) 'which will hardly endure in our gardens' (p. 246).

Under 'Pulmonaria maculosa', the next plant to be described, comes the note 'found out by John Goodier,[3] a great searcher and lover of plants dwelling at Mapledurham in Hampshire' (p. 248)—which should surely belong to his other species *P. angustifolia*. 'Limonium Rauwolfii' (*L. sinuatum*) is placed among the Buglosses, and the Evening Primrose 'out of Virginia' (*Oenothera biennis*) among the Stocks. Of Campions, Colum-

1 Tulips had first been flowered in western gardens in 1559: see Gesner, *Cordi Hist. Stirp.* p. 213, where it is called Tulipa Turcarum, and said to have been grown from Byzantine seed. Crispin van de Passe (Chrispin Paas of Utrecht, the engraver), in his *Hortus Floridus* (Arnhem, 1614), has pictures of 36 different Tulips mostly called by the names of Dutch growers. The tulip-mania in Holland reached its height in 1636–7, when as much as 13,000 gulden was paid for a single bulb. English travellers, e.g. William Crowne, 1636, and Peter Mundy, 1640, comment vividly upon the craze.

2 Onorio Belli had sent seeds and plants of this to De l'Ecluse from Crete in 1594; cf. his letter in *R.P.H.* p. ccc.

3 See below, pp. 291–4.

bines and 'Larkes heeles' he gives many abnormalities; and with the last he joins the 'Indian Cresses' or 'Nasturtium indicum' (*Tropaeolum majus*) which 'was first found in the West Indies and from thence sent into Spain unto Monardus [Nicolas Monardes, whose book as *Joyfull Newes*[1] had been translated into English by John Frampton in 1577], from whence all other parts have received it' (p. 281). And so after a number of other species, Violets, Poppies, Camomiles, the Red Adonis, Sun-flower and Marigolds, he comes to the 'Carnations and Gilloflowers' in which he specially delighted. The 'great' doubles, 'Hulo', 'Grimelo', 'Granado', 'Chrystall', 'Savadge' and the rest, number about fifty named kinds; one, 'the Poole flower', is said to grow wild 'upon the rocks near Cogshot [Carisbrooke] Castle in the Isle of Wight'; several come from Mr Ralph Tuggy of Westminster, who was already making a name for himself by his skill with them (pp. 306–12). Pinks, Sweet Johns and Sweet Williams were less esteemed. Thrift (*Armeria maritima*), already freely used as an edging plant, is included with the Pinks.

Then come the Composites—some like the 'Friars' Crown' (*Carduus eriophorus*) rather curious plants for the garden—the Legumina, the Peonies, Hellebores and Helleborines including the famous Ladies' Slipper (*Cypripedium calceolus*)[2] which 'groweth in Lancashire near upon the border of Yorkshire in a wood or place called the Helkes, three miles from Ingleborough, the highest hill in England and not far from Ingleton as I am informed by a courteous Gentlewoman, a great lover of these delights, called Mistris Thomasin Tunstall, who dwelleth at Bullbanke near Hornby Castle' (p. 348), 'Lilly Convally' (*Convallaria majalis*), the Gentians (*G. lutea*, *asclepiadea*, *cruciata*, *acaulis*—called by him *verna*- and *pneumonanthe*), the Campanulas, including *Lobelia cardinalis* 'neer the river of Canada where the French plantation in America is seated' (p. 357), *Convolvulus*, *Datura* whose seeds he suggests as an anaesthetic for those whose limbs have to be amputated (p. 362), *Nicotiana*, 'Mervail of Peru' (*Mirabilis jalapa*),[3] 'Hollihockes and Floramours' (*Amaranthus*).

Then he discusses the Mandrake of which 'counterfeit roots' were apparently still being sold (p. 377), 'Love Apples' (*Lycopersicum esculentum*—Tomato) which 'grow naturally in Barbary and Ethiopia yet some report them to be first brought from Peru' (p. 380), the 'Maracoc' (*Passiflora coerulea*) of which he gives 'the Jesuites figure', showing the instruments of the Passion, for comparison with his plate (p. 395), and a

1 The Nasturtium is called by Monardes 'Flower of Blood' (cf. *Joyfull Newes*, p. 92), and is said to have been brought from Peru.

2 Gerard in his *Herball*, p. 359, had pictured it, described it as growing in Germany, Hungary and Poland, and stated that he had a plant of it in his garden received from his friend 'Master Garret Apothecarie'.

3 For a discussion of the use and danger of this plant by William Mount in 1580, cf. Gunther, *Early British Botanists*, p. 262.

large variety of shrubs and flowering trees including twenty-four different Roses, such as the double-yellow 'first brought into England by Master Nicholas Lete from Constantinople which, as we hear, was first brought thither from Syria; but perished quickly...afterwards it was sent to Master John de Franqueville, from which is sprung the greatest store' (p. 420). He has interesting notes on the Indian Fig (*Opuntia ficus-indica*), denying that it is the same as Pliny's Opuntia, which he identifies with a seaside plant (?*Salicornia* sp.), and on the Yucca, quoting Gerard's account of its introduction and stating that Gerard sent a plant to Jean Robin and that Vespasian, his son, sent one back to De Franqueville which 'now abideth and flourisheth in my garden' (p. 434). Other shrubs complete the Garden of Pleasure and 'now lastly (according to the use of our old ancient Fathers) I bring you to rest on the Grasse'—three ornamental kinds being described.

His second book, dealing with the Kitchen Garden, is of special interest as the first detailed account of the vegetables commonly grown for 'sallets' and other dishes. William Harrison, in the edition of 1587, had commented upon the disuse of herbs in England during the century between Henry IV and Henry VIII, but said that they were then coming into favour again. The bills of fare under Elizabeth and James I hardly bear out this claim;[1] and when in 1614 Giacomo Castelvetro, an Italian refugee, wrote his *Brieve Racconto di tutte le Radici*, he wished to show to the English how many more vegetables might be grown and eaten. So Parkinson's account, seventy years earlier than John Evelyn's *Acetaria, a Discourse on Salads*, is worth study.

Evidently he felt that the subject had not yet received enough attention; for in the introductory chapters not only does he insist upon the need to prepare and manure the ground properly and give careful directions for culture and planting, but he complains that seed is not produced in sufficient quantity and has still to be brought from beyond the seas. Nevertheless his list of plants is much longer than might be expected and shows signs of real advance both in the variety of species and in the method of treatment. Radishes, Lettuces, Carrots, Parsnips, Turnips, Cabbages and Leeks of various kinds come first; then Artichokes, Melons, Cucumbers and 'Pompions'; then a large number of herbs, Mints, Clary, Parsley, Fennel, Borage, Marigold, Orache, Beet, 'Patience' (or Monks Rhubarb, *Rumex alpinus*), Chives and Garlic; and then 'Sallet' herbs, Asparagus, Lambs' Lettuce, Purslane, Spinach, Endive, Succory, Chervil, Rampion (for its roots), Rocket and Cresses, Horse Radish as sauce for fish, Burnet used in wine, Skirrets, and Clove Gillyflowers to be pickled in powdered

1 Cf. an article in *Italian Studies*, II, No. 5, by the Mistress of Girton, Miss K. T. Butler, which contains much information on the subject as well as a full account of Castelvetro.

sugar. Castelvetro's *lattuca romana* he had received from John Tradescant, who 'first brought it into England' (p. 498) and grew it as Castelvetro recommended by tying up the outer leaves so as to whiten the heart (p. 469). Smallage (Celery) he knew but despised; 'his evil tast and savour doth cause it not to be accepted into meats' (p. 492). 'Cole-flower' (Cauliflower) was still rare—'it is hard to meet with good seed' (p. 469)—but was 'of the more regard at good men's tables' (p. 504). 'Sweet Parsley or Sweet Smallage' which degenerates in our soil till 'it becometh no better than our ordinary Fennel' he first saw in the 'Venetian Ambassadors Garden in the Spittle yard near Bishops gate Street' (p. 491). 'Potatoes' (now so-called) are of three sorts, the Spanish or sweet, the Virginian or 'solanum tuberosum esculentum' of Bauhin, and the Canadian or Artichoke of Jerusalem:[1] the latter are 'grown to be so common here with us at London that even the most vulgar begin to despise them' (p. 518). He brings the book to an end with Strawberries and a few medicinal herbs, Angelica, Winter Cherry (*Physalis*), Asarabacca and Liquorice (p. 533).

The third part deals with the Orchard and like its predecessors begins with introductory chapters, discussing the situation and lay-out, the making of a nursery both of seedlings and of stocks, the 'divers manners of grafting', the care and propagation of all sorts of trees, and particularly the culture of the vine. He deals in detail first with soft fruit including Grapes; then with stoned fruit including Almonds; then with 'Oranges', 'placed in a close gallerie for the winter time' (p. 584), Apples of which he distinguishes more than fifty named kinds, Quinces, six kinds, and Pears thirty-two; and finally Walnut, Horse Chestnut and Mulberry. This is followed by a 'Corollarie' or account of twenty-two other trees such as are planted for use or ornament or rarity: it includes several American novelties including the 'Buckes horne' Sumach (*Rhus typhinus*) and the 'Virginian Ivie' (*Vitis hederacea*).

Of the thousand or more plants thus described 780 are depicted on 109 full-page plates. Copied as many of these are from pictures already published, their arrangement involves a measure of redrawing; and in the main this is done with skill, though the cutting is coarse and the rendering of foliage often stiff and clumsy. A large number, for example the abnormal Primroses of p. 243, are original, and if not of much artistic merit are at least recognisably accurate. There is a full-face portrait of the author, a pleasant bearded man holding a nosegay, prefixed to the first edition; and it is interesting to compare this with the similar but older and more worried features of his likeness in 1640.

The book is full of allusions to contemporary gardeners and botanists, some of which have been already quoted. Many of these men supplied

1 Cf. above, p. 242 for their introduction by De Franqueville.

Parkinson with plants—Francis le Veau,[1] 'the honestest roote-gatherer that ever came over to us' (p. 88); Mathias Caccini, the 'nobleman of Florence' who had corresponded with De l'Ecluse and after his death sent a Daffodil to Christian Porret at Leyden (pp. 90 and 410); Vincent Sion of Flanders, who had a garden on the Bankside in London until 1620 and supplied another rare Daffodil to 'George Wilmer of Stratford Bowe Esquire'[2] and to Parkinson (p. 104); 'Mr Doctor Flud' (can this be Robert Flud the famous Rosicrucian?) who gave him seed of a Daffodil from the University Garden at Pisa (p. 92);[3] 'Mr Humfrey Packington (Pakington) of Worcestershire Esquire at Harvington' who gave a double-blush Anemone (p. 213); Master Brian Ball, Apothecary of Coventry (p. 357); 'Master Doctor John More' (Moore for whom see Munk, *Roll of R.C.P.* I, pp. 174–5) who sent seeds from Italy and Spain (pp. 372,[4] 401) and received others from Parkinson in exchange while he was at Padua (*Theat. Bot.* p. 612); Master William Ward, chief keeper of the King's Granary at Whitehall who had a house at 'Boram' (Boreham) in Essex (pp. 593, 610); and 'Master John Millen, dwelling in Olde Street who from John Tradescante and all others that have had good fruit hath stored himself with the best' (p. 575).[5]

In addition there are a number of references to literature. One to 'Master Gerard in his Herball' refers to his statement that he had raised from seed sent him from Strasburg a double white Clematis, and is somewhat sceptical; 'Clusius doubteth whether there be any such'; 'I doe much doubt whether the double will give any good seed' (p. 292). Others mention 'Basilius Beslerus that set forth the great booke[6] of the Bishop of Eystot [Eichstadt] in Germany his garden' (Basil Besler, *Hortus Eystettensis*, 1613) (p. 410); 'Prosper Alpinus in his booke of Egyptian plants' (Prospero Alpino, *De Plantis Aegypti*, Venice, 1592) (p. 410); 'the Collectour who is thought to be Joannes Molineus of the great Herball or History of plants and generally bearing Daleschampius name

1 From him De l'Obel also received seeds, cf. MS. *St. Ill.* II, p. 129; and a Grass, l.c. I, no. 68. He describes Veau as 'Salmuriensis' (of Saumur in Anjou) and says that he 'made botanical journeys into Portugal and other parts' (*St. Ill.* p. 32).

2 Cf. *Theat. Bot.* p. 1551, for his raising and losing a West Indian Locust-tree: Johnson, *Gerard*, p. 135, recording this Daffodil speaks of him as 'Mr Wilmot late of Bow' and p. 1543 says that he saw the Yucca flowering in his garden. 'Willmer's dowble Narcissus' is mentioned in *The Garden Book of Sir Thomas Hanmer*, p. 26, as grown at Bettisfield *c.* 1660. George Wilmer took his degree at Trinity College, Cambridge, in 1600–1 and came of a well-known Stratford family. He died in 1626 and was succeeded by a son of the same name.

3 Cf. also *Theat. Bot.* p. 531, where seed is of a Daisy.

4 Cf. also *Theat. Bot.* p. 1523.

5 Goodyer visited 'Millaine' and his garden in 1628 (Gunther, *E.B.B.* p. 54) and Johnson refers to his garden in Old Street in *Gerard*, p. 481.

6 It has very large pages and admirable plates, and was reissued in 1640.

because the finding and relation of divers herbs therein expressed is appropriate to him and printed at Lyons' (*Historia generalis plantarum* commonly called *Historia plantarum Lugdunensis*, published anonymously, but assigned to Jacques d'Aléchamps with assistance from Jean Desmoulins and Jean Bauhin, 1586–7) (p. 603); 'Fabius Columna in his Phytobasanos unto whom Clusius giveth the greatest approbation' (Fabio Colonna, *Phytobasanos*, Naples, 1592) (p. 272); 'Spigelius in his Isagoges' (Adrian Spieghel of Brussels and Padua, *Isagoges in rem herbariam*, Padua, 1606) (p. 372); 'Bauhinus upon Matthiolus' (Gaspard Bauhin who edited and enlarged the Works of Pierandrea Mattioli, Frankfort, 1598) (p. 516); 'Cordus, as it seemeth in his History of Plants' (Valerius Cordus, 1515–44, whose *Historia Stirpium* was edited by Conrad Gesner, Strasburg, 1561) (p. 425 and frequently).

These are sufficient to prove that Parkinson had at least done his best to master the literature of the subject; and this gives his work a distinction lacking to most other English writers. He is capable of bad blunders; 'Joachimus Camerarius saith in his *Hortus Medicus* [Frankfort, 1588] that in Borussia which is a place in Italy as I take it' (p. 202); for he had not travelled or had much education. But his knowledge is not merely that of a working druggist and gardener: he has real botanical science.

The *Theatrum Botanicum*, the largest of British Herbals and in John Ray's judgment a book fit to be named alongside the great continental histories of plants, was published in 1640 when Parkinson had given up his apothecary's business and devoted himself to his garden and his books. It acknowledges obligation on the title-page—'with the chiefe notes of Dr Lobel, Dr Bonham and others inserted therein'. It is dedicated to Charles I. A rather discursive preface 'To the Reader' states the losses and gains of its postponement, glances ironically at Johnson's speed and youth, denies envy of his achievement and urges age and experience and the importance of the subject as his excuse for writing. Commendatory letters in Latin from Sir Theodore de Mayerne and Sir Simon Baskerville called 'The Rich', the royal physicians, and from three Oxford doctors, Thomas Clayton, the Regius Professor and Master of Pembroke College, John Bainbridge, first Professor of Astronomy, and John Speed, son of the cartographer, two of whom use Parkinson's book as an excuse for praising their newly founded Botanical Garden; and eulogies in verse from Sir Matthew Lister; from John Maurice, possibly John Morris, the Regius Professor of Hebrew at Oxford or probably the 'Mr John Morrice Gentleman of Isselworth beyond Braindford [Isleworth and Brentford]' from whom he obtained a prescription for pleurisy (*Theat. Bot.* p. 790) and who got seeds from Virginia (l.c. p. 858), a list of which came into Goodyer's possession (Gunther, *E.B.B.* pp. 74, 370); from J. D. Leet (Jan de Laet) of Leyden; and from John Harmar of Oxford, afterwards

Professor of Greek, ring changes on the theme 'No night of Age shall cloude bright Parke-in-sunne'. Then without prelude save a list of the 'Classes or Tribes'—a very poor arrangement considering the date of its appearance—the author begins his first tribe, the 'Sweete smelling Herbes' with the Hyssop.

Of this arrangement the principle is not botanical but pharmacological; and pharmacology in Parkinson's day was still based upon the four elements and their corresponding tempers. His classes are seventeen in number, and as typical of the age of transition between herbalism and science they are worth notice. The 'Sweete smelling' serve to link this book with its predecessor; for these are garden-favourites, fragrant and culinary. The next four tribes, Cathartic, Narcotic (Poisons and Antidotes), Nephritic (Saxifrages—plants which 'break the stone' in two senses), and Vulnerary, are primarily for the Apothecary; and the selection shows how powerful was the belief in 'correspondences' or the still more precise 'signatures', and in the traditions which reinforced them. Thus 'a wreath of Periwinkle [*Vinca minor*] worne about the legs defendeth them from the crampe' (p. 384); 'the dried herbe Paritary [*Parietaria officinalis*] made up with hony into an Electuarie is a singular remedy for any old continuall or dry cough' (p. 437); 'Wintergreene [*Pyrola rotundifolia*] is very cold and drying and exceeding astringent and glutinous withall whereby it is a most singular remedy for greene wounds' (p. 510); and 'Muscus ex cranio [the mildew on human skulls in charnel-houses] is in our times much more set by to make the Unguentum sympatheticum which cureth wounds without local application of salves' (p. 1313)—every sentence invites research into the reasons which suggested it. Tribes 6 and 7 are the cool and the hot plants—such of them as have not already appeared. Then follow six tribes which show the beginnings of real classification—Umbelliferous; Thistles and Thorny Plants (including *Acanthus*, *Ononis*, *Rhamnus*, *Rubus*, *Juniperus* and *Salsola kali*); Ferns; Pulses; 'Cornes' (including Maize, Millet and Buckwheat); and 'Grasses, Rushes and Reedes': in this last there is a real advance in the number and arrangement of species. Then ecology gets a share and Class 14 is labelled 'Marsh, Water and Sea plants with Mosses and Mushromes': as it includes sea-weeds, corals and sponges, its contents are decidedly mixed. Class 15, 'the Unordered Tribe', is frankly 'a gathering Campe to take up all those straglers that have lost their rankes' (p. 1325). Class 16 is 'Trees and Shrubs' from the Oak to the Black Currant. Class 17, 'Strange and Outlandish Plants', is an alphabetical list of drugs and perfumes including, besides a number of foreign herbs and spices, ambergris, 'Blatta Byzantina' (apparently an ointment made from the operculum of the 'Purpura or purple Periwinkle'), 'Mumme' (embalmed Egyptians),[1] 'Spermaceti'

1 This recalls the appearance of 'mummy' in the *Grete Herball* of 1526.

('found in the head of one onely sort of Whale fish', p. 1607) and 'Unicornes Horne' (of which, after describing the Indian wild ass and the value of the horn against poison, he adds, 'it is somewhat probable that even all those hornes formerly mentioned both in France, Venice or elsewhere and that also of our Kings kept at Windsor or the Tower, is but of the Sea Unicorne, for even such was brought home by Sir Dudley Digges[1] found on the shore and cast up by the sea', p. 1611). There follow a short Appendix (pp. 1672–88) and a competent index of Latin names.

His method within the Tribe is to group into chapters plants which bear a similar name or are related by shape of leaf or of flower or by similarity of structure. On occasion there are curious muddles: thus for example on p. 99, among the Wormwoods, there appears *Senecio cineraria*; in the chapter headed Germander, pp. 104–7, we have *Teucrium botrys*, *Dryas octopetala*, *Paederota bonarota* and *Veronica chamaedrys*; on p. 524 three colour varieties, blue, white and blush, of the flowers of Bugle (*Ajuga reptans*) are treated as distinct species; under Sanicle, pp. 532–4, are *Sanicula europaea*, *Pinguicula vulgaris*, *Saxifraga geum* and *Cortusa matthioli*; and in the chapter on 'Mountaine Chickweede', p. 762, the first three pictured are *Stellaria aquatica*, *Lamium amplexicaule* and *Veronica hederaefolia*. But in the main his 'genera' are more successful than his larger groupings. Indeed his chief defect is that he is almost wholly uncritical in accepting distinctions. Not only are flower-colour, size and other purely superficial differences made sufficient for a new heading, but in many cases where different authors have described the same plant, especially if from different places, these are set out as separate. In some chapters, as in Book IX, chap. 25, where there are four different 'Knotberries', all of them *Rubus chamaemorus*, these mistaken distinctions are numerous; and the book is swollen by them beyond all reason. A list of these duplications is given by How as an appendix to his edition of De l'Obel, *Stirpium Illustrationes*.

Within the chapters his method is generally to number each 'species' and give a short description of it. This is often, indeed perhaps generally, his own and usually indicates how the particular plant differs from others next to it. He has neither the fulness nor the precision of Thomas Penny; is indeed not very accurate or scientific: but he has a good vocabulary and a pleasant habit of using homely and effective comparisons. Many of his accounts are certainly derived from others, and this is often indicated by the mention of a name in the title of the plant. But he seldom or never gives any direct reference to the source of his record. Indeed, the only approach to this that I have noted is on p. 984: 'This Teasell whereof

1 From his journey to Russia in 1618 when John Tradescant accompanied him: see above, p. 244.

I have no knowledge and but follow mine author [an unnamed German apparently] whom I will not so farre mistrust as to say there is none such, for who knoweth all the diversities that other countries doe produce?' But it is not fair to make this the basis for a charge of plagiarism. He is doing what everyone of his contemporaries did—even if a few like De l'Ecluse were more careful to acknowledge the origin of their specimens. His woodcuts, like his descriptions, are seldom original. Very many of them are identical with those of Plantin, though careful comparison shows that they are not printed from the same blocks. Some (e.g. nos. 4 and 5 on p. 509) are misplaced.

Then follow sections labelled 'Place', 'Tyme', 'Names' and 'Vertues'; and under these the various species are sometimes grouped together and never very clearly separated. Parkinson apologised in his introduction for not distinguishing each plant, but urged rightly that his method saved space. But the result is that locality is not always or often exactly defined and that generalisations are unpleasantly frequent.

There is indeed a curious feature about his treatment of Place. On certain occasions, few in all and most of them worth noting, he gives details with great precision. For example, whereas his ordinary note is illustrated on p. 273 by the clause 'Most of those Centories are found in our owne country in many places', on this occasion he adds, 'The first yellow Centory [presumably he means *Blackstonia perfoliata*] groweth in many places of Kent, as in a field next unto Sir Francis Carew his house at Bedington neare Croydon[1] and in a field next beyond Southfleete Church towards Gravesend'—the first of which may be compared with Gerard, *Herball*, p. 1066, where he says of 'Lande Caltrops', 'I founde it neare the Parke of Sir Fraunces Carewe neere Croidon'.

Here are some of Parkinson's longer notes: of *Rubia peregrina*, p. 275, 'it groweth in many places in our owne land, as at a place called Hodhill in Dorsetshire, on that side is next to the river, in the parish of Stompaine [Stourpaine] two miles from Blandford; at Warrham likewise in the same shire on a mud wall in the same towne; and at a place called Somerpill, neere to a Chappell, which is by the landing-place as ye come from Astferry to Chesell[2] in great abundance'. Of Scurvygrass (*Cochlearia alpina*) (p. 286): 'It hath been also found growing upon Ingleborough hill in Lancashire, assured me by a worthy Gentlewoman, Mrs Thomasin Tunstall, remembered in my former booke [*Paradisus*, p. 348]...I heare

1 This branch of the family had been at Beddington since the time of Edward III. Sir Nicholas, master of the horse to Henry VIII, beheaded 1539, was Francis' father. Francis died childless in 1607.

2 On p. 500 he speaks of 'the marshes neere Ast ferry in Glocestershire', that is, Aust on the Severn. Chesell Pill is marked on the map of Gloucestershire in Camden, *Britannia*, as the inlet running into the Severn south-west of Aust, near the modern tunnel. Somerpill cannot be traced.

also that it groweth nigh unto a Castle in the Peake of Derbishire which is 30 miles distant at the least from the sea, and that the late Earle of Rutland had some brought from thence.' On p. 348, *Atropa belladonna*: 'At Framingham [Framlingham] in Suffolk, under Jesus Colledge wall in Cambridge, at Ilford in Essex, at Croydon among the elmes at the end of the town: at Moore Parke[1] in the Parke of Sir Percivall Hart at Lellingstone[2] in Kent on the Conny burries [Rabbit warren], in Burling Parke likewise, as also in the way that leadeth from S. Mary Cray to Footes Cray over against the gate of a great field called Wendell.' On p. 429 *Parnassia palustris*: 'In the Moores neere Lynton[3] and Cambridge at Hesset and Drinkestone in Suffolke, in the Butchers close thereby; in a meadow close on the backside of the Parsonage house of Burton, and at the bottome of Barton hills in Bedfordshire [five miles north of Luton] as also in the middle of the great Townefield of Hadington [Headington][4] which is about a mile from Oxford, and on the other side of Oxford in the pasture next unto Botley in the high way.' On p. 621 *Cardamine bulbifera*: 'At Mayfield in Sussex in a wood called Highreede, and in another wood there also called Foxholes, both of them belonging to one Mr Stephen Perkhurst at the writing hereof.'[5] On p. 699 *Polygonatum multiflorum*: 'Beside those that Gerard hath named it groweth in a wood two miles from Canterbury by Fishpoole hill, as also in a bushie close belonging to the Personage of Alderberry [Alderbury] neare Clarindon two miles from Salisbury, the next close is called Speltes, and in Chesson wood on Chesson hill betweene Newington and Sittingbourne in Kent.' On p. 729 *Sedum roseum*: 'Upon the mountaines of Pendle and Ingelborough oftentimes on the very raggiest places and most dangerous of them scarce accessible and so steepe that they may soone tumble downe that very warily do not looke to their footing, from whence hath beene sent me some rootes for my garden' (perhaps by Mistress Thomasin

1 Is this the Herts locality in which Goodyer found *Cicuta virosa*?

2 Lullingstone, 7 miles east of Foot's Cray: cf. also p. 407 where the fields about Sir Percival's house are given for *Gentiana pneumonanthe*—a locality added to those copied from Gerard.

3 From Linton he had reported *Galega officinalis* 'growing wilde in the Medowes' (p. 418).

4 Given as Herington, p. 534.

5 Evidently derived though without acknowledgment from Goodyer, who after a tour in 1634 noted bulbifera 'at Mayfield in a wood of Mr Stephen Penckhurst called Highwood' (Gunther, *E.B.B.* p. 74, or 'Highreed', l.c. p. 186). Presumably Parkinson got this record from him direct: it was not published by Johnson. The original MS. note printed by Gunther is in Goodyer's hand but very illegible (Magdalen Coll. Libr. *Goodyer Misc. MSS.* p. 61). The Pankhursts were a Mayfield family; and 'reed' or 'rede' (a clearing in the forest) is a common Sussex place-name. Foxhole Farm is in the neighbouring parish of Hadlow Down (*Place-names of Sussex*, p. 396).

Tunstall aforesaid; and perhaps she wrote this charming note of the locality—which is surely not by Parkinson!). On p. 1045: *Ceterach officinarum*: 'About Bristow and other the West parts plentifully as also on Framingham Castle, on Beckensfield Church in Barkshire[?], Strowde in Kent and elsewhere', and p. 1050 *Asplenium ruta-muraria*: 'In many places as at Dartford and the bridge at Ashford in Kent, at Beckonsfield in Buckinghamshire, at Wolley [Woolley] in Huntingtonshire, on Framingham Castle in Suffolke, on the Church walls at Mayfield in Sussex, and on the rocks neare Weston super mare in Sommersetshire.'[1]

The confusion about Beaconsfield is typical of a weakness in geography which we have noted in his former book. 'At Buckworth, Hamerton and Richwersworth [? Rickmansworth in Herts] in Huntingtonshire', p. 553, and 'in Porteland which is an island belonging to Cornewall not far from Plimmouth', p. 849, are perhaps his most notable blunders. But the record of three plants which were reported for him from the Wye valley is similar: on p. 425 of *Chrysosplenium oppositifolium* 'about Tidnaham in the Forrest of Deane, at Ashford and Iden in Kent, at Chepstow in Essex'; on p. 455 of *Asparagus officinalis* 'in the marshes of Tidnam neare Chipstoll [*sic*] and in Apleton medow in Glostershire which is about two miles from Bristow from whence the poore people doe gather the buddes or young shootes and sell them in the markets of Bristow much cheaper than our garden kinde is sold at London'; on p. 538 of *Alchemilla vulgaris* 'in the pastures nigh Tidnam and Chepstow'—Tidenham stands in the angle between Wye and Severn and south of the Forest of Dean.

Presumably these and other localities in the north and west were supplied to him by friends, some whose identity can be guessed,[2] some doubtless with illegible handwritings—which may explain the indecipherable names under *Rubia peregrina* and near Chepstow. Parkinson himself almost certainly wrote a script very difficult to decipher;[3] and he was an elderly man copying carelessly and inexpert in the correction of proofs. He had himself evidently travelled very little, although, as has been indicated, he shows some knowledge of Kent, especially the north coast, Dartford and Gravesend and Rochester, and the salt marshes by Faversham (for *Peucedanum officinale* there see p. 880) and Canterbury. He must surely have explored the Romney Marsh district: for he reports the Aldborough Peas (*Lathyrus maritimus*) 'at Rie at Pemsey [Pevensey] in Sussex, at Gilford [Guldeford] in Kent over against the

1 Of these two ferns many localities are given by Gerard: but only Bristow and Dartford coincide with Parkinson's.

2 There seems no clue to the source of his important record of *Arbutus unedo* 'of late dayes found in the west part of Ireland', p. 1490.

3 Gunther, *E.B.B.* p. 266, printed a letter signed by him, and on pp. 358–71 lists supposedly written by him among the Goodyer MSS.: these are very hard to read.

Comber [Camber Castle]', p. 1060; *Urtica pilulifera* 'naturally growing time out of minde, both at the towne of Lidde and in the streetes of the towne of Romney in Kent where it is recorded Julius Caesar landed'—hence the name Romania and the sowing of the Nettle as an antidote to the cold of Britain, p. 441; *Mercurialis annua* 'by a village called Brookeland in Rumney Marish in Kent', p. 297; *Solanum dulcamara* with white flowers 'by Saint Margates Church in Rumney Marsh', p. 350; and the Christmas-flowering Thorn (*Crataegus oxyacantha*) 'in High street or Whey street in Rumney Marsh',[1] as well as at Glastonbury, p. 1025: 'in Rumney Marsh' also he reports the occurrence of what seems to be *Malaxis paludosa*, p. 505.

But it is naturally in London that his localities are most detailed; and here both sides of his interest are active. Thus on p. 1039 of *Osmunda regalis*, 'I tooke a roote thereof for my garden from the bogge on Hampsteed Heath not farre from a small cottage there'—this being the locality reported by Gerard, *Herball*, p. 969;[2] and on p. 1173 of *Scirpus sylvaticus*, 'we have sometimes found it in our simpling walkes betweene London and Kentish Towne in the bottome of a field'. First 'in my garden in Long Acre' he discovered *Scrophularia vernalis* 'growing naturally, not in any of my sowen beds but straglingly in wast places', p. 609; then in the neighbourhood of Gray's Inn—*Saxifraga granulata*, p. 423, 'in grassie sandy places on the backeside of Grayes Inne, where Mr Lambes[3] Conduit heade standeth'; *Barbarea vulgaris*, p. 820, 'in the next pasture to the Conduit head behind Grayes Inne that bringeth water to Mr Lambes Conduit in Holborne'; and *Lactuca serriola*, p. 814, 'on a high banke by the footeway going downe Grayes Inne lane unto Bradford bridge'.[4]

Further afield *Brassica rapa*, p. 864, 'I found going from Shorditch by Bednall Greene to Hackney'; *Stachys palustris*, p. 588, 'by the path sides in the fields going to Chelsey and Kensington, by Hackney in the ditch sides of a field called the shoulder of mutton field, and in Surry in S. Georges fields, and in the medowes by Lambeth and under that tree by Stangate over against Westminster bridge which standeth alone upon the banke and none else: in Kent by Southfleet and on the backside of the Churchyard of Nettlesteede, hard by Sir John Scot's house, and in the middle of the next field to the Lime Kilne at the foot of Shooters hill'; *Damasonium alisma*, p. 1245, 'I have gathered in the ditches on the left hand of the Highway from Halloway to Highgate'; *Lathyrus nissolia*,

1 From Goodyer, Gunther, *E.B.B.* p. 74.

2 Johnson, *Gerard*, p. 1131, may have been mistaken in saying that it was all destroyed there, for he reports it in his list of Hampstead plants (ed. Ralph, p. 44).

3 William Lamb, a London merchant, rebuilt the conduit in Holborn in 1577: it was taken down in 1746.

4 For this, cf. p. 453: 'Bradford Bridge at the lower end of Grayes Inne Lane by London.'

p. 1079, 'about the hedges and bushes towards Highgate, Pancras church,[1] etc.'; *Vaccinium myrtillus*, p. 1458, 'Hampsteede Heath, Fincheley and Saint Johns wood'; *Alchemilla arvensis*, p. 449, 'neare unto the meere-stones by Lambeth which divide the liberties of London from Surrey'.

A man so sedentary as Parkinson obviously relied for his plants upon more-travelled friends.[2] Chief of these were the two already familiar: John Tradescant, whose son John the younger was out in Virginia and brought home with him the 'Berry-bearing Fern' 'in this present year 1638, presently after the death of his father' (p. 1045); and Wilhelm (Guillaume) Boel the Frieslander, who, trading with Portugal, Spain and North Africa, brought to England a large supply of seeds and plants, particularly a consignment from Spain in 1608 of which some came to De l'Ecluse at Leyden and were noted in his *Curae posteriores*, pp. 32–9. He is described as 'Doctor Boel who is now resident at Lishborne [Lisbon]' (p. 483): but otherwise we are told little of his history—though according to De l'Obel, *Stirp. Ill.* p. 51, he had done some botanising in London, and found a Sedge (perhaps *Carex hirta*) at Highgate. He was apparently sufficient of a botanist to be capable of recognising the importance of grasses and sedges as well as flowering species. Boel's relationship with Parkinson seems to have become a little difficult. The two passages on pp. 707 and 1064 in which his complaint is expressed do not wholly clear it up. They are as follows: 'This [*Erodium gruinum*] and the Third [*Geranium?*] among a number of other seeds were brought to me by Guillaume Boel which he gathered in Spaine upon my charge; however, Mr Goodier getting the seeds from Mr Coys caused it and divers other things to bee published in his name: notwithstanding I told him the charge was mine that procured it and many other'; and 'All these Cichelings grow in Spaine and from thence were brought with a number of other rare seedes besides by Guillaume Boel and imparted to Mr Coys of Stabbers [Stubbers] in Essex, in love, as a lover of rare plants, but to me of debt, for going into Spaine almost wholly on my charge hee brought mee little else for my mony, but while I beate the bush another catcheth and eateth the bird: for while I with care and cost sowed them yearely hoping first to publish them, another that never saw them unlesse in my garden, nor knew of them but by a collaterall friend, prevents me whom they knew had their descriptions ready for the presse.' This rather cryptic complaint is made clear by reference to Goodyer's records published by R. T. Gunther, *Early British Botanists*, pp. 138–41, 146, and to Johnson's *Gerard*, appendix, pp. 1627–8. John Goodyer saw the plants grown from

1 Johnson, *Gerard*, p. 1250, 'in the medow grounds about Pancridge Church'.

2 Gunther, *E.B.B.* p. 358, suggests that the lists of exotic plants among the Goodyer MSS., written apparently by Parkinson, were given by him to travellers or merchants, whose business took them overseas, as a guide to them in selecting what to bring back.

Boel's seed in Parkinson's garden in 1616. Four years later he received seed of them from William Coys. Descriptions of the plants grown from this were sent by him to Johnson, who, knowing their history and that Parkinson was preparing a book, added them to his edition of Gerard as an appendix. It is hardly surprising that Parkinson felt ill-used. But Boel was hardly to blame; nor was Goodyer, who sent the notes to Johnson before Parkinson had written anything.

Another botanist to whom he refers freely is 'Mr George Bowles, a young Gentleman of excellent knowledge in these things' (p. 390), who not only supplied him with a large number of localities near his home at Chislehurst (as in the passage cited for *Paris quadrifolia*) and in Shropshire (p. 954), and 'by the shadie woods sides of the mountaines and their vallyes in Wales' (p. 297), where he found *Impatiens noli-tangere*, but on occasion gave him seeds. Bowles was, as we shall see, already acquainted with Johnson, and was then and later a man of 'excellent knowledge in Herbarisme' (p. 1241).

John Goodyer, already mentioned in the *Paradisus*, reappears as the discoverer of *Sagina nodosa* 'on a boggy ground below the red well of Wellingborough in Northamptonshire' (p. 428) (cf. Johnson, *Gerard*, p. 568), of *Geranium lucidum* 'in our owne Country' (p. 708, *Gerard*, p. 938); of *Carum segetum* 'in the fields among the corne in divers places' (p. 931, *Gerard*, pp. 1017–18); of 'Capons taile Grasse' (? *Panicum viride*)[1] (pp. 1162–3, *Gerard*, pp. 30, 29), and of the 'Christtide Greene Oake' at Malwood Castle in the New Forest (p. 1646). Parkinson claims to have had the knowledge of these plants from him direct and not through Johnson's *Gerard*:[2] he pays a glowing tribute to his zeal and skill.

In addition there are a number of less known helpers: John Newton, 'a chirurgion of Colliton in Somersetshire',[3] who brought from New England an Avens (p. 136) and *Lobelia cardinalis* (p. 596); Dr Anthony Sadler, 'a physician in Exeter, son to the elder Sadler an apothecary there', from whom he got the double-flowered *Hesperis matronalis* (p. 628); 'Master [Walter] Stonehouse[4] a reverend minister of Darfield in Yorkshire' who found *Viola palustris* (p. 755); William Quick 'a worthy apothecarie in his time' who found Alexanders [*Smyrnium olusatrum*] 'wilde in some of the Iles about our owne land' and sent its seed to Coys and Parkinson (p. 930);[5] Zanche Silliard 'an apothecarie of Dublin' who sent him *Drosera anglica*[6] and 'Dr Coote' [William Coote of Johnson,

1 Druce identified this as *Festuca myurus*: this is doubted by D. Stapf (cf. Gunther, *E.B.B.* p. 171).

2 As presumably he did of *C. bulbifera*; see above, p. 259.

3 Doubtless Colyton in Devon is meant.

4 Cf. *Alumni Oxon.* and Gunther, *E.B.B.* pp. 271–3, 348–51 and below, pp. 300–2.

5 Cf. Coys' Garden List in Gunther, *E.B.B.* p. 318.

6 Cf. below, p. 300, for Richard Heaton's comment on this.

Gerard, p. 305][1] who reported the same from Ellesmere (p. 1053); and 'Master Edward Hassellwood' who found a 'browne kinde of gumme growing on okes and sent some of it' (p. 1391). Houses belonging to Sir John Leveson at Rochester (pp. 517, 944), to 'Sir Thomas Lucees [a misprint for Lucas]'[2] near Colchester (p. 640), and to 'Mr Gouch a divine' near Southampton (p. 640) are also mentioned as localities for rare species. 'Master Doctor Matthew Lister', author of the commendatory verses, 'being in Venice got three or four seedes which he sent me' of the Rhubarb (*Rheum rhaponticum*) which had been brought from Thrace to Prospero Alpino at Padua, 'whence some apothecaries at Venice had it': from these '4 seeds' 'I and many other my friends as well in England as beyond sea have bin furnished' (p. 157)—a record to which Thomas Johnson had already referred in his edition of Gerard's *Herball*, p. 394.

Two Welsh references are of interest. The first is the list of localities for *Meconopsis cambrica* 'in many places of Wales in the valleys and fields at the foote of the hils and by the water sides, about a mile from a small village called Abbar [Aber], and at the midway from Denbigh to Guider [Gwydir] the house of a worthy gentleman Sir John Guin [Wynn][3] as also neere a woodden bridge that giveth passage over the river Dee to a small village called Balam which is in North Wales, and in going up the hill that leads to Banghor, as also nere Anglesey in the way to the said Sir John Guin his house' (p. 370); this is explicitly stated to have been 'found by Lobel in his life time'. The second is for *Oxyria digyna*, 'of this sort (if it be not the same with the second [*Rumex scutatus*] whereunto it is very like) no author ever made mention before now, and scarce is it knowne to any but the gentleman of Anglesey called Mr Morris Lloid of Prislierworth that found it on a mountaine in Wales and shewed it to Dr Bonham in his life' (p. 745). This last has an additional importance because although Parkinson couples Bonham's name with De l'Obel's on the title-page as one from whom he has drawn much matter, this seems to be the only reference to Bonham in the text. The place, now called Pres Iorwerth or on the maps Prys Iorwerth (in each case a shortened form of Preswyl Iorwerth), is a farm near Cerrigceinwen and not far from the Capel Mawr crossing on the road from Llangefni to Bodorgan; that Bonham has reproduced its name in so easily identifiable a form is testimony to his accuracy. Morris Lloyd, grandson of the poet, herald and antiquarian of that name, was murdered by Cromwellian soldiers in his barn in 1647; and the family seems to have died out with his son William.[4]

1 Chaplain to Lord Herbert of Cherbury: see below, p. 282.

2 The locality is drawn from De l'Obel, cf. *St. Ill.* p. 97, where the name is spelled correctly.

3 Cf. *D.N.B.* He died in 1626.

4 Cf. J. E. Griffiths, *Pedigrees of Anglesea and Carnarvonshire Families*, 1914; *Archaeologia Cambrensis*, III (1903), p. 280. I owe this information to my friend Mr G. M. Ll. Davies.

Of English books to which acknowledgment is made one of the most interesting is the hortus siccus of Simeon Foxe, youngest son of the martyrologist, and a very distinguished London doctor. On p. 265 Parkinson writes, 'both these sorts [of Thalictrum] I saw in a Booke of dried herbes belonging to Doctor Foxe, President of the Physitians Colledge of London', and on p. 941, 'two figures taken out of Dr Foxe's booke of dryed herbes which he had from Padoa garden'. Foxe was President of the College in 1634–40, but his stay in Padua where he took his M.D. degree must have been in the last decade of the sixteenth century. The two pictures are of dried leaves, exactly similar to those sent out by the garden in the books of pressed samples which were used as a sort of catalogue advertising their plants. Francis Willughby when he and Ray visited Padua in 1664 brought back a similar book dated 1620 and signed by Domenico Zanetti who was custodian of the garden, 1617–28;[1] and this is still preserved by Lord Middleton at Birdsall. So, too, John Evelyn, visiting Padua in August 1645, 'gave order to the gardener[2] to make me a collection for a *hortus hyemalis* by permission of the Cavalier Dr Veslingius [Johann Vesling] who was then [1638–49] Prefect and Botanic Professor as well as of Anatomie' (*Diary*, ed. Bray, p. 170).

To his English predecessors there are naturally a number of allusions, though less than we might have expected. Of Turner he only speaks some ten times and on most of these merely to cite his name along with other botanists as an authority for nomenclature—thus pp. 59, 162, 495, 613, 856, 988, 1026. On p. 893 of *Conopodium denudatum* he says, 'Dr Turner tooke it to be Apios', a reference to *Names*, p. Bi*b*; but probably derived from Gerard, *Herball*, p. 906. On p. 1311 of *Lycopodium alpinum*, 'Turner and Tabermontanus call it Chamaecyparissus and so doth Gerard, but they did not mean Lavender Cotton as Gerard doth, but a kinde of Mosse which Turner fitly Englished Heath Cypresse'—this alludes, of course, to the famous blunder of Gerard, and refers to Turner, *Names*, p. Ci*b*. The notable thing about these allusions is that no one of them implies any acquaintance with Turner's *Herbal*. All Turner's detailed discussions, for example on the identity of the various Wormwoods or of Britannica and Bistort (*Theat. Bot.* p. 392), and all his discoveries as of Scurvygrass (*Theat. Bot.* p. 286) or *Trinia glauca* at Bristol (*Theat. Bot.* p. 880) are wholly omitted. Considering the great length at which Parkinson discusses *Names* and the fulness with which he quotes authorities, it is very disappointing to find him almost entirely ignoring the greatest in this respect of his fellow-countrymen.

Penny is almost equally, but much more naturally, omitted. There are

1 Cf. P. A. Saccardo, *La Botanica in Italia*, 1901, II, p. 115.

2 Giovanni Macchion, custodian, 1631–94.

references to the plants which De l'Ecluse named for him—the *Anemone* (p. 334), *Swertia* (p. 404), *Hypericum balearicum* (p. 666), *Coronilla* (p. 1091) and *Cornus suecica* (p. 1461). But *Rubus chamaemorus*, although Penny's picture is reproduced and his plant is kept separate as 'Chamaemorus Anglica' from those described by Hesketh and after him by 'Mr Bradshaugh, a Gentleman of the country',[1] and from the Norwegian types, is inserted without any mention of him. To Hesketh there is one interesting reference, in a passage derived from De l'Obel's *Stirpium Illustrationes*,[2] concerning the plant which Parkinson says 'hath not beene related by any before' and names Lancashire Buglosse: it 'groweth in one of the Iles about Lankashire, there found by Mr Thomas Hesket' and is obviously *Mertensia maritima*. To Lyte there is perhaps only a single reference and that in regard to the name 'Aisweede or Axeweede' for *Aegopodium podagraria* (p. 943).

Gerard is, of course, very freely mentioned and his mistakes (as e.g. in the misplacement of figures) are noted (e.g. p. 846). 'His corrector's' emendations are also noted (e.g. p. 173) as is, on occasion, 'Mr Johnson's' failure to emend (e.g. p. 361). In the main Parkinson's words on p. 990 are a fair expression of his attitude: after pointing out a mistake of Gerard's he adds, 'many such faults have passed Mr Johnson's correction which I am loth in every place to exhibit knowing that none of us all can publish anything, but there may bee slippes and errours in many places thereof'. Considering that Gerard's book was by this time exposed as a piece of plagiarism and that Johnson had accepted a purely commercial proposition, to rush through a revision of Gerard for the express purpose of forestalling Parkinson's life's work, his generosity is very remarkable. Johnson was a much younger man; he admitted that his work was unduly rushed: there would have been every excuse for Parkinson if he had felt and written with bitterness. Yet apart from one complaint that Johnson has claimed originality where in fact the *Paradisus* had anticipated him (p. 1341) and another about Coys and Boel with its covert reference to those who have stolen his thunder, he keeps a worthy silence. Even in the matter of the Barnacle Tree his comment is without malice. It deserves quotation as an indication both of his temper and of his judgment. 'To finish this treatise of sea plants, let me bring this admirable tale of untruth to your consideration, that whatsoever hath formerly beene related concerning the breeding of these Barnackles, to be from shels growing on trees, etc., is utterly erronious, their breeding and hatching being found out by the Dutch and others in their navigation to the Northward as that third of

1 No doubt the 'Roger Bradshaghe' who sent John Redman's note on Chamaemorus to Johnson (cf. *Gerard*, p. 1629 and below, p. 278).

2 MS. (Magdalen College), II, p. 10. De l'Obel adds 'we tried to grow it at Hackney, but in vain: it died'.

the Dutch in Anno 1536 doth declare' (p. 1306).[1] It is pardonable that he prints alongside this a reduced copy of the famous picture of the tree!

Indeed, considering the vigour with which his contemporaries criticised one another, Parkinson devotes very little space to comment upon mistakes. Thus, for example, expressing his doubts about the true character of the 'Stoechas purpurea odorata' on pp. 70–1, he describes the controversy between Desmoulins and De l'Obel over the Chrysocome of Dioscorides at great length and, though himself disagreeing with Desmoulins, gives a full and unbiassed record of his contention. So, too, on p. 318, reproducing Mattioli's 'fained Leopards bane with flowers added by Lugdunensis [Desmoulins' Herbal]' and describing Gesner's attack upon it, he gives Desmoulins' reply and tentatively suggests a solution of his own. On occasion he can point out a mistake without camouflage: 'Our London dispensatorie or Pharmacopoeia Londinensis[2] in the description of Unguentum marciatum maketh Camphorata to be Abrotanum which is utterly untrue', but he adds, 'yet I think it may very well be the substitute or succedanium thereof for that oyntment' (p. 569). Similarly, in acknowledging his own mistake in following De l'Obel's opinion that *Daphne mezereum* might be the Chamaedaphne of Dioscorides, he concludes with a wise saw, 'Let the criticke carper examine this animadversion, but let the judicious convince me and I will yeeld' (p. 701).

The most important of such problems is, of course, that of his relationship with De l'Obel; for here, as we have already noted, William How, whose *Phytologia Britannica* was published anonymously in 1650 and will be considered in detail below, published in 1655 a small volume drawn from De l'Obel's unfinished *Stirpium Illustrationes*; and added to it a spiteful preface and notes drawing attention to Parkinson's plagiarisms from it. A short list of 'errors' by Parkinson is appended—but these are relatively few considering the huge size of the *Theatrum* and the age of its author: nor do they illuminate the question of plagiarism.

Parkinson's own account of his debt to De l'Obel—a debt acknowledged on his title-page—is given on p. 1060 after his story of the famous Aldborough peas. It runs, 'Mr John Argent,[3] Dr of Physicke of the Colledge in London, brought from thence also the whole plant such as you see is here figured, which he gave to Dr Lobel in his lifetime, to be inserted in his Workes, but he prevented by death failing to performe it,

1 This comment is obviously derived from Johnson, who prints Gerard's tale (*Gerard*, pp. 1587–9) and adds his own repudiation of it, citing J. I. Pontanus, *Rerum Amstelod. Hist.* II, chapter 22 (Amsterdam, 1611, p. 134), and Colonna, *Phytobasanos*, pp. xiv–xix. For this latter reference cf. De l'Obel, *Adv.* (ed. 1605), p. 456, where after quoting it he adds that barnacles had been sent to him from Scotland by J. Cleve, pharmacist to James I.

2 First published in 1618.

3 Of Peterhouse, Cambridge, President of the R.C.P. eight years, 1625–7, 1629–33.

I have by purchasing his Workes with my money here supplied'. How, in *Stirpium Illustr.* pp. 164–5, comments violently upon this, stating that the whole of this chapter of Parkinson (two pages including four pictures) is 'plundered from this Treasury and but one ravish'd Plant acknowledged', and then adds as interjections, 'whose Volumes were compleat The Title! Epistle! and Diploma affixed!' and after quoting Parkinson 'and murdered his genuine scrutiny in treacherous oblivion erecting upon this despoyld Fabrick that Theater on which is personated the great Rhapsodist'. The suggestion is that De l'Obel's book was finished, was bought and suppressed by Parkinson, and that in fact we have in him a repetition of Gerard's behaviour to Dr Priest. This is wholly untrue.

In the absence of any clear knowledge of the size and completeness of the 'Workes' left by De l'Obel, it was till lately not easy to estimate the justification for such a charge. Now that Gunther has found the whole of the *Stirpium Illustrationes* as De l'Obel left it, we can see that though How in his volume picked out only a relatively small part of the material for publication, the MS. containing some 835 plants in all is by no means even a major source of Parkinson's work. How printed the best and most interesting notes; and even these Parkinson hardly plagiarised. The remainder was merely a random series of even shorter and more scattered material than appears in How's volume: and it is not the fact that Parkinson inserted most or all of it. He has taken very little, for the simple reason that very little was worth taking. His tome is avowedly a compilation: he copies whole sections almost word for word from his own *Paradisus*; he is constantly mentioning and quoting from De l'Obel; and he makes no pretence that the bulk of his work is original. But, even if he had incorporated all that De l'Obel had left, certainly the vast majority of his chapters would have remained unaffected.

Indeed, there is no secret about the sources of the *Theatrum*: John Ray described them adequately in a letter to John Aubrey (*Further Correspondence of J.R.* p. 162). The book was based upon Gaspard Bauhin's *Pinax*, published at Basel in 1623: to this were added some of Jacques Cornut's Canadian plants, Paris, 1635 (p. 333); some from Giovanni Battista Ferrari, Rome, 1633 (p. 1526); from Jan de Laet (p. 600); from Castor Durante, Rome, 1585 (pp. 31, 867); from Paul Reneaulme, Paris, 1611 (p. 1400); and from Tobia Aldini, physician to the Cardinal Odoardo Farnese and reputed author of the account of rare plants in the Farnese garden, Rome, 1625—a work really written by Pietro Castelli (pp. 183, 238, 1488): there were allusions (p. 1207) to work as late as Johann Vesling's edition of Prospero Alpino issued at Padua in 1638. Most of these books were published after De l'Obel's death, so that the vast amount of Parkinson's work in which they are cited and discussed could not be derived from the *Stirpium Illustrationes*.

Nor with every respect to De l'Obel is his contribution to the *Theatrum* always an advantage. We have already noted some of the plants which he reported erroneously as English; and his own descriptions and localities, though often more exact than Parkinson's, are not on a high level of accuracy. As a botanist he is not at all in the same class as either of the Bauhins, and their prominence gives the *Theatrum* its chief claim to merit. It is from Gaspard Bauhin's *Prodromus* and *Pinax*, the latter being by far the most complete catalogue of synonyms ever published, that most of the quite excellent discussions of nomenclature and the references to other and less known European writers are derived. Certainly Parkinson knew them well, even if he translated and printed them badly—as witness his reference to *Trientalis europaea*, on p. 510, 'in the Beeche wood [a mistake—in betuletis] in Scotland as it is recorded by Bauhinus, who saith Dr Craige [a mistake—Bauhin wrote D. Cargillus, *Prodromus*, p. 100] sent it him from thence'; to Myagrum paniculatum (*Neslia paniculata*), on p. 870, 'Bauhinus saith it grew with his brother, John Bauhinus, at Mount Belgrade (in horto Montembelgardiaco) [Montbéliard] by the name of a Myagrum' (cf. *Prodromus*, p. 52); and to the abundance of Gratiola[1] 'without St Justines gate at Padoa' (*Theatrum*, p. 221), which is in *Prodromus*, p. 108, 'ad portam S. Justinae', where Bauhin himself had gathered it.

Of these other sources the most important have been already mentioned, either as found in the *Paradisus* or in the account of more recent books. It was from Gaspard Bauhin, *Pinax* (e.g. pp. 222, 456), that he got knowledge of the plants raised 'in that curious garden of that Venetian Magnifico Signior Contarini'[2] to which he freely refers (e.g. pp. 28, 209, etc.). It was from De l'Ecluse, *Exotica*, that he learnt of 'Honorius Bellus [Belli, whose letters to him from Crete were published in his *Rariorum Plantarum Historia*, pp. ccxcix–cccxiiii] a famous physitian living long in Candie' (pp. 238, 293), and 'attesting that in the mountaine Ida in Candy there groweth in great plenty a white Peony' (p. 1380): from De l'Ecluse, too, that he drew his knowledge of the *Observationes* of Petrus Bellonius (Pierre Belon) the great French traveller and writer on birds and fishes[3] (pp. 101, 254, 293). So, too, it was 'in Mr Purchas his fourth booke of Pilgrimes'[4] that he found the records of 'Mr William

1 Bauhin's picture is plainly *Lythrum hyssopifolium*, but Parkinson attaches it to the 'True Hedge Hyssop' (*Gratiola officinalis*).

2 Bauhin, who mentions him on many occasions in his *Prodromus*, e.g. pp. 26, 70, 76, 103, calls him 'Niccolo Contarini a Senator of Venice'. Parkinson, l.c. p. 28, speaks of him as if his garden was at Padua, but the only member of the famous family to whom this could apply is Vincenzo who held a special professorship there for a time.

3 De l'Ecluse translated the record of his travels from French into Latin, and this was published by Plantin in 1589.

4 Samuel Purchas, *Pilgrimes*, published 1625: Parkinson's references are to this edition, those to Finch on the same page, Vol. 1, p. 429.

Finch an English merchant' as set out on pp. 376, 601, 1248, and dealing with plants at Agra in the Mogul's country seen in 1610; in book III, p. 236, of the same volume, the report of Mr Joseph Salbanke written to Sir Thomas Smith, Governor of the East India Company, in 1609, and describing the Indigo at Bianie (p. 602); and in volume iv, p. 1369, the story of 'Master Lewis Jackson dwelling in Holburne' who saw the 'Fountain Tree' at Ferro [1] (p. 1645). But some of his contacts are more immediate; as in regard to Javanese plants, 'Dr Justus Heurnius, both divine and physician,[2] for the Dutch factory in the Kingdom or Ile of Java sent into Holland a small booke or collection of certain herbes, etc. growing in that country with the vertues and uses (which booke as I understand by my good friends, Dr Daniel Heringhooke[3] [Herringhoeck] and Dr William Parkins, both English, is kept in the University Library at Leyden in a close cupboard having a glasse window before it)' (p. 376); or 'I had the knowledge thereof [of a *Trillium*] given me from Mounsier Loumeau of Rochell, preacher,[4] who had it out of Canada' (p. 390); or of a Cretan *Sedum*, 'it was sent me from Hieronymus Winghe[5] (Jerom van Wingen] a Canon of Tournay in Flanders, who received it from Candy' (p. 722); or of *Buxus sempervirens*, 'Fernelius [Jean Fernel, the eminent French physician] onely doth number the leaves hereof among things that doe purge, but the practice thereof is worne out of use: yet I remember that Doctor Smith [Richard Smith of St John's College, Cambridge, President R.C.P., died 1599], that was one of Queene Elizabeth's Physitions, appointed the decoction of an ounce of the leaves of Boxe for a purging medicine' (p. 1429).

Here are four pleasant reminiscences. The first is of 'English Tabacco' (*Nicotiana tabacum*), 'I have knowne Sir Walter Raleigh, when he was prisoner in the Tower, make choise of this sort to make good Tobacco of which he knew so rightly to cure' (p. 712). The second is of the Banana (*Musa paradisiaca*), 'I having tasted of one that Doctour Pay gave me did thinke I had tasted of an Orris roote preserved with sugar' (p. 1496).

1 The most south-western of the Canary Islands. Parkinson gives the page in Purchas wrongly as 1639.

2 Second son of Jan van Heurn, Professor of Medicine at Leyden. He wrote *De Legatione Evangelica ad Indos capessenda admonitio* (Leyden, 1618), an exposition of the need and character of missionary work, and was the translator of the Acts and reviser of the translation of the Gospels into Malayan: this version with an English preface and dedication to Robert Boyle was published at Oxford in 1677.

3 Incorporated at Cambridge from Leyden, 1639 (Venn, *Al. Cantab.* II, p. 359).

4 S. de Lommeau, Pastor of the Reformed Church at La Rochelle, and Moderator of the Synod in 1599.

5 De l'Obel, *Stirp. Illustr.* p. 131, says that he received seeds of Melons from 'Hieron: de Winghen'. The painter of the same name, though a contemporary, can hardly be the person meant; for he spent most of his life at Frankfort.

The third is of the Sensitive Plant (*Mimosa pudica*),[1] 'I have seene of the living plant as it grew in a pot at Chelsey in Sir John Davers [Danvers] garden[2] where divers seeds being sowne therein about the middle of May 1638 and 1639, some of them sprang up to be neare halfe a foot high...with severall branches and paires of winged leaves, containing some eight some ten small leaves on a side...foulding themselves upward close to the middle ribbe upon any touch thereof: this I proved in those two severall yeares' (pp. 1617–18). And here, finally, is another which illustrates the professional interests of Ludgate Hill—interests which were gradually replaced by those of Long Acre: 'The greater Gentians...are by their bitternesse so availeable against putrefaction, the plague or pestilence, that the Germans did formerly make a Treakle therewith at Iena which was transported into our country and we thereupon called it Iene Treakle...which Treakle was bitter and therefore the more likely to worke good effects; but that Ieane Treakle, which hath since crept into the place of it among the vulgar, because it is sweet and pleasant is for that cause greedily sought after and for the cheapnesse: there is nothing in it that can doe them good, being nothing but the drosse and worst part of Sugar taken from it in purifying...yet in London it hath beene upon occasion both censured and condemned by a Jury, and many hundred weights thereof beene publikely burned in the open streetes before their doores that sold it,[3] as a just witnesse to all that it is not a thing tollerable in a Common-wealth. I have thus farre digressed...' (pp. 407–8).

The *Theatrum* is an old man's book, diffuse and often ungrammatical in its long and ill-penned sentences, unequal in the quality of its contents, uncritical in its inclusion of species imperfectly defined, lacking the precision, the arrangement, and the accuracy of detail that mark the trained mind. It is a pandect, perhaps the last 'omnigatherum', and its collector is not a man of first-rate ability or profound scholarship, so that his experience and reading do not compensate for his lack of field-work and of scientific insight. Indeed, the secret of his worth is that, like Mistress Tunstall, he is 'a great lover of these delights'—that he has the authentic

1 Johnson reported that James Garret had received a dry plant from the Earl of Cumberland, who brought it from San Juan de Puerto Rico; Garret sent it to De l'Ecluse in October 1599, by whom it was figured and described. Then Johnson had seen and drawn a dried sprig in November 1632 'with Mr Job Best', *Gerard*, pp. 1599–1600.

2 ''Twas Sir John Danvers who first taught us the way of Italian gardens', says John Aubrey of his later garden at Lavington, Wilts. (*Nat. Hist. of Wilts*, ed. Britton, p. 93). His Chelsea garden, next to the mansion formerly belonging to Sir Thomas More, was laid out between 1610 and 1628: he died there in 1655. Aubrey (*Lives*, II, p. 226) records that Francis Bacon 'much delighted in his curious garden at Chelsey'.

3 For this method of punishing the sellers of bad medicines cf. Barrett, *History of Apoths*. p. 5.

passion for a garden and the quiet wisdom of a gardener, than which there are few things more precious. His lore of 'pleasant plants and their vertues' represents more truly than the work of any other early naturalist the ingrained tradition of folk who have lived near to the earth and learnt something of its ways. And this attitude of appreciation, an attitude as much aesthetic as scientific, an attitude widely spread among our fellow-countrymen, is more largely responsible than is commonly recognised for the achievements of our scientists. In the recent enthusiasm for his younger contemporaries, Goodyer and Johnson, his reputation has suffered even as it did at their hands in his life-time. So it is fitting that without disputing their superiority in botanical science or exploratory zeal, a tribute should be paid to the patient and painstaking author of what Ray described as 'the most full and comprehensive book of that subject extant'.[1]

Moreover, Parkinson had come to fill the position which Hugh Morgan had made in the previous century. Both men were essentially gardeners: their pharmacology led them to their gardens; their gardens turned them into botanists. Between them their lives cover the period which transformed England into a country of gardens; and in that change they were typical, and perhaps the most effective, agents. Gerard with his 'Blew Pipe the later physicians do name Lilach', his Judas Tree and his Laburnum; Coys, the first as Wilmer was perhaps the second to flower the Yucca in England, and who apparently acclimatised and let loose upon us *Linaria cymbalaria*; Richard Shanne growing some forty varieties of Tulip in his garden at Methley;[2] John Tradescant introducing Spiderwort and Snakeweed[3] and Scarlet Runners[4] and growing what was perhaps the first English Horse Chestnut;[5] John Millen stocking his grounds in Old Street with the best of fruit trees; Ralph and 'Mistresse' Tuggy making a reputation for Gillyflowers and Auriculas[6] at Westminster; these all, amateurs and professionals alike, represent the founders and benefactors of our horticulture. John Gerard had failed in 1596 to persuade the University of Cambridge to let him lay out for it a physic-garden:[7] but in 1622 Henry, Sir John Danvers' elder brother, afterwards Earl of Danby, had granted to the University of Oxford the site ever since appropriated to its Botanical Garden, and by 1640 Jacob Bobart had begun the work which was to immortalise him and his sons, and in which a great succession of students, Walter Stonehouse of Darfield; William Coles, author of

1 Letter to Aubrey in *Further Correspondence*, p. 159.

2 His list, like Coys', is printed in Gunther, *E.B.B.* pp. 301–21.

3 *Aristolochia serpentaria*: so Johnson, *Gerard*, p. 848.

4 Johnson, l.c. p. 1213.

5 Johnson, l.c. p. 1443.

6 Cf. *The Garden Book of Sir T. Hanmer*, p. 80, 'the first sower of them was one Tuggey in Westminster about thirty yeares since'.

7 Cf. the letter quoted by Gunther, *Early Science in Cambridge*, pp. 371–2, from Jackson, *Gerard's Catalogue*, p. xii.

The Art of Simpling in 1656 and of *Adam in Eden* in 1657; Robert Lovell of Christ Church, in whose *Pambotanologia* in 1659 'all plants not in the Physick Garden are noted with asterisks'; William Browne, chief compiler of the *Catalogus Oxoniensis* in 1658; William How, author of the *Phytologia Britannica* in 1650; and Christopher Merret, who replaced it with his *Pinax* in 1666, were to get their training.[1] Here was the beginning of a real scientific department—a pioneer which should show the way to similar or analogous development in other fields.

CHAPTER XVI. THOMAS JOHNSON AND HIS FRIENDS

Before Parkinson's *Theatrum* had been finished, its publication was forestalled by the decision of the successors of John Norton to rush through a new edition of Gerard's *Herball*. John Norton had died in the same year as Gerard, but the rights had been preserved and in 1632 were assigned by his cousin, Bonham Norton, to three partners, Adam Islip, Joice Norton and R. Whitaker. They commissioned Thomas Johnson to prepare a new edition on condition, as he tells us, that the work was done within a year (*Gerard*, p. 1591); he undertook the task, and presented a copy of his book to the Society of Apothecaries on 28 November 1633.

H. Wallis Kew and H. E. Powell in their admirable book, *Thomas Johnson, Botanist and Royalist*, admit that 'the position for Johnson cannot have been altogether a comfortable one' since he was a friend of Parkinson and was printing his translation of Ambroise Paré (the great French surgeon) with Thomas Cotes, Parkinson's publisher. This is to put the matter very mildly. Johnson knew that Gerard's book was very faulty: this was by then common knowledge; for De l'Obel had spoken with violence on the subject, even though his remarks had not yet been printed. He knew that Parkinson had announced the production of another work in 1629 and himself added, 'which I thinke by this time is fit for the Presse' (*Gerard*, pref.). He also knew that his rival was an old man for whom the delay arising out of his being forestalled might well be disastrous. Moreover he was on his own showing immensely busy, and was according to his biographers planning a full-scale British flora. It is difficult to reconcile these facts with the wholehearted eulogies of his

1 Coles in the preface to *Adam in Eden* gives a list of the 'very eminent Botanicks in the University of Oxford' who helped him: 'Mr Steevens, Principall of Hart-Hall in Oxford, Mr Lydall, Mr Brown, Mr Wit, Mr Hawley, Mr Beeston, Mr John Crosse, the apothecary', for the last of whom cf. Gunther, *E.B.B.* p. 304.

character which they and others have perhaps too generously heaped upon him. His behaviour may have been commercially legitimate: it is hard to justify by any other code—unless we recognise that he was a young man of remarkable energy and considerable ability with a career to make and that he had just got his feet on the ladder. The opportunity to show his fitness and the element of competition and a race against time were evidently too attractive to leave room for other considerations.

Of his gifts there was already excellent evidence. Born at or near Selby probably in 1604, educated perhaps at Pontefract, apprenticed in London to William Bell, an original member of the Society of Apothecaries, in 1620, he completed his 'servitude' in 1628, became a Free Brother of the Society and set up in business in Snow Hill at the sign of the Red Lion.[1] In 1626, a date mentioned several times in his *Gerard*, e.g. pp. 450, 630, 676, he had visited his native county, finding *Calamintha acinos* 'a little on this side Pomfret [Pontefract]', p. 676, *Campanula latifolia* 'upon the bankes of the river Ouse', near Selby (p. 450), *Anagallis tenella*, of which he drew a rather stiff and conventional picture, 'in the bishopricke of Durham and in two or three places of Yorkshire' (p. 630), *Empetrum nigrum* 'on the tops of the hills of Gisbrough, betwixt it and Rosemary-topin (a round hill so called) [Guisborough and Roseberry Topping in Cleveland]' (p. 1383), *Stratiotes aloides* 'about Rotsey [Rotsea near Cranswick] a small village in Holdernesse' (p. 825), and *Gentiana pneumonanthe* at 'a place called Netleton Moore' near Caistor (p. 438), this last presumably the result of a visit to a Lincolnshire relative. By this time too he had begun to explore the neighbourhood of London; for in August of the same year he discovered *Stachys arvensis* in a cornfield 'not far from Greenhive in Kent' (p. 699) and succeeded in identifying it in G. Bauhin's *Prodromus* —a note of some importance as proving the range and care of his studies.

Thus equipped, in 1629 he made his first simpling voyage into Kent, and printed a racy and very readable account of it.[2] The narrative of the ten gallant apothecaries, 'Jonas Styles, William Broad,[3] John Buggs,[4]

1 So *The Diary of Thomas Crosfield* (ed. F. S. Boas), p. 20.

2 This and Johnson's other small books were reprinted by T. S. Ralph in 1847. Full résumés are given in Kew and Powell, *Thomas Johnson*.

3 He contributed commendatory verses to Parkinson's *Paradisus*, and had evidently done some independent botanising; for Johnson (*Gerard*, p. 825) records that he had found *Stratiotes aloides* 'in the fennes of Lincolnshire'.

4 In 1632 Buggs was imprisoned as an 'empirick', or apothecary practising physic without license, at the instance of the Royal College of Physicians. The Apothecaries agreed to pay his fines. He was released and apparently went to Leyden and obtained his M.D. On his return he was almost appointed Physician at Christ Church Hospital: but the College succeeded in preventing this, and he was again forbidden to practise: cf. Goodall, *Historical Account*, pp. 413–19. He seems to have been a lively and popular person and is described in a petition of the Royal College of Physicians to the Lord Chamberlain as 'one of the Queen of Bohemia's players, sometime an apothecary'.

Leonard Buckner,[1] Job Weale, Robert Larking [Lorkin], Thomas Wallis, two Edward Browns, one being Broad's servant and myself', meeting at break of day at St Paul's and embarking in boats for Gravesend; of the storm, for which nothing but a quotation from the first Aeneid will suffice; of the separation and successful reunion of the boat-loads; of the visit to the *Royal Prince* with her sixty-six guns at Chatham; of the crossing to the isle of Sheppey and the grand scene of their appearance under suspicion before the Mayor of Quinborow (Queenborough);[2] of Buggs' speech in their defence, of Styles' comic interlude and of the draughts of beer in which any unpleasantness was drowned; these and the subsequent adventures are told with a heroic ardour and a version of the Latin tongue which are highly diverting. The lists of plants interjected between the various episodes serve to remind the reader that these are not the records of an earlier John Gilpin, nor of the Pickwick Club in its infancy. But botanising has seldom been more epically described.

Moreover both here and in the visit to Hampstead[3] a fortnight later, it is evident that some at least of the party and eminently its historian were very competent workers. If none of the plants noted are of outstanding interest, the records are an excellent beginning of a local flora, and seem to be accurate and trustworthy. The difficulty with them, as with all early lists whether of wild or garden plants, is that the names are often given without any botanist's authority and being accompanied by neither picture nor description are very hard to identify. 'Sedum medium vulg.', 'Viola purpurea', 'Chamaecyparissus' (in the garden of the Bull at Rochester), 'Hypericon ubiq.: in aridis' and such like cannot be defined with much confidence.[4] Moreover the printing (especially in this first book) and the Latin are not inerrant. It is generally easy to tell what Johnson means, but his grammar is often ill-calculated to express it.[5]

1 Renter Warden of the Society 1647, Upper Warden 1651, Master 1656: he died during his year of office: cf. Barrett, *History of Apoths.* pp. 63, 66, 68.

2 'Quinburrough consists off one smalle streete, accompted auntientt, having a Mayor and certaine priviledges'; so P. Mundy who visited it in 1640 (*Travels*, IV, p. 55).

3 For the fame of Hampstead, cf. Drayton, *Poly-olbion*, XVI, ll. 249–52:

'But Hampsted pleads himself in simples to have skill
And therefor by desert to be the noblest hill;
As one that on his worth and knowledge doth rely
In learned physick's use and skilful surgery.'

4 It is necessary to add that many of the identifications from the *Mercurius* accepted by W. A. Clarke, *First Records*, and G. C. Druce, *Comital Flora*, seem to me highly precarious.

5 A good instance of this is to be found in the name of the apothecary whose shop he visited at Sandwich. He refers to him as Caroli Anati, Carolum Anatum and Carolo Anato (*Opuscula*, ed. Ralph, pp. 25, 32). Was his name Charles Duck?

But in spite of its defects this first publication must certainly have drawn attention to the 'Socii itinerantes' and their leader, and made them known to other students. Johnson must have seen much of Parkinson during his apprenticeship; for as we have noted he contributed commendatory verses to the *Paradisus*. But in 1630 he met George Bowles of Chislehurst, the most enthusiastic and useful of the younger botanists whose work has hardly received the attention that it deserves, and found with him *Radiola linoides* 'on a heath not farre from Chislehurst' (*Gerard*, p. 569); and in 1631 he took wine with John Goodyer on one of his visits to London.[1] These two supplied him with much help, and share with him the credit for initiating a thorough exploration and description of the British flora. It is probable that the need and possibility of this were first fully brought home to him by the experience of his little book's reception and by the discovery that these new friends shared his interest in field-work.

In 1632 by the advice and at the charges of Thomas Hicks, Warden of the Apothecaries' Society, a longer excursion was planned.[2] Broad, Buckner, Lorkin and James Clarke[3] with Hicks and Johnson set sail on 1 August for Margate. Richard Pollard, the inn-keeper there, and Simon Rose, the local doctor, joined them in their search and a long list of plants was compiled, including the 'Plantain with spoky tufts' of *Gerard*, p. 420 (*Plantago major* var. *polystachya*) and *Alisma ranunculoides*, found by Johnson in company with Broad and Buckner 'in a ditch on this side Margate' (*Gerard*, p. 418), which is, however, not a first record (despite W. A. Clarke, *First Records*, p. 151 and G. C. Druce, *Comital Flora*, p. 310), since Penny had shown it to De l'Obel some fifty years earlier.[4] They visited Nash and its garden and Queakes (Quex Park), another house in the neighbourhood, formerly the seat of Sir Henry Crisp and well known to Gerard.[5] Next day they moved on to Sandwich and Sandown Castle, inspected the garden of a Belgian, Gaspard Niren, and (crowning wonder) were shown in the druggist's shop the stuffed skin of a sea-serpent fifteen feet long which had been killed on the sandhills while it was eating rabbits. The journey next day from Sandwich to Canterbury yielded a fine list of plants—five pages of his report. Sunday was spent in rest and worship. On Monday they started at dawn, broke their journey

1 Gunther, *E.B.B.* p. 56.

2 The account was published as *Descriptio Itineris Plantarum Investigationis ergo suscepti* by Thomas Cotes, 1632.

3 He and Thomas Smith had already accompanied Johnson on a simpling ride through Windsor Forest; cf. *Gerard*, p. 30.

4 Cf. above, p. 168. It is noteworthy that Johnson's picture of the plant is copied from De l'Obel's.

5 Cf. e.g. *Herball*, p. 192. For Sir Henry Crisp, grandson of the Sir Henry of Henry VIII's time, see *Visitation of Kent*, 1619, p. 74.

at Faversham to see the herb-garden of Nicholas Swayton, 'the honest and skilful apothecarie' of *Gerard*, p. 303, but even so got to Gravesend in time to get a boat back to London that night. To this record as to its predecessor is attached a list of Hampstead plants, now much more complete, but not arranged in any sort of order. There are a few synonyms added and in general the plants are easy to identify.

During July of this year he had visited the garden of John Tradescant in South Lambeth and seen there the 'Snake-weed' (*Aristolochia serpentaria*) and the 'Indian Painted Storks-bill' newly imported. In November 'being with Mr Job Best[1] at the Trinity house in Ratcliffe' he was shown and drew a dried specimen of the 'Sensitive herbe' (*Mimosa pudica*) of which stories were beginning to arouse interest. The picture is not a great work; Johnson had not Penny's skill with the pencil: but it testifies to his zeal for exploring and recording in all possible quarters.

So we come to the *Herball* and the remarkable change which Johnson made in it in the year 1632–3. In the main this falls into three types, correction, improvement and addition; and in each type he put in a very large amount of work.

First he had to deal with the errors of Gerard's book, the misplaced pictures, the confused species, the blunders of fact. The illustrations were in themselves a large undertaking; for the old plates obtained by Norton were not available and a very numerous set from the house of Plantin was used in their stead. Many of these were in fact similar; for Bergzabern had copied from his predecessors, and the same designs originating with Fuchs or even Brunfels reappear in all these collections.[2] But Johnson was more familiar with the continental authorities, especially De l'Ecluse, and was able to remove most of Gerard's mistakes and to add a number of new species, either unillustrated or unmentioned, in the *Herball*. He included his own pictures, the four that appeared in the *Descriptio*, the Sensitive Plant, and his masterpiece, the excellent drawing of a large bunch of Bananas which he received on 10 April 1633 from John Argent, then President of the College of Physicians: these came from Bermuda, were hung up in Johnson's shop, 'became ripe about the beginning of May and lasted until June: the pulp or meat was very soft and tender and it did eate somewhat like a Muske Melon' (*Gerard*, p. 1515)—a verdict more complimentary than that of Parkinson.

Of Gerard's confusion of species the chapters on the Gentians are a sufficient example. Here Johnson rewrote 'Chap. 107 Of Bastard Felwoort' recording four 'species', *G. acaulis*, *G. verna*, and two annuals: over these latter he is himself confused, for he describes the smaller as

1 Presumably son of Thomas Best, senior warden of Trinity House at this time.

2 B. D. Jackson wrote on this subject in *Transactions*, *Herts. Nat. Hist. Society*, 1906; and Mrs Arber, *Herbals*, treats fully of it.

having a flower divided into five parts, and says that the flowers of the larger are like it in shape; *G. amarella* and *G. campestris* were not clearly distinguished till Ray's time: but Johnson at least ascribes to them the localities in Kent and Sussex which Gerard had given to *G. verna*. He also queries Gerard's localities for *G. pneumonanthe*—'I suspect that our author knew it not...this seldome or never grows on chalkie cliffes but on wet moorish grounds' (p. 438): but leaves unchallenged Gerard's claim to have found *G. cruciata* in Essex (p. 434). Of 'Gentiana concava', the strange soapwort found by Gerard, he adds a note, '[Gaspard] Bauhine received this plant with the figure thereof from Doctor Lister, one of his Majesties Physitions and he refers it unto Saponaria...in his *Prodromus* [Book VI] pag. 103'[1] (*Gerard*, p. 435).

On occasion his corrections are due to the advice of others. Thus in pointing out that Gerard had inserted a Cress (probably *Nasturtium amphibium*, but Jackson, *Gerard's Catalogue*, p. 51, says *Sisymbrium strictissimum*) in place of *Senecio saracenicus* as the famous simple called 'Saracen's Consound' which involves placing the Cress in a new place on p. 274 and altering p. 428, he notes that 'Mr John Goodyer was the first, I thinke, that observed this mistake' (p. 275) and states on p. 428 that 'the true Solidago here described and figured was found Anno 1632 by my kinde Friends Mr George Bowles and Mr William Coot in Shropshire in Wales...as one goeth from Dudson in the parish of Cherberry to Guarthlow'.[2] So, too, on p. 1629 he prints a letter from Mr Roger Bradshaghe (Bradshaugh of Parkinson)[3] enclosing notes made by John Redman on some of Gerard's northern plants. Most of these deal with abnormal specimens and merely point out that they are not now to be found in the localities named: but a few criticise his ignorance of northern speech!

Emendations and improvements are marked with a dagger if Johnson regards them as substantial, and additions or rewritten passages are marked with a double cross. This is an admirable system and makes the authority for any part of the book easy to discover. Indeed, the care and ingenuity with which this part of the work has been done deserve all the praise that has been bestowed on it. If Gerard's book was to be kept as the main source, it could hardly have been 'emaculated' on a better system. There are, indeed, plenty of cases in which mistakes have been kept, and as we shall see some in which Johnson has made matters worse. But considering the difficulty of revision and his avowed reluctance to assume too easily that his predecessor was wrong, we can only applaud the judicious and thorough way in which he has performed his editorial duties.

1 'Saponaria concava Anglica...in comitatu Northampton in nemore quodam Spianie dicto, a Gerardo Chirurgo Londinensi reperta...et Londino a D Listero una cum figura suis coloribus illustrata missa fuit', *Prodromus*, p. 103.

2 Cf. below, p. 295.

3 Cf. above, p. 266.

In certain places—not deemed sufficiently important to be marked—he has made improvements on other than botanical lines. One such which seems to have been overlooked by Kew and Powell is on p. 194. Here he has inserted (under Red Lilies) the quotations from Ovid's *Metamorphoses*, from Theocritus, Vergil and Nemesianus which Gerard had copied from Camerarius[1] and printed under Hyacinths (*Herball*, p. 101). But for the Ovid, in place of a rather crude rendering in Gerard, he has printed a heading 'Which lately were elegantly rendred in English by Mr Sands' and then a quotation from George Sandys' translation published in 1626. This is testimony not only to the thoroughness with which he improved Gerard, but to his own range of culture. For Sandys, youngest son of the Archbishop of York, was a competent scholar and something of a poet, and his rendering is both faithful to the original and dexterous in its management of the heroic couplet. There are a good many similar if less striking cases where changes only intended to improve the literary quality of the book have been made.

By far the most important part of Johnson's work is his additions; and these range from his admirable survey of the history of botany 'To the Reader' and his 'Catalogue of Additions' to his Appendix of species omitted in the main body of the work. The history is the first sketch of the subject to be written in English and, as a brief introduction to the chief names in plant-lore and study from Solomon to Parkinson, something of a masterpiece. There is, for example, a very interesting discussion of the correct name of the 'unknowne Author to whom the printers have given the title of Apuleius Madaurensis', in which Johnson claims to have seen four manuscripts of him and heard of a fifth—'the first[2] I saw some nine yeares agoe with Sir Robert Cotton' the great collector and antiquary —but to have found him called 'Plato Apoliensis'.[3] And the records of many of the more recent botanists are models of precision and sound judgment. There are mistakes: he gives the *De Proprietatibus* to Bartholomew Glanvill and the date as 1397; he omits all mention of those who have not left books behind them; and he ignores altogether Jean Bauhin although Cherler had produced the first version of his *Historia universalis* in 1619. But as evidence of Johnson's knowledge of the relevant literature and as a preliminary to his account of Gerard's work and of his own method and sources both literary and personal, it is excellent.

1 They occur in his *Hortus medicus*, p. 46, published in 1588: Camerarius was a great classical scholar—though not so great as his father.

2 Presumably that listed by T. Smith, *Catalogus Biblioth. Cottonianae*, p. 85, as 'Herbarium Appulei Platonici quod accepit ab Aesculapio Saxonice'.

3 From one of these MSS. of 'Apoliensis Plato' Goodyer copied and sent the picture and brief account of a Saxifrage printed on *Gerard*, p. 604, in a chapter added by Johnson.

This description of Gerard has been already noted and discussed; as has his statement of the limitations of time imposed on him, of the character of the changes proposed, and of the method by which they are introduced. His claim that his additions are largely from De l'Ecluse is fully borne out by the book itself: the new chapter on Ranunculi (pp. 964–5), for example, consists entirely of matter from him; and his *Curae posteriores*, the posthumous notes published in 1611, is freely used. Johnson ends with a dignified apology for the conditions under which his edition has been produced, and with generous tributes to his helper, John Goodyer, to George Bowles and to seven named members of the Society, his 'loving friends and fellow travellers in this study'.

In the book itself the additions are of three main sorts: from other writers, from his own observations and from his friends. Of the first many, probably the majority, come from De l'Ecluse and De l'Obel; others from Gaspard Bauhin, especially his *Prodromus*, Giovanni Pona's Catalogue of the plants on Monte Baldo, and Prospero Alpino's of Egyptian Plants and Rhubarb; one from Pierre Belon[1] on the gathering of Ladanum (p. 1291) and one from William Turner's *Names* which he corrects on the identity of *Lathyrus montanus* (p. 1236). He also refers his readers (on p. 145) 'to the Florilegies of De Bry, Swertz, Robine, or to Mr Parkinson'—that is the *Florilegium novum* of Johann Theodor von Bry, 1612, which he quotes in detail on p. 122; the *Florilegium* of Emanuel Sweerts of Breda, Frankfort, 1612, and Amsterdam, 1620; the *Catalogus Stirpium Lutetiae* of Jean Robin, Paris, 1597; and of course the *Paradisus terrestris* to which and to Parkinson's garden he refers at least a dozen times and always with respect. Twice he reports a plant on the authority of Jan Dortmann the apothecary of Groningen—the *Lobelia dortmanni* on p. 105[2] and what appears to be a mistaken rendering of *Saxifraga hirculus* on p. 1284[3]—but the records come by way of De l'Ecluse; as does that from 'Dr Nicholas Colie' a Dutch physician on p. 1133.

Of Johnson's own observations a large number are naturally drawn from his expeditions into Yorkshire and the north in 1626 and into Kent and Hampstead as printed and already described. There are, however, a good many new references to London and its neighbourhood. All three Mustards (*Brassica sinapis*, *B. alba* and *Sisymbrium officinale*) may be found 'on the banks about the backe of Old Street and in the way to Islington' (p. 245); *Damasonium alisma* 'I found a little beyond Ilford

1 He translates a passage for comparison with Gerard's version of it.

2 This record he doubtless derived from De l'Ecluse, *Curae post.* p. 40, who gave its picture and description.

3 The seed-vessel, growth and flower colour are correct, but it is said to be six-petalled, and the leaves are much too large.

in the way to Rumford' (p. 418); *Scutellaria galericulata* 'growes by the ponds and waters sides in Saint James his Parke, and in Tuthill Fields' (p. 479); *Centaurea scabiosa fl. albo* 'in a field neere Martin [Merton] Abbey in Surrey' (p. 729); *Caucalis nodosa* 'floures in June and July and grows wilde upon the banks about S. James and Pickadilla' (p. 1023); *Lycopsis arvensis* 'upon the dry ditch banks about Pickadilla' (p. 799); and *Oenanthe crocata* 'groweth amongst oysiers against Yorke House a little above the Horse ferrey against Lambeth' (p. 1060). Of this last he has a warning: 'Pernitious and not excusable is the ignorance of some of our time that have bought and (as one may probably conjecture) used the roots of this plant instead of those of Peionie; and I know they are dayly by the ignorant women in Cheape-side sold to people more ignorant than themselves by the name of Water Lovage; *Caveat Emptor*. The danger that may ensue...'[1] (p. 1060). Johnson shows his professional interest by a number of references to the herb-women. Thus of *Calamintha acinos* 'brought to Cheapside market, where the herbe women called it Poley mountaine' (p. 676); and of *Melilotus caerulea* 'Gardiners and herbe women in Cheapside commonly call it by the name of Balsam' (p. 1191).

The help that he received from his friends is always generously and fully acknowledged. We have noted his many references to Parkinson and that some of these related to his garden. As an example, and because the problem of its status as a British plant is of some interest, we may cite his addition of the 'Great Tower Mustard' (*Arabis turrita*—still found in the grounds of St John's College, Cambridge) to Gerard's chapter: after an adequate description of the plant he writes, 'It is a stranger with us, yet I am deceived if I have not seen it growing in Mr Parkinson's garden'[2] (p. 272). So, too, John Tradescant's garden yielded him several species besides the 'Virginian Spiderwort', for example *Spiraea ariaefolia* (p. 1043) and *Aesculus hippocastanum* (p. 1443). Ralph Tuggy had lately died (p. 589), but his garden besides the 'Gillofloures' for which it was famous gave him two Myrtles (p. 1413) and the Passion flower (*Passiflora coerulea*) then 'in good plenty' and bearing many blossoms (p. 1592).

Other gardens are new to fame. Thus of *Cucubalus baccifer*, which De l'Obel had reported as growing wild in England, Johnson says, 'Yet I have not seen it growing but in the garden of my friend, Mr Pemble, at Marribone' (p. 615) and again of Hesketh's Cloudberry, which Gerard separated from *Rubus chamaemorus*, there is a note, 'My friend, Mr Pimble,

1 Twenty-five years before Plat in the second part of his *Garden of Eden*, p. 131, had written 'Cheapside is as full of these lying and forswearing Huswives as the Shambles and Gracechurch Street are of the shameless crew of Poulters wives'.

2 It may be noted that Parkinson (*Theat. Bot.* p. 853) gives a redrawn version of the same picture and a different description, referring to Bauhin as well as De l'Ecluse, but no note of personal acquaintance with the plant or with Johnson's statement.

of Maribone received a plant hereof out of Lancashire; and by the shape of the leaf I could not judge it to differ from the Chamaemorus formerly described', that is from Penny's original plant (p. 1420). More identifiable than 'Mr Pemble'[1] is Sir John Tunstall, 'Gentleman Usher unto her Majestie' (p. 161),[2] in whose garden at 'Edgcome by Croydon' Queen Anne's flowers were at that time kept and among them 'that curious Colchicum which is called by some Colchicum variegatum Chiense', *C. variegatum*, a Greek species already pictured in Parkinson, *Paradisus*, which Johnson drew again (p. 163). On his visit he found growing in great plenty 'close by the gate of the house' *Dipsacus pilosus* (p. 1168). A further reference to gardens and to a friend whose interest became important later is on p. 996, where it is noted that 'the red floured mountaine Avens (*Geum rivale*) was found growing in Wales by my much honoured friend, Mr Thomas Glynn,[3] who sent some plants thereof to our Herbarists in whose gardens it thriveth exceedingly'.

Of other friends the most interesting is William Coote, already mentioned by Parkinson as 'Dr Coot' for a report of *Drosera anglica* from Ellesmere, and by Johnson along with George Bowles in connection with the finding of *Senecio saracenicus* in 1632 near Cherbury. Of a Hawkweed (probably *Hieracium boreale*) Johnson writes on p. 305: 'This I had from my kinde friend Mr William Coote who wrot to mee that hee found them growing on a hill in the Lady Briget Kingsmills ground at Sidmonton not far from Newberry in an old Roman camp close by the Decuman port, on the quarter that regards the West-South-West, upon the skirts of the hill.'[4] From this it is clear that William Coote was a man of education, familiar with the lay-out of Roman camps and capable of bringing the compass into descriptions of localities. His identity has never been discussed, but from the three references which we have cited it is more than probable that he was the William Coote who was elected to a fellowship at Trinity College, Cambridge, in 1610, took his doctorate in Divinity in 1635,[5] and then as Aubrey informs us became one of the chaplains of Lord Herbert, living with him at Cherbury.[6] Unfortunately, apart from

1 A Mr Thomas Pimble was punished for selling bad medicines in 1622: cf. Barrett, *History of Apoths.* p. 17: but he lived at the sign of the Crane in Southwark.

2 His pedigree is given in *Visitation of Surrey* (Harl. Soc. XLIII), p. 189.

3 Of Glynnlhivon in Carnarvonshire. He also sent *Diotis maritima* from the Welsh coast, cf. *Gerard*, p. 644, and *Acorus calamus*, l.c. p. 63, in May 1632.

4 Camden, *Britannia*, ed. 1607, p. 196: 'Sidmonton the seat of the family of Kingsmils Knights and Burghclere that lies under a high hill on the top of which there is a military camp with a large trench.'

5 *Trinity Coll. Admissions*, II, p. 238. There is no record of his parentage, birthplace or career.

6 *Lives* (ed. 1813), II, p. 390: 'Dr Coote, a Cambridge scholar and a learned, was one of his chaplains.'

the dates of his degrees at the University, little is known of him, but he must have been known to George Herbert, fourth son of Lady Magdalen, who became a fellow of Trinity in 1614, was in residence until 1627 and as we have seen shared his elder brother's interest in 'botanics' and gardening. If he was the younger brother of Thomas Coote who came up from Westminster to Trinity in 1602 and became a fellow in 1608, the connection with Herbert, also of Westminster, would be more certain.[1] That Coote was at Cherbury at the time of George Bowles' visit is clear from their joint discovery of *Senecio saracenicus*; that they both were familiar with Ellesmere, and presumably with Richard Herbert's place there, is also evident from references by Parkinson (*Theat. Bot.* p. 1053) and Johnson (*Gerard*, p. 250). But though this may suggest another link between the students of botany and the Herbert family, it does not tell us much.

Of his companions in the journeys into Kent only Leonard Buckner seems to have made any further contribution to the *Herball.* He had been in Oxford in 1632, almost immediately after their August expedition, and had brought back an Equisetum (probably *E. sylvaticum*); 'hee had it three miles beyond Oxford, a little on this side Euensham [Eynsham] ferry, in a bog upon a common by the Beacon hill neere Cumner wood in the end of August 1632' (p. 1115). He also found *Stachys germanica* 'growing wilde in Oxfordshire in the field joyning to Witney Parke a mile from the Towne' (p. 697)—this being a first and very interesting record.

A friend of whom there seems to be only one mention is named in connection with *Anagallis femina*: 'I also being in Essex in the company of my kinde friend Mr Nathaniel Wright found this among the corne at Wrightsbridge being the seat of Mr John Wright his brother' (p. 618). Wrightsbridge, named after the family, is near Hornchurch; and John Wright who died in 1644 is buried in the Church of Southweald (so Morant, *History of Essex*, I, p. 122). His brother Nathaniel, who had graduated at Emmanuel College, Cambridge, in 1623–4, took his M.D. at Bourges and was incorporated at Oxford in 1638. He was physician to Oliver Cromwell in Scotland (Venn, *Alumni Cantab.* IV, p. 475).

But of all his helpers the two whom he selects for special mention in his preface, John Goodyer and George Bowles, are by far the most important. Goodyer's descriptions are always inserted as signed paragraphs and thus form a distinct and easily recognisable feature of the edition. Bowles is mentioned some fifteen times and usually in connection with new observations. Both these will be discussed later.

The skill and thoroughness with which these very varied contributions are worked into the text at once preserve the general form and contents

1 I owe this suggestion to Dr F. E. Hutchinson, editor of *The Works of George Herbert.*

of Gerard's book and 'emaculate' it and bring it up to date. As a year's work, and even so combined with other interests both literary and professional, it is a very remarkable achievement. But the success of it, and the great improvement made in the *Herball*, must not blind us to the three facts, that Johnson has often been content to leave very poor sections unreformed, that his 'improvements' are not always for the better, and that in his own work there are on occasion bad blunders. A few samples of these blemishes will illustrate their character.

Here for example is the remark which introduces the chapter 'Of Rushes': 'I do not here intend to trouble you with an accurate distinction and enumeration of Rushes; for if I should, it would be tedious to you, laborious to me, and beneficial to neither'—which may well be the case considering the confusion which he found and which, by adding a paragraph about Rushes used for making 'mats and bottom chairs' and another introducing *Scirpus palustris*, he made worse confounded (pp. 34–5). It is difficult to believe that a man with real scientific quality could have written that sentence; and certainly his treatment of Grasses, Sedges and Rushes compares very poorly with Parkinson's.

A few pages later, on pp. 46–7, he leaves Gerard's chapter 'of Stitchwort', *Stellaria graminea*, inserted between that on *Typha latifolia* and that on the Spiderworts and *Tradescantia*—merely because Stitchwort is called 'Gramen leucanthemum'. So too he fails to revise Gerard's placing of *Paris quadrifolia* and *Botrychium lunaria* (p. 405) in the same chapter, or of *Myosurus minimus* among the Plantains (p. 426). But it may perhaps be urged that radical changes of Gerard's order were impossible, and that allowance must be made for the difficulty of fitting the new work into the old frame.

Unfortunately some of his own additions are equally blameworthy. For instance, on p. 767 he arbitrarily inserts a cluster-headed Lychnis into a chapter of Labiates; on p. 271 he includes what De l'Obel found at Plymouth and named 'Thlaspi hederacium' among the 'Wooddy Mustards' although both picture and description are obviously, as Ray pointed out in *C.A.* p. 296, those of *Cochlearia officinalis* and should go to p. 401; and on p. 96 he adds a description and picture of the 'Lancashire Asphodill' (*Narthecium ossifragum*) to those already given by Gerard though the plants are identical and Gerard's account is actually better. So, too, on p. 1183 he confuses De l'Obel's *Carduus pratensis* with Penny's *Carduus heterophyllus*, insists that they are the same species, and therefore, having seen *pratensis* but not *heterophyllus*, condemns Penny's picture and description as bad. Further—and this is a sad revelation of his incompetence—writing on p. 858 of our friend the Virginian Creeper (*Vitis hederacea*) he says: 'There is kept for novelties sake...a Virginian by some (though unfitly) termed a Vine, being indeed an Ivie.'

His interest is in fact not so much in botany as in pharmacology. For his profession he has both pride and enthusiasm. We have seen how frequently and vigorously he denounced those unauthorised practitioners, the herb-women of Cheapside. The same loyalty to his calling made him add a note to Gerard's story of the herbalist Cornwall 'dwelling by the Bagge and Bottle' who sold to the Apothecaries Ammi for Wormwood; 'certainly our Author was either misinformed, or the people of these times were very simple, for I dare boldly say there is not any Apothecary or scarce any other so simple as thus to be deceived now' (p. 1101)—though it may fairly be asked whether he could reconcile this confidence with the indifference to the knowledge of herbs which he described a year later in the first pages of his *Mercurius*.

But though his practical and professional interest gave a bias to his studies and, as these examples show, must make us hesitate to bestow upon him the unbounded praise of his recent biographers, he was manifestly a man of real ability. Characteristic and conspicuous is his treatment of Gerard's chapter 'Of the Goose Tree'. He prints the picture and the whole of Gerard's description; and then adds between double crosses a brief paragraph beginning 'The Barnakle whose fabulous breed my Author here sets downe, and divers others have also delivered, were found by some Hollander to have another originall, and that by eggs as other birds have'. He then quotes the record and the reference—'*Pontani rerum et urb. Amstelodam., hist. lib. 2. cap. 22*'; and adds a reference to Fabio Colonna's proof in his *Phytobasanos* (*Plantarum Novarum Historia*, pp. xiv–xix)[1] that the shells are 'a kind of Balanus marinus'.

Johnson was a keen, well-read, energetic man whose work upon Gerard's *Herball* immensely increased its value, and made it so popular that a second edition was demanded in 1636. To this he added an Advertisement on the last page in which he set down first an apology that no alterations or additions except the correction of a few errata have been made; secondly that this is due to his determination 'to travell over the most parts of this Kingdome...for I judge it requisite that we should labour to know those plants which are and are ever like to be inhabitants of this isle; for I verily believe that the divine Providence had a care in bestowing plants in each part of the earth fitting and convenient to the foreknowne necessities of the future inhabitants; and if we thoroughly knew the vertues of these we needed no Indian nor American drugges'—a superstition very widely spread and which throws light on the motives of his botanising; thirdly that this new material resulting from his travels and from his friends 'gone into forreigne parts' will be published separately and serve as a supplementary volume to the existing *Herball*. It was no doubt wise to let the *Gerard* alone: to replace it by a full and independent

1 Published at Naples, 1592; cf. above, p. 267, *note*.

work would have been the only satisfactory method. But even if he did not wish to revise his edition, he might have taken the pains to correct the errors in pagination (three successive leaves numbered 29 and 30 for example), which are far too frequent.

Before going on to Johnson's further botanical works, mention must be made of 'The Workes of that famous Chirurgion Ambrose Parey, Translated out of Latine and compared with the French by Th: Johnson'. This was published by T. Cotes (the publisher of Parkinson's *Theatrum*) and R. Young in 1634, although, as Kew and Powell suggest, it may well have been finished earlier. It is of special interest not only as showing Johnson's energy and range of ability, but as throwing light on his contacts with people of importance. Here there are two points to be noted. First, the book had been 'entred for the copie' in 1629 as a translation by George Baker from the French; and in fact the last part of it, 'The Apologie and Voyages', was according to Johnson's preface 'translated into English out of French by George Baker, a Surgeon of this City, since that time, as I heare, dead beyond the seas'.[1] Now George Baker,[2] sometime surgeon to Queen Elizabeth, whom we have encountered as the friend of Gerard and who had died in 1600, was the grandfather of Johnson's friend George Bowles; for Frances, Baker's eldest child, had married Robert Bowles, Groom and Yeoman of the Tents to the Queen. It was perhaps from George Bowles that Johnson got the invitation to complete the translation.

The second point is the dedication of the book to Edward Lord Herbert of Cherbury. Kew and Powell remark that 'we do not know in what manner they were acquainted' (l.c. p. 75). It is at least possible that here again George Bowles was responsible. For Bowles' younger brother, afterwards Sir William, married Margaret youngest daughter of John Donne, Dean of St Paul's. Donne was the close friend of Herbert's mother, the Lady Magdalen, who after his father's death had married Sir John Danvers. George Bowles was certainly at Cherbury in 1632 and he may then have obtained permission for his friend Johnson to dedicate to Lord Herbert his forthcoming translation.

In the same year, 1634, Johnson and his fellow-travellers undertook a longer tour, planned as it seems partly because he was himself staying at Bath in medical attendance on a lady of means, Mistress Ann Walter, and partly for the exploration of the famous Avon gorge at Bristol. The full story of the tour—how all the great ones of the Apothecaries, Richard Edwards, the Master, Edward Cooke,[3] the Senior Warden, Thomas Hicks,

1 Cf. Kew and Powell, l.c. p. 34.

2 He had written a commendation of Gerard printed in the *Herball* in which he compared him with a French herbarist, probably Jean Robin, cf. above, p. 208.

3 Master, 1640.

Master Elect,[1] Roger Harry Yonge,[2] William Broad, Robert Lorkin and James Clarke[3] with three servants, rode from London to Reading; spent the night there with Dr John Bird and the druggist John Watlington; went on to Marlborough where Johnson met them; and next day with John Buggs, now M.D.,[4] Philip Parsons, Principal of Hart Hall, and two more friends William Jackson of Oxford and Richard Kendall of London, went on to Bath; how they saw George Gibbs' garden,[5] called upon Philip Langley at Mangersfield,[6] met John Price the Bristol druggist and were guided by him to the St Vincent's rocks—is printed in the *Mercurius Botanicus*, the record and full catalogue which Johnson compiled of their tour. The return, the visit to Matthew Chock at Keynsham; the Sunday at Bath; Trowbridge and the Rev. Thomas Pelling;[7] Salisbury and Dr Richard Haydock's *Opus Anatomicum* with its copperplate engravings,[8] Dr James Lake, and William Wilson the druggist; Southampton and John Cropp; Cowes and Captain Humfrey Turney; Newport and Thomas Juning; Carisbrooke Castle and Thomas Basket and Captain Andrew James;[9] Stonyplace and William Kitchin; Ryde and Portsmouth; Chichester and Anthony House and Isaac Thornbury; Petworth, Godalming ('Godliman'), Guildford and so to London: these are more briefly chronicled. The plants except those in Gibbs' garden are listed alphabetically and all together, with localities in black letter and English for the rarer species, in Latin and general terms for the common. As the plants previously found in Kent are also included, the catalogue forms a substantial preparation for a flora of southern England. *Phyteuma orbiculare* 'betweene Selbury Hill and Beacon Hill in the way to Bathe', *Geranium sanguineum*

1 Elected Master, 18 August.

2 Cf. Barrett, l.c. p. 22.

3 Renter Warden, 1652: died that year.

4 According to Goodall (l.c.) of Leyden. Foster, *Al. Oxon.* I, p. 206, says of Padua and quotes Wood, *Fasti*, I, 479: but this is an error; Wood, l.c., recording Buggs' incorporation at Oxford on 16 June 1635, says a Londoner, doctor of Leyden.

5 Of this Johnson printed a catalogue in *Mercurius Bot.* Goodyer also left a list of his plants; cf. Gunther, *E.B.B.* pp. 346–8. Parkinson, *Theatrum*, p. 133, reports that in 1640 Gibbs had lately returned from Virginia 'with a number of seeds and plants'.

6 Gunther, l.c. p. 281, quotes How's MS. Records, 'at Mangersfield in Mr Langton's yard': cf. also *Merc. Bot.* p. 77 of *Cotyledon umbilicus* 'upon the walls at Mangersfield'.

7 Of Magdalen College, Oxford, B.A. 1619–20, Rector of Trowbridge from 1621: Foster, *Al. Oxon.* III, p. 1139.

8 He translated from the Italian *The Arts of curious Painting, Graving and Building*, Oxford, 1598: cf. Wood, *Ath. Oxon.* I, c. 678. His *Anatomy* does not seem to have been published.

9 In *Description of a Journey*, p. 59 (Camden Misc. XVI, 1936) there is a list of the castles and captains in 1635, 'Cowes Castle, Captain Turry; Cashocke Castle, Captain James'—thus confirming Johnson. The Captain of Cowes is given as Tourmy by Oglander, *A Royalist's Notebook*, p. 99.

from St Vincent's rocks and *Ornithogalum pyrenaicum* 'betweene Bath and Bradford, not far from little Ashley' are the best finds of the tour.

The book is interesting partly because the many names of doctors and apothecaries mentioned in it are evidence of the rapid growth of interest in 'simpling', and partly from the fact that in the account of the exploration at Bristol there is not only a description of the Hot wells ('fons aquae tepidae gustuique gratae') but an account and discussion of the crystals ('adamantes sexanguli et quadranguli') dug from the upper part of the cliffs. It has also a dedication to Sir Theodore de Mayerne, Dr Matthew Lister, Dr Ottwell Meverell and Dr Laurence Wright, in which apology is made for these botanising journeys and a list given of prominent doctors who have studied herbs. The list of these is striking—'Anguillara, Rondelet, Mattioli, Belon, Pena, De l'Obel, De l'Ecluse, Camerarius, Penny, Bauhin, Spieghel and many others; Conrad Gesner in regard to his account of Mont Pilatus; Francesco Calzolari and Monte Baldo; and finally Giovanni Pona and the same hill'. It is not quite the list that we should have expected.

With the *Mercurius* was bound up an account of the *Thermae Bathonicae*, the baths and city with a plan and elaborate details of the various treatments, no doubt composed while Johnson was looking after his patient before the tour began.

His final journey is the long and important visit to North Wales in 1639 described in the Second Part of the *Mercurius* published in 1641, dedicated to Thomas Glynn[1] of Glynn-lhivon (Glynllifon), and containing a list of plants supplementary to that issued in 1634.

The journey was undertaken by Johnson and Paul Sone,[2] apparently not an apothecary; and being ignorant of Welsh they took with them Edward Morgan, perhaps a connection of Hugh, and the man who was soon after appointed to supervise the Physic Garden in Westminster. They set out on 22 July, spent the first night at Aylesbury and the second at Stratford-on-Avon; then stayed two days with Robert Lee[3] at Billesley; then through 'Bremicha' (Bromwich) to 'Vulcani municipium' (Wolverhampton), where they found *Salix pentandra*; next day to Newport, Shackerforth Mill and Chester, where Walter Stonehouse who had come across from Yorkshire met them, having spent a night in squalor at

1 See above, p. 282; described in Foster, *Al. Oxon.* II, p. 574, as son and heir-apparent of Sir William. He was brother of Sir John Glynn the judge (for whom see *D.N.B.*).

2 Possibly brother of Thomas Sone, 'citizen and grocer' of London, whose daughter Elizabeth died in 1670 and is buried in Hackney Church.

3 Doubtless the grandson of Sir Robert Lee, Lord Mayor of London in 1602–3 and himself knighted in 1642. He died in 1659. Cf. *Visitation of Warwickshire* (Harleian Soc. LXII), pp. 70–1. He may be the Robert Lee of Hart Hall, Oxford, who was created M.D. in 1642–3, cf. Foster, *Al. Oxon.* III, pp. 894, 898.

Stockport and left a Latin epigram in revenge on his chamber door. They stayed at Chester with Dr Samuel Bispham, went on through Flint and Holywell to 'Ruthlanda' (Rhuddlan), and the next day to 'Garth-gogo' (Pen y Gogarth or Great Ormes Head), where they botanised, and thence to Aberconway and 'Bodskalan' (Bodysgallen) the hospitable house of Robert Wynn. Then by the narrow and terrifying track over Penmaenmawr they came to Bangor and Caernarvon and to Thomas Glynn's home four miles beyond and near Llandwrog with its wonderful view of the sea and the Isle of Anglesey and Ireland in front and of the British Alps behind. On 3 August they took horse again for 'Widhfa' or, as the English call it, Snowdon. The terrors of the ascent and the cloud-capped grandeur of the summit are worthily described (how our ancestors did hate mountains!): but arrived at the top the party nevertheless spread out its plants and consumed its lunch as prosaically as if they had not been making a first ascent. Their specimens deserve remembrance. There is first the famous 'Nasturtium petraeum' which in later years was distinguished by the name 'Johnsoni'[1]—*Arabis petraea*; then *Oxyria digyna*, *Viola palustris*, *Thymus serpyllum*, *Sedum roseum*, *Silene acaulis*, *Saxifraga nivalis*, *S. hypnoides*, *S. tridactylites*, *S. stellaris*, 'Gentianella Bavarica' (?), *Festuca ovina*, *Lycopodium alpinum*, another Moss not easily identifiable, *Saussurea alpina*, *Salix herbacea*, *Allosorus crispus*, 'and to our surprise, except that Pena and De l'Obel had noted some such thing in *Adversaria* [presumably referring to p. 189]', *Armeria maritima* and *Silene maritima*.

After a full day's rest the party crossed the strait to Abermenai; visited the house in which Dr Kyffin, Dean of Bangor in Richard III's time, had acted on behalf of the future Henry VII; found a number of plants in the sand-hills and rocks; and then went on from Newburgh to 'Bellum-Mariscum' (Beaumaris) where they stayed with Richard Buckley (Bulkeley) at Baron Hill, the splendid house built by his father in 1618, and after dinner rode on to Penmon point 'opposite a little island'. Then hearing great stories of the plants on the Carnedhs they went back to Bangor, but ran into such rain and cloud on Carnedh-Llewelyn that their guide lost his nerve, told them fearsome tales of eagles, and brought them back empty-handed to the hospitable roof of Thomas Glynn. Next day the return began. 'Penmerfa' (Penmorfa near Tremadoc), Harlech, Barmouth and the crossing of another perilous hill 'Altan-ownreth', 'Llanguerel' (Llwyngwril), Machynlleth and so to Montgomery, where they spent a night enjoying the munificence of Edward Herbert, Baron Cherbury.[2] Thence to 'Guerndee' (Wernddu) where they found the two plants that

1 Cf. e.g. Ray, *Synopsis Brit.* II, p. 174.

2 It is probable that he was himself absent with the king: no doubt his chaplain, William Coote, Bowles' friend, entertained them.

George Bowles had already discovered, *Impatiens noli-tangere* and *Senecio saracenicus*. There Walter Stonehouse left them, making for Shrewsbury on his way to Yorkshire. The three went on by Ludlow, 'Lemsterem' (Leominster), Hereford, Gloucester and Oxford.

The catalogue only contains such plants as had not previously been listed: it is in fact a supplement to the previous volume. But, as such, it contains not only the plants of the new Welsh tour but those found elsewhere by his friends Walter Stonehouse, George Bowles and John Goodyer, and a certain number found by De l'Obel and (in some cases) noted in his unfinished book, *Stirpium Illustrationes*, which had now (in 1641) been included in Parkinson's *Theatrum*. With his usual care Johnson had marked with an asterisk all plants derived from Parkinson and had prefixed another sign to such records as he regarded with uncertainty. It is necessary to add that these signs are not always inserted where they should be; and that the localities even of rare species are often given in purely general terms. One or two of the records, for example *Mertensia maritima*,[1] 'by the salt pans betweene Barwicke and the Holy Island', are new, unexpected and without any discoverer's name.

The purpose of these lists is made quite clear in the introduction to the record of the tour. It is to prepare for a complete and illustrated description of all British plants. By the publication of these lists the field already explored would be displayed: attention would therefore be drawn to areas and species still unknown: and so he and his friend John Goodyer could together look forward to producing a complete Flora. Gaspard Bauhin had done something of the sort for general botany in his *Pinax*; he had done it for an area in his *Catalogus plantarum circa Basileam*, 1622; they are the necessary preliminaries to any proper knowledge of the country's medicinal resources.[2]

It is one of the tragedies of the Civil War that this plan was not carried through by its originator. Two years after its announcement Johnson had left Snow Hill and joined the London Royalists at Oxford. Almost on his arrival he was recommended by the King for the degree of Bachelor of Physic and four months later was created Doctor. But although in the main these 'Caroline creations' bore no relation to academic or intellectual worth, Ray who was no lover of the Stuarts said in all sincerity of Johnson that 'on account of his notable skill in Medicine and especially in botany, the University of Oxford very deservedly honoured him with the degree and title of Doctor' (*Historia Plantarum*, I, Introd.).

Soon after, he joined Colonel Rawdon at Basing House; and was active during the long months of the siege until in a sortie in September 1644

1 Parkinson, *Theat.* p. 765, had reported this 'in one of the Iles about Lankashire. There found by Mr Thomas Hesket', Gerard's correspondent.

2 *Merc. Bot.* II, p. 3.

he was wounded by a shot in the shoulder and, fever setting in, died a fortnight later.

John Goodyer, Johnson's 'onely assistant' in his edition of Gerard, is in these days no longer as Canon Vaughan described him in 1901[1] a 'forgotten Botanist'. Since then Miss M. E. Wotton in 1917,[2] G. C. Druce in several of his works, and R. T. Gunther in 1922 in his *Early British Botanists* have repaired the neglect. Indeed, the last-named volume, being based upon the study of Goodyer's library bequeathed by him to Magdalen College, Oxford, gives a very full account of his career and a complete collection of his botanical descriptions. There is indeed a risk, with him as with Johnson, that the enthusiasm of his discoverers has led to an exaggeration of the importance of their discovery. To describe Goodyer as 'an incomparable botanist of sound judgement and of immense industry' is to suggest an achievement much more remarkable than he in fact accomplished.

Born in 1592 at Alton he spent almost the whole of his life within a few miles of his birthplace. In 1616 when he began his collection of botanical books he seems to have been in the service of Sir Thomas Bilson, son of the Bishop of Winchester who died in that year, and to have lived at West Mapledurham near Buriton. Then he had a house and garden at Droxford some eight miles away in the Meon valley. Finally, on or shortly after his marriage to Patience Crump, he settled in a house in the Spain at Petersfield, where he spent the rest of his days. He may have been often in Oxford during the Civil War—though the evidence is not very convincing; and no doubt was occasionally in London. But he seems to have become very stay-at-home and to have devoted himself increasingly to local affairs and to the production of an interlinear translation of the *Materia Medica* of Dioscorides. There he died in 1664. In 1657 William Coles of New College wrote of him as 'the ablest Herbarist now living in England, a man well stricken in years'.[3]

Of the excellence of his early work, his success in discovering new plants and his skill in description, Johnson's additions to Gerard and his own manuscripts at Magdalen printed in sequence by Gunther give ample proof. If *Pulmonaria angustifolia* on 25 May 1620, 'flowering in a wood by Holbury House in the New Forest' (*Gerard*, p. 809, *E.B.B.* p. 115), is his best known find, *Ludwigia palustris* on 19 August 1645, 'growinge in the rivulett on the east side of Petersfield' (l.c. p. 188), a spot in which it was last seen two centuries later in 1848, is of equal interest. *Veronica hybrida* at St Vincent's Rocks (as recorded by Johnson, *Mercurius Bot.* II, p. 36, cf. l.c. pp. 75–6), *Thesium humifusum*, 5 July 1620, 'on the side of a chalkie hill as you goe from Droxford to Poppie Hill' (l.c. p. 36),

1 In an article in the *Cornhill*.

2 In *Hants and Sussex News*, 11 April.

3 *Adam in Eden*, p. 161.

and the four British Elms (*Ulmus campestris*, *minor*, *montana* and *glabra*) first distinguished by him (l.c. pp. 37–43) are hardly less proof of his competence. Gunther indeed credits him with some forty additions to the British flora. This is certainly an overstatement since, for example, *Frankenia laevis* had already been found by De l'Obel at Portsmouth, perhaps in the same locality as Goodyer's 'Haylinge' (cf. Parkinson, *Theatrum*, p. 1485), and *Epipactis palustris* by Robert Abbot the 'learned preacher and diligent Herbarist' at Hatfield (cf. Gerard, *Herball*, p. 175):[1] yet the list, considering that most of it belongs to a short spell of searching, is creditable both to his energy and to his flair for discriminating novelties. Of these and very many other plants both native and exotic he has left descriptions, carefully drawn and as a rule sufficient to make the identity of the species clear. If they have not quite the excellence of those by Penny, and are occasionally involved or defective, the most and best of them follow a regular sequence and are expressed in clear and, for the time, technical language. It seems evident that Goodyer realised how unsatisfactory was the attempt to accommodate the English flora to continental pictures and descriptions; and set himself in the years 1616–21 to replace this method by another.

That the intention was fully justifiable cannot be questioned. Turner had done most of his botanising abroad, and had seen his work as primarily that of identifying the plants already described by classical authors or described and pictured by recent herbalists. He did a great work in fixing the names of British species, but was hardly interested in those which could not be equated with forms already known and named. Lyte and Gerard had imported continental books and given them as much relevance to our flora as their very insufficient knowledge of it made possible. But there was as yet no independent study of our plants, and no certainty that the effort to attach them to the species of Dioscorides or of Clusius was always possible of fulfilment. A fresh exploration and a fresh system of establishing discoveries by description or by picture or by preservation were urgently needed. Goodyer deserves the credit of realising this and of trying to meet the situation.

But he was heavily handicapped, at first probably by lack of the necessary education and knowledge, and then when he had collected and studied his books by a lack of scientific insight. This latter should be noted. In 1620 he found 'Acinos odoratissimum' (*Ocimum basilicum*) growing in the garden of his brother-in-law's father, William Yalden, at Sheet near Petersfield among Sweet Marjoram (*Origanum majorana*): on

1 Unfortunately Gunther, who claims Abbot along with Hugh Morgan as friends of Johnson (*E.B.B.* p. 274), has not recognised that they belong to an earlier generation. Abbot was known to Gerard: he entered Christ's College, Cambridge, in 1573, and died in 1613.

11 October 1621 he wrote, 'it is to be considered whether the seeds of sweet Marjerome degenerate and send forth this herbe or not' (Johnson, *Gerard*, p. 677). Turner, as we have noted, made the same guess about Cornflower (*Centaurea cyanus*): but by Goodyer's time a student of botany should surely have tested his fancy by experiment before propounding it. Still more serious is the lapse about the ear of wheat in which he and Johnson were both involved. On p. 65 of his *Gerard* in the chapter on 'Corne', Johnson inserts 'a rare observation of the transmutation of one species into another, in plants; which though it have beene observed in ancient times, as by Theophrastus...and by Virgil...yet none that I have read have observed that two severall graines, perfect in each respect, did growe at any time in one eare: the which I saw this yeare 1632, in an eare of white Wheat, which was found by my very good friend Master John Goodyer, a man second to none in his industrie and searching of plants, nor in his judgement or knowledge of them. This eare of wheat was as large and faire as most are, and about the middle thereof grew three or foure perfect Oats in all respects.' Presumably he mistook some abnormality of awn, being predisposed to interpret it as a change of species by his classical authorities, of whom Theophrastus speaks of Spelt into Oats and Vergil of Barley into Darnel and Oats. But the observation thus emphasised did much harm. In the preface to the *Catalogue of Foreign Plants* bound up with his *Observations* and written probably in 1669 John Ray, discussing the specific differences of plants, declared that 'a true transformation of species cannot be denied unless we set aside the evidence of first-hand and reliable witnesses', and proceeded to quote, along with a case from Olaus Worm's *Museum* of barley and rye in a single ear, this case observed by Goodyer of whom he thought highly.

It is indeed tempting to suppose that it may have been Goodyer's consciousness of his own lack of scientific training that slowed down his field-work, led him to the collection and annotation of his very complete botanical library, and eventually set him on to the rather futile task of producing an English rendering of Dioscorides. From the facts available it is evident that his interest, though it never died out, swung from plants in the field and garden to plants in the study; and that if he had for a few years the hope of producing a descriptive flora of England, this did not survive the death of his friend Johnson and the upheavals of the Civil War.[1] His output in 1621 was so large and on the whole so good as to promise great things, but this splendid year was never again equalled.

1 It is necessary to add that Gunther overestimates the work of Goodyer's later years, because of his greatly exaggerated view of the merit of William How's *Phytologia Britannica*—which was in fact nothing but a slightly increased reprint of the lists in Johnson's *Mercurius*.

Perhaps it was the lack of local encouragement or of friends and correspondents who could do for him what Willughby and Skippon, Robinson and Sloane did for Ray. Perhaps the number of books which he collected made the task seem overwhelming and put him out of conceit with his own efforts. Perhaps he was never more than a country land-agent who for a few years had an enthusiasm for collecting wild flowers which passed into a kindred enthusiasm for collecting books. In any case it is difficult not to feel that Gunther's eulogies of him are out of proportion to his achievement.

Of the English botanists of the first half of the seventeenth century Johnson and Goodyer have lately been fully studied. Their colleague and contemporary, George Bowle or Bowles, is still without any adequate recognition. Yet as the helper of Johnson in his youth and of Ray in later life, as a doctor with a large practice in London, and as a member of a family which had held positions of responsibility under the Crown, he would seem to deserve more attention than he has received. His influence in promoting the emergence of the modern scientific outlook—an emergence which took place during his life-time—may well have been far greater in his own day than that of many whose writings have survived.

Of his connections something has already been said. William Bowle of Bromley, his grandfather, was Groom of the Tents to Queen Elizabeth and died in 1609. His son, Robert of Chislehurst, inherited the office and married Frances, daughter of George Baker surgeon to the Queen and the friend of Gerard.[1] George, their eldest son, was born in 1604, as is shown both by the *Visitation of Kent* (Harleian Society, XLII, p. 114) and by his entry at the Queen's College, Oxford, 'in 1620 at the age of 16'.[2] He took no degree until 1640 when he became M.D. at Leyden on 26 September. In 1639 he married Abigail, daughter of William Booth a London merchant; and probably after his return from Holland settled in his house in the Strand 'in a court on this side of the May Pole on the left side'—that is near the present site of St Clement Danes Church. He was admitted an Honorary Fellow of the College of Physicians in 1664; died on 4 April 1672; and was buried at Chislehurst.[3]

His connections in later life seem to have been mainly with London and its neighbourhood. His uncle, Alexander Baker, was a magistrate in Middlesex whose daughter Anne married Thomas Pakington of Edgware. His brother, Sir William Bowles, became Groom of the Tents and Gentleman of the Privy Chamber to Charles II, and as we have noted had married Margaret, youngest daughter of John Donne, Dean of St

1 See above, p. 286.

2 Foster, *Al. Oxon.* I, p. 160. On his marriage in 1639 he was stated to be 28, but marriage ages both for men and women are notoriously inaccurate.

3 Cf. E. A. Webb, *History of Chislehurst*, pp. 227–8.

Paul's. His family connection with Chislehurst certainly continued: but we hear little of any close contact with the place after his marriage, except that, according to Merret, *Pinax*, p. 94, he planted *Pinguicula* there.

From the fragments of knowledge preserved to us it seems probable that he, like Goodyer, spent a few years in eager study of our flora, but afterwards, though not losing his interest altogether, found other claims upon him too exacting to leave him much leisure. As a medical candidate (and this is how Johnson describes him) he had the means and the will to seek for plants; and it is in that stage of his career that he made his name as a 'botanographist'. We have referred to his principal records in other connections. Here it is worth describing them in more detail. In 1632 he seems to have done a considerable tour. Perhaps its first discovery is of *Lythrum hyssopifolium* 'at Dorchester in Oxfordshire at the backe side of the enclosed grounds on the left hand of the town in the grassy places of the champian corne fields' (*Gerard*, p. 582); then we find him at Bath recording *Caucalis latifolia* 'in the corne fields on the hils' about the town (p. 1023) and at Bristol in July noting *Trinia glauca* from St Vincent's Rocks as Turner and De l'Obel had already done, and sending Johnson plants and seed of it (p. 1053), and noting *Rubia peregrina* in the same place (p. 1120). After this he seems to have paid the visit to Cherbury, where William Coote was chaplain to Lord Herbert, on which they found *Senecio saracenicus* 'in Shropshire in Wales in the hedge in the way as one goeth from Dudson [Dudston] to Guarthlow' (p. 428) and Bowles added *Impatiens noli-tangere* 'in Shropshire on the banks of the river Kemlet [Camlad] at Marington in the parish of Cherberry under a gentlemans house called Mr Lloyd: but especially at Guerndee [Wernddu] in the parish of Cherstock [Churchstoke] half a mile from the foresaid river, amongst great Alder trees in the highway' (p. 446). Fuller detail is given in the account of *Campanula hederacea* (p. 452): 'This pretty plant was first discovered to grow in England by Master George Bowles, anno 1632, who found it in Montgomeryshire, on the dry banks in the high-way as one rideth from Dolgeogg [Plas Dolguog near Penegoes], a worshipful gentleman's house called Mr Francis Herbert [great-grandfather of the first Earl of Powis and a connection of Lord Herbert], unto a market towne called Mahuntleth [Machynlleth] and in all the way from thence to the sea-side.' Clearly at this time he visited Aberdovey, where he discovered *Matthiola sinuata* 'upon the rocks' (p. 461) and again saw *Rubia peregrina* 'out of the cliffes' (p. 1120).[1] The tour seems to have ended with a visit to Richard Herbert at Ellesmere, where he found *Teesdalia nudicaulis* 'in the fields about Birch in the parish of Elesmere, in the grounds belonging to Mr Richard Herbert and that in great plenty'

1 Cf. Ray, *Memorials*, p. 172: 'At Aberdovy lives the lady Lloyd who informed me that Rubia tinctorum was found growing wild on the rocks there by Dr Bowles.'

(p. 250) and *Cicuta virosa* 'in the ditches about Ellesmere and in divers ponds in Flintshire [the part of Flint nearby]' (p. 257). One more plant, apparently a more cut-leaved variety of *Heracleum sphondylium*, is recorded by Parkinson, *Theatrum*, p. 954, and by Johnson, *Mercurius*, II, as found by Bowles in Shropshire: but no details are mentioned. Another, *Andromeda polifolia*, of which his record is not given by Johnson, is found in the *Phytologia Britannica*, p. 106: 'In great quantity at Birch in the moores by Ellesmeere in Shropshire—Dr Bowle.' A third is a 'grass' possibly *Carex pendula* 'in Ellesmere moores', *Phyt. Brit.* p. 54. Dr Coote reported to Parkinson from Ellesmere some years later, and no doubt was often there on visits from Cherbury. Bowles cannot himself have stayed very long: indeed, his whole tour can hardly have been leisurely: for he returned to Chislehurst in time to find the Equisetum already reported by Leonard Buckner at the end of August (p. 1115). He was evidently living there at this time; for the plant-records, *Radiola linoides*, July 1630 (p. 569), *Oxalis acetosella fl. rubro*, April 1633 (p. 1201), *Nardus stricta*, August 1633 (p. 1630), show a variety of dates. From Kent, too, 'the highway between Crayford and Dartford', and at some time between 1634 and 1641 came his most striking species *Orchis hircina*, which Johnson, *Mercurius*, II, does not assign to him, but which is recorded with his name by Ray, *Catalogus Angliae*, p. 342. From Kent, 'between the two parkes at Eltham on the mud', and before 1650 came his record of *Cochlearia anglica* in *Phytologia Britannica*, p. 5.

His records in the *Phytologia*, with the exceptions already mentioned, seem all to come from a single visit. They are not many and may well be noted.

The first is of *Geum rivale* 'in a pasture in great quantity neer Lynn belonging to Mr Thursby', p. 23—this being somewhat remote from the rest. There follow—'Wild Purslane' 'About Ramsey Meare in the foot pathes by the Rope Mills', p. 97 (this being the Ramsey in Huntingdonshire on the western edge of the fens); 'Solidago saracenica' where in addition to his 'Dudson and Guarthlow' record he gives 'on the five mile banke neere Whitlesea', p. 115—an interesting statement which must surely mean that he has confused *S. saracenicus* (assuming this to have been his Welsh plant) with the famous fenland species *S. paludosus* for which Whittlesea is a good locality. There remain three, all from the same area: 'Gnaphalium odoratum flore albo elegans pusilla planta' (a plant which is not easily identifiable) 'on a goodly heath by Barneck', p. 48; 'Pulsatilla rubra' evidently a dark form of *Anemone pulsatilla* 'on a heath towards Barneck three miles from Stanford where there are tenne thousand of these plants'; and 'Orchis Arachnitis', our *Ophrys sphegodes*, 'a brave plant and flowers betimes, I was much taken having never seen it before, it grows upon an old Stone pit ground, which is now green, hard by Walcot a mile from Barneck, as fine a place for variety of rare plants as ever I beheld', p. 82.

This last is a record of much interest, not indeed in itself—for the Spider Orchid has a way of appearing in such places and often in numbers —but for its subsequent history. Ray inserted a shortened version in his *Catalogus Angliae*, 1670, p. 224—'In an old stone-pit-ground hard by Walcot a mile from Barneck. Dr Bowles'; but before the book was published he received from Bowles along with a large number of herbal prescriptions records of his three rarest finds, *Orchis hircina*, *Senecio saracenicus* (only the Welsh locality) and what is now called 'Orchis forficulum referens. The Earwig-Orchis'. Of this it is said: 'This with the Andrachnitis, Arachnitis and Sphegodes maximus holosericeus praecox [all synonyms of *Ophrys sphegodes*] on a kind of little common with bushes on it between Blatherwick, the seat of the Staffords and Finchett [Fineshade Abbey] the seat of the Kerkhams [Kirkhams]:[1] also in a dry barren ground with hills that hath been left from digging of stones, as one goes from Wansford-bridge, after you have passed the next wind-mill, towards Upton on the left hand, half a mile towards Wallcot in Northamptonshire.' This record, the latter part of which almost certainly refers to the same place and plant as that on p. 224, Ray printed in his Appendix on p. 342.

It is also of importance because it is quoted by Thomas Fuller in the introductory paragraphs of his account of his native county in *Worthies* (ed. 1811, II, p. 157): 'Know, Reader, that Doctor Bowle my worthy Friend and most skilful Botanographist, hath taken notice of a heath in this county nigh to Stamford whereof he giveth the commendation "as fine a place for variety of rare plants as ever I beheld" (*Phytologia Britannica*, p. 82); who, I am sure, hath seen in this kind as much both here and beyond the seas, as any of his age and profession.' This friendship of Bowles with Fuller may perhaps date back to Johnson's lifetime; for Fuller and Johnson were certainly together in the siege of Basing House in 1644; but more probably it arose through Sir John Danvers with whom Fuller spent some considerable time in his house at Chelsea. In any case, it is welcome testimony to the reputation of Bowles as a botanist.

1 Either Robert or his son Walter must then have been head of the family.

CHAPTER XVII. WILLIAM HOW AND CHRISTOPHER MERRET

The death of Johnson and the absorption of Goodyer and Bowles in other interests might have been a greater disaster than was actually the case. The development of interest in botany and indeed in scientific studies generally was now becoming so rapid that there was little risk of ground once gained being lost. Progress which had depended upon individual pioneers was now the concern of organised groups and institutions. The 'New Philosophy', if not yet formally acknowledged, had found its herald in Francis Bacon; and an increasing number of supporters gave attention consciously or unconsciously to the programme and method which he had proclaimed. We have seen that the actual business of exploring the country's flora had been begun in England by gardeners and herbalists before any advocacy of observation and induction had been heard. There was still a space of twenty years before the foundation of the Royal Society gave that exploration a permanent and official backing. But in the interval Johnson's work was carried on and his *Mercurius* was made the basis first of William How's *Phytologia Britannica* and then of Christopher Merret's *Pinax Rerum Naturalium Britannicarum.*

William How is a man of whom little is known. Born in London in 1620 and going to Merchant Taylors School in 1632, he went to St John's College, Oxford, in 1637, graduated in 1641 and took his M.A. in 1644. Then he joined the King's forces and was apparently put in command of a troop of horse: but with the waning of the royal cause he returned to London, began to practise in St Laurence Lane and soon moved to a house in Milk Street, Cheapside. It seems clear that he had not much experience of plants in the field, but that he had got at least one of Johnson's books, the Description of the Kentish tour, with Johnson's own notes in manuscript. It is, of course, very possible that he met Johnson in Oxford and he may well have had close contact with him.

In any case, as Kew and Powell have pointed out, his work was begun by pasting up into a single list the two alphabetical catalogues in the two parts of the *Mercurius*. These he augmented by a number of records from 'Dr Johnson's MS.', by others without any name attached, by very many from Walter Stonehouse, some from George Bowles and Richard Heaton, and a few from 'Mr Loggins' and 'Mr Sare'. To the list thus completed he prefixed a flamboyant dedication, 'Viris Apollineis', written in a Latin which is like a nightmare parody of the worst kind of euphuistic extravaganza, a dedication the point of which seems to be that the medical man need not go searching the tropics for his drugs, since every disease has by divine decree its native remedy[1]—though, of course, the quacks who sell Mithridatic Lovage for Spignel deserve to be shown up. The book was printed by Richard Cotes and published anonymously by Octavian Pulleyn with the imprimatur of John Clarke the President, and Sir Maurice Williams, George Ent, John Micklethwaite and Ahasuerus Regimorter the Censors of the College of Physicians in 1650.

It is a very hasty and defective piece of work. Johnson's records are printed verbatim so that his notes, for example that *Gentiana bavarica* is 'found on mountain tops if my guess is not mistaken', *Merc.* II, p. 22 or *P.B.* p. 46, or that 'Millefolium umbellatum' is 'most frequent according to Park. pag. 1258 yet I have not seen nor heard of it by any other', *Merc.* II, p. 26 or *P.B.* p. 76, are still in the first person. One or two of the additions from his MS., for example *Habenaria viridis*, 'by Barkway', p. 82, are useful. Most are bare names without locality and scarcely identifiable. One addition at least is a pure tautology: he not only prints from Johnson the 'Saponaria folio concavo sive Gentiana concava' on p. 109, but adds 'Gentiana concava' (the same plant from the same place) on p. 46. He multiplies varieties extravagantly: there are thirty-one supposedly different Orchids, pp. 82–4: there are two 'Genistella sive Chamaespartum' (presumably *Genista anglica*, though Chamaespartum is usually applied to *G. pilosa*), one 'from South Sea Castle' without authority, the other 'towards Portsmouth' with a reference to the margin of De l'Obel's 'Historia plantarum Teutonice edita': there are a large number of white-flowered forms and other trivial varieties, treated as distinct. Moreover there are a great many entries which classify as British obvious casuals or garden escapes. Such are *Anagyris foetida* 'at Collisdon [Coulsdon] two miles from Blechentree [corrected among errata on last page to Blechenligh, i.e. Bletchingley] in Surry', p. 7: *Peonia corallina* 'in Mr Feilds well close in Darfield...I believe it came first out of a garden with some dung', p. 96: *Physalis alkekengi* 'observed by Mr Jervase Dickson a physitian in a hedge by Bently near Doncaster in Yorkshire and by Mr Parker of Stockport in Cheshire in severall places of that country wilde', p. 115: or *Verbascum phoeniceum* 'neer Oxford. Mr Thomas', p. 128. These are new in How's book: in addition he has some thirty others mainly derived from Gerard like *Campanula media*, *Doronicum pardalianches* and *Soldanella alpina* which were accepted as British until Ray in the preface to his *Catalogus* twenty years later listed and rejected them. There are also several 'Corals', many sea-weeds and a few fungi.

1 This sentiment was repeated a few years later by How's friend, William Coles, on the title-page of his *Adam in Eden*, London, 1657. This book quotes the *Phytologia*, e.g. on p. 63; but its localities are largely drawn without acknowledgment from Gerard and Parkinson.

Of the unnamed records only one or two seem to be both new and interesting. One is the addition on p. 75 to Johnson's record of *Stratiotes aloides* of the note 'and in the new ditches of the Dutch workes of Hatfeild [in Lincolnshire] within three or four yeares after they were made'—an allusion to the drainage scheme carried out there by Cornelius Vermuyden in 1626 and described by William Dugdale, *History of Imbanking*, 1662, pp. 144–5. Another is on *Teucrium scordium*, p. 110: 'On the bankes in the Isle of Ely you may mow it.' Another is on *Peonia officinalis*, p. 95: 'In a close belonging to Mris Anne Steavenson at Sunningwell in Barkeshire of above fifty yeares standing.' Another under 'Asarum', p. 12, and added to De l'Obel's record states that Asarabacca (presumably *Asarum europaeum*, though De l'Obel's plant was *Antirrhinum asarina*) was also found at Eynsham near Oxford.

There is one very curious medical note, that under 'Typha palustris', p. 126. Here he quotes from *Gerard emac.* p. 46, Lit. B, a sentence describing, in detail, an inguinal hernia, and adds that the description, of the fall of the gut, is mistaken. Gerard, not Johnson, is dealing with the supposed value of the seeds of *Typha latifolia* as a cure: How's correction is a mere piece of medical pedantry.

But the value of the book is not in its editing of Johnson or in How's own contributions, if any, but in the material which he collected from his friends. Goodyer, who is only quoted once or twice, and Bowles we have already discussed. They belong to an older generation. So in fact does Walter Stonehouse. But unlike the others he was hardly known as a botanist until How received from him the great mass of material here published. As R. T. Gunther demonstrated in the *Journal of Botany* for 1920,[1] Stonehouse was a Londoner, born in 1597, and a relative of Sir William Stonehouse, Bart., of Radley, whose daughter married William Langton, President of Magdalen College, Oxford, 1610–26. He went up as one of the first scholars to Wadham College in 1613, was elected a fellow of Magdalen in 1617 and seems to have remained in residence until his marriage in 1629. He was presented by the University to a living near Canterbury in 1630 and may have made acquaintance with Johnson there. In 1631 he moved to Darfield[2] in Yorkshire, some five miles east of Barnsley, at the invitation of John Savile[3] of Methley, a dozen miles to the north. There he became an enthusiastic gardener, his catalogue in manuscript being still preserved in Magdalen College (MS. No. 239) and having

1 Cf. also his *Early British Botanists*, pp. 271–3, 348–51.

2 A living worth at that time the adequate sum of £250 a year according to Walker, *Sufferings*, l.c.

3 Son of Sir John, Baron of the Exchequer, by his second wife: succeeded his half-brother, Sir Henry, in 1632: cf. *Yorks. Archaeol. Assoc. Records*, IX, pp. 150–2. Sir John's first wife had been Jane, daughter of Richard Garth of Morden (see above, pp. 170, 243, etc.).

been printed in the *Gardeners' Chronicle*, 1920. He was ejected and imprisoned according to Walker, *Sufferings of the Clergy* (ed. 1714), pp. 123, 373, in 1648, returned in 1652 and died in 1655. His only daughter, Anna, married William Cave, the Church historian and Canon of Windsor: she died at Islington in 1691. His commendatory verses containing anagrams of his and Tradescant's names were printed in the Tradescant Catalogue of 1656. Presumably during the years of his deprivation he had come to London and frequented the garden and museum at South Lambeth. In 1648 also he may have supplied How with the data which he has used so freely.

These records are in fact very numerous. Many refer to Darfield: 'Netherwood Hall grounds', p. 11; 'a close of Sir Edward Rodes [Rhodes][1] at Haughton [Houghton] in Darfield parish', p. 17; 'at Wombwell Head', p. 45; 'in mine own Rie', p. 88; 'in Mr Feilds well close', p. 96; 'in divers hedges', p. 105; 'in mine owne orchard', p. 124: others to the neighbourhood: one to the west 'by Uden Lodge in Pennyston (Penistone) Moore', p. 83; several to the east, Brodsworth Wood, pp. 15, 83; 'Scosby leas neer Donkester' (Scawsby), pp. 33, 48; 'next Hampoole wood as one goes from Hutton-Pagnell' (Hooton-Pagnell), p. 44; three to the north, 'in Womersley field', p. 87; 'betwixt Pontfract and Ferry-bridge', p. 87; 'in Mr William Pickering his parsonage house at Swillington', p. 29; and a few more remote represent more serious expeditions: 'neer Bridlington on the shoare', p. 43; 'most plentifully mixt with Pinguicula in a very low and squallid meadow near Knasborough' (Knaresborough)—this of *Primula farinosa*, p. 98; 'neere Stockport, the country people there call it the Merry-tree. Whence I should think it the Merasus of the Hungarians (mentioned by Clusius Pannon, lib. cap. 24) had not hee said that hath black berries whereas this hath them of a delayed red'—of *Prunus avium*, p. 25; 'neer Eldenhole in the Peake and about Buckstones' (Buxton), p. 61; 'betwixt Peasly [Pleasley] and Mansfield', p. 56; 'neere Nottingham between the town and the gallows', p. 24. The plants reported are in the main not rare; many of them are the white varieties so freely chronicled by How; it is not a great botanist's list: but he was certainly observant and interested. Only one of his notes comes from Oxfordshire, 'neer Gosworth bridge' (Gosford over the Cherwell), p. 11; two are from Berkshire, 'in Duckleton lottes', p. 28, and 'in a lane going from Tubny [Tubney near Fyfield] to Newbridge', p. 115. The two most interesting come from Wiltshire, on pp. 41–2, 'a kinde of sweet smelling Female Ferne [? *Aspidium aemulum*] somewhere about the Marquesse of Hartford [Hertford—William Seymour Earl 1621, became Marquis in 1641 and Duke of Somerset in 1660] his forest of Savernake; but so long since that I have forgotten both the exact place and whether it any way

1 Cf. Dugdale's *Visitation of Yorks*. 1665–6, Surtees Soc. p. 266.

differ in shape from the common, only this I remember, that the old Earl took so much notice of it, that hee caused a fair inscription to bee made in his Garden pond at his house of Totnam [Tottenham] neer it, to direct to it'—a fascinating sign of the interest which was being taken in such subjects; and on p. 51, 'the inquiry of the Grass some nine miles from Sarisbury by Mr Tuckers at Maddington wherewith they fat hogges, and which is four and twenty foot long, which may happily be a kinde of Gramen caninum supinum though Gerard English it ill Upright Dogges-grass pag. 26'—a reference which as Ray noted in his *Catalogue*, p. 142, is derived from Fuller, *Worthies*, II, p. 438 (ed. 1811). Fuller has a paragraph headed Knot Grasse, and states that 'it is a peculiar kind; and of the ninety species of Grasses in England is the most marvellous'. He gives details of its length, of the peculiar hollow in which it grows, of its reaping with sickles, and of the proposals for growing it elsewhere. In Thomas Davis, *General View of the Agriculture of Wiltshire*, drawn up for the Board of Agriculture in 1794, there is a discussion of this grass which he, like Fuller, declares is found only in 'two small meadows at Orcheston, six miles north-west of Amesbury, usually called the Wiltshire Long Grass Meads', p. 41. His opinion, quoted from 'Mr Sole of Bath', is that the grass is 'nothing more than the "black couch or couchy bent" (Agrostis Stolonifera) one of the worst grasses in its native state the kingdom produces,...but in these meadows, when abundantly fed with water, it is of a juicy, succulent, nourishing quality as grass and makes the most desirable hay in the district...the stalk is frequently eight or ten feet in length from the original root;...the crop...is perhaps not eighteen inches high'. At this period the identity of this grass and the circumstances producing its abnormal development were keenly debated both in the *Proceedings* of the Bath Agricultural Society and by botanists. Withering, *Arrangement* (6th ed. 1818), II, p. 187, has a long note on the subject: so has T. A. Preston, *Flowering Plants of Wilts*. 1888, pp. 393–4. The species appears to be *Agrostis alba*, though specimens in the British Museum labelled Orcheston or Maddington Grass are *Agropyron repens*.

The other botanist of importance who contributed to How's book is Richard Heaton. He has always been well known as one of the earliest students of Irish plants; and his records of *Gentiana verna* 'betwixt Gort and Galloway [Galway]' (p. 46) and of *Dryas octopetala* 'in the mountains betwixt Gort and Galloway: it makes a pretty show in the winter with his rough heads like Viorna [*Clematis vitalba*]' (p. 120) are familiar. He noted for How a number of other Irish species: *Juniperus nana* 'upon the rocks by Kilmadough [Kilmacduagh near Gort]' (p. 64); *Scilla verna* 'at the Rings end of Dublin' (p. 60); *Pyrola minor* 'in a bogge by Roscre [Roscrea] in the Kings County' (p. 100); *Epipactis atrorubens* 'found by Lysnegeragh' ('probably near Roscrea'—*Cybele Hibernica*, 2nd edition,

p. 342) (p. 57); and two others from Edenderry in King's county, 'Soon-a-man-meene: in English the juyce of a faire Woman: in a wood near Eddenderry: I referred it to Rubus saxatilis, but the berries of this plant were yellow' (according to *Cybele Hibernica*, p. 110, *R. saxatilis*) (p. 116) and *Drosera anglica* 'plentifully in a bogge by Edenderry. The leaves are above a span long. I gave some of the plant to Zanchie Sylliard Apoth. of Dublin, which he sent to Mr Parkinson who in his description mentions the said Zanchie as if he had found it' (p. 105)—a reference to Parkinson, *Theatrum*, p. 1053.[1] That he was a distinguished member of Trinity College, Dublin, took his D.D. there in 1661 and then in 1662 became Dean of Clonfert, has also been freely stated, although Pulteney, *Progress of Botany in England*, II, p. 194, says 'I cannot collect any anecdotes of him', and even Dr R. L. Praeger[2] only notes his connection with Ireland and with How and Merret.

But what is plain from How's volume is that Heaton was not always and only an Irishman, that on the contrary he did a good deal of field work in England and that several of his finds here are of importance. He found white Pimpernel (*Anagallis arvensis*) 'in a wood on the south side of Chislehirst heath over against Scadbury Park' and 'Anemone tuberosa radic.' (probably *Anemone pulsatilla*) 'upon Cotswald hills neere Black-Burton' (p. 8). Then there are 'Branched Maidenhaire' (presumably *Asplenium adiantum-nigrum*) 'in Devonshire' (p. 121) and *Melittis melissophyllum* 'in Mr Champernon's wood[3] by his house on the hill side neere Totnes' (p. 74). And finally there are two from the north: *Senecio palustris* 'a stones cast from the east end of Shirley Poole neere Rushie Moore belonging to Mr Darcy Washington[4] in Yorkeshire' (a place two miles to the south of Askern near Doncaster, according to F. A. Lees, *Flora of West Yorks*. p. 293) (p. 30) and a variety of *Lysimachia vulgaris* (so Lees, l.c. p. 378), 'Lysimachia lutea minor' as Heaton calls it, 'by Shirly Pool in Yorkeshire' (p. 72). Plainly the man who supplied these notes had spent a considerable time in England after he had gained some knowledge of plants.

From the fact that he is only known in Dublin as obtaining the D.D. in January 1661, it seems clear that he graduated previously in England. The *Alumni Dublinenses* says 'See *Venn*', which presumably means that the editors identified the Dean of Clonfert with the Richard Heaton sizar of St John's College, Cambridge, who matriculated there in 1620, took his B.A. in 1623–4 and his M.A. in 1627. Venn accepts this identification,

1 See above, p. 263.
2 *Flora of West Ireland*, p. 31.
3 Presumably Henry Champernowne of Dartington, cf. *Visitation of Devon*, 1620 (Harl. Soc. VI), p. 53.
4 Son of Richard Washington of Adwick-le-Street (cf. Dugdale's *Visitation*, p. 273); matriculated University College, Oxford, 1605 (Foster, *Alumni Oxon*. p. 1578).

which had previously been suggested by C. C. Babington,[1] and adds that he was in Ireland as Prebendary of Killaloe in 1633, and died there in 1666. It is confirmed by the entry in the Register of St John's. There Mr F. P. White has found the words, presumably in Heaton's own hand, 'Ego Richardus Heaton Eboracensis'. The double connection with Yorkshire should make it certain that the Johnian and the botanist are one and the same. The only other evidence of his career seems to be that contained in a petition by Richard Heaton, D.D., in August 1664 to the King for a regrant of lands of Balliskenagh. There it is stated that he had served Charles I as chaplain to the regiment of foot under Colonel James Strangways and afterwards to Lord Wentworth's Life Guard of horse, and that he had been ruined by the rebellion in Ireland (*Calendar of State Papers, Ireland*, Addenda, 1625–70, p. 503).

Presumably it was after the appearance of the book that How received 'some notes and a box full of the fresh juli' from Dr Thomas Browne of Norwich. Browne refers to this in his letter of 13 July 1668 to Christopher Merret (G. L. Keynes, *Works of Sir T. Browne*, VI, p. 362) when he claims to have sent *Acorus calamus* to Goodyer 'about 25 years ago' and 'more lately' to Dr How. How's letter, printed in Wilkin, *Sir T. Browne's Works*, I, p. 417, acknowledging a 'rare present' which may well be these notes, is dated from Milk Street, 20 September 1655. It is written in a style less turgid and involved than that of his Latin prefaces, but obscure and not too grammatical. He deals with the Westminster Garden ('Botanotrophium Westmonasteriense') for horticulture, medicine and perfumery, of which he was apparently the founder or proprietor,[2] and for which he seems to be planning a catalogue; of this garden Edward Morgan, Johnson's companion, was superintendent, or as he calls himself 'herbarist'.[3] How, of course, died before he could make use of Browne's material.

The *Phytologia*, bad though it certainly was, seems to have sold well. How's premature death in 1656, a few months after the publication of his attack on Parkinson, prevented the production of the revised edition for which he had made a few preparatory notes;[4] and the popular demand

1 Cf. *Al. Cantab.* IV, p. 527; the knowledge that Babington identified the Dean with Heaton of St John's I owe to Mr F. P. White of that College, who also drew my attention to the Irish State Papers.

2 William Coles, *Adam in Eden*, p. 582, speaks of it as 'Dr Howe's garden at Westminster' and in his Preface speaks of 'the late noble Institutions of some Physick Gardens: amongst which Mr Morgan of Westminster hath one in his tuition which by the noblenesse of Dr How is already very full fraught'. The lessee of the ground was apparently a Mr Gape: cf. H. Field, *Memoirs of the Garden at Chelsea* (London, 1820), p. 8.

3 Cf. the 'Approbation' prefixed to *Adam in Eden* and signed by him as 'Herbarist to the Physick Garden at Westminster'.

4 These were written in his copy of *P.B.*: cf. Gunther, *E.B.B.* pp. 279–93.

for his book had to be met elsewhere. So Cave Pulleyn, son and successor of Octavian, How's publisher, approached his friend Christopher Merret, who was living in the house belonging to the College of Physicians[1] on Amen Corner in charge of William Harvey's Museum and Library which were kept there, and invited him to produce a complete Catalogue of British fauna, flora and fossils. Merret sought the help of a skilled botanist, John Dale the friend of Goodyer, whose work has been described by Gunther in *Early British Botanists*, pp. 294–8. He bought the eight hundred engravings of plants which Johnson had prepared for the history which he and Goodyer had planned. Dale's death did not turn him from his purpose. He secured the services of Thomas Willisel, an old soldier of Lambert's corps, a man of little education but highly skilled in field-work; and sent him every summer for five years into various parts of England. He got some assistance from his younger son Christopher, and also, as appears from *Pinax*, p. 214, from his elder boy Robert; and got the loan of Goodyer's manuscripts from his nephew and executor Edmund Yalden. He preserved as many specimens as possible in his 'hortus siccus' and was ready to show these to any genuine student. Birds and fishes he got by enquiry from the London salesmen: stones and metals from his own previous experience in chemistry and technology. His museum was available: so was the Royal Society. He certainly did his best as a sedentary and relatively inexpert naturalist to get all the help that was obtainable.

Nor was he ill-fitted for the task that he thus undertook. Born in 1614 at Winchcomb, he had gone up to Gloucester Hall, Oxford, in 1631, moved to Oriel two years later, graduated in 1635, began to practise in London about 1640[2] and proceeded to his M.D. in 1643. In 1651 he had been elected to the College of Physicians, in 1654 he had been Gulstonian Lecturer; and in the same year had been nominated by Harvey as curator of his Museum and Library. When the Royal Society was founded in 1662 he became an original member of it. He was a man who had lived and worked in Oxford long enough to understand the scope and character of his undertaking, and who, unlike How, wrote reasonably good Latin without solecisms if without much distinction. He had made a fine start in his medical career, had secured a position of some importance and was beginning to be well-known and well-esteemed. There must have seemed every prospect that he would throw himself keenly into work so relevant to his profession, and would bring to botanical and zoological

1 This was first situated in Knight Rider Street: after the Fire it was moved to Ivy Lane.

2 In the opening of his *Short View of Frauds by Apothecaries*, published 1670, he describes himself as having had '30 years not unsuccessful practice in this great City'.

studies what they had not had in England since Penny's time, a man with a good mind and a good training.

And since How's work a further advance in cataloguing had been made by the appearance in 1656[1] of the *Musaeum Tradescantianum*. We have already noted John Tradescant's establishment of a garden and a collection of curiosities in South Lambeth; and that his son, John Tradescant junior, had carried on the double venture. As explained in the preface of his catalogue John junior had undertaken with the help of two friends a survey and list of all his rarities. Then his son died, his friends were busy, William[2] Hollar kept him waiting ten months for the two portraits prefixed, and publication was delayed. When issued the two lists of the museum and the garden were bound up together, and the former was divided into two main sections, natural and artificial. The natural falls into seven categories: Birds with their eggs, beaks, feathers and claws, including the egg of a Cassowary or 'Emeu' (with a reference to Harvey, *Generatio Animalium*, p. 61), 'the claw of the bird Rock who as the Authors report is able to trusse an Elephant', the 'Dodar from the Island of Mauritius'[3] and 'Barnacles four sorts' (pp. 1–4); Four-footed Beasts including 'a natural Dragon above two inches long'; 'a Doe's head and horns from Saint James's Parke neer London' and 'Elkes hoofes' (still preserved at Oxford) (pp. 5–7); Fishes including 'a Phocaena's head given by T.W.' (probably Thomas Wharton)[4] (pp. 8, 9); Shells (pp. 10–14); Insects and Serpents (pp. 14–17); Fossilia (that is anything 'dug out of the earth'), metals, earths, 'succi concreti' (including corals, salts, 'alumen Cantabrigiense', etc.), various organic fossils ('Ophytes from Whitby', marble from York, etc.) and gems (including British emeralds, English and Scottish pearls and the Lapis Bezoar) (pp. 17–26); and Exotic Fruits, leaves, wood, seeds, etc. (pp. 26–34). In the second section along with curiosities usual and unusual ('Flea chains of silver and gold with 300 links a piece and yet but an inch long' being the most unexpected) are pieces of amber with flies and spiders, 'splene-stones', 'a booke of Mr Tradescant's choicest flowers and plants exquisitely limned in vellum by Mr Alex. Marshall',[5] a piece of a log from 'Bagmere in Cheshire near

1 A second title-page was issued by the same printer John Grismond and publisher Nathaniel Brooke along with a dedication to Charles II in 1660. It had previously been dedicated to the President and Fellows of the College of Physicians.

2 An error for Wenceslaus: he was in Antwerp from the fall of Basing House until 1652.

3 Destroyed by order of the Vice-Chancellor at Oxford in 1755, its head and foot being still preserved there.

4 A list of Tradescant's 'Benefactors' is appended to the volume. Wharton (1614–73) was a noted London doctor and anatomist, author of *Adenographia*.

5 Is this the Mr Marshall whose additions to the 'curious booke of flowers in miniature' Evelyn saw at Fulham (*Diary*, 1 Aug. 1682)? For his pictures of flowers on vellum cf. *Brit. Mus. Cat. Drawings*, III, pp. 92–4.

Breereton [between Congleton and Middlewich]', 'blood that rained in the Isle of Wight attested by Sir Jo. Oglander',[1] and one souvenir of old John's exploit, 'a copper Letter-case an inch long taken in the Isle of Ree with a letter in it which was swallowed by a Woman and found'. With such a list before him—and as we shall see he made some use of it—Merret had guidance for the parts of his task outside How's field.

Less immediately relevant though bearing closely upon his intention is the new edition of Sir Hugh Plat's[2] *The Jewel House of Art and Nature* issued 'by Elizabeth Alsop at her house in Grubstreet' in 1653 with an appendix on Minerals, Stones, Gums and Rosins by 'D. B. Gent' (Arnold de Boate or Boot). Plat's book first printed in 1594 is a collection of 150 'new and conceited experiments' ranging from 'how to brew good and wholesome Beer without any Hops at all' (no. 9—in which he refers to 'Doctor Turner' along with other authors of 'learned Herbals') to a 'perspective Ring that will discover all the Cards that are near him that weareth it' (no. 2), and the familiar poacher's trick of catching birds in hoods of bird-limed paper (no. 56). The book is a forerunner of the *Century of Inventions*, published in 1663 by Edward Somerset, second Marquis of Worcester,[3] and of the experiments with which Hooke was required to amuse and instruct the virtuosi of the Royal Society. In 1653 De Boate added to it a long descriptive catalogue which contains, among much that is trivial and at second hand, notes on Bristol Stones and their difference from the true Adamant or Diamond; on the taking of Margarites (pearls) from the 'Horse-Mussel' (presumably *Unio margaritifer*) found in Buckinghamshire 'and so plentifull in some parts of the river of Clun that they do more than cover the bottome of that river....I have some few which I took out of the shel myself' (p. 221); and on the reddish stone 'like unto a dull or coarse Ruby' which a jeweller friend riding into Suffolk found 'about four miles from Sudbury' (p. 222).

That Merret used Plat's book is improbable; no direct citation or allusion seems to occur. The case is different in regard to a later work, Joshua Childrey's *Britannia Baconica*[4] *or the Natural Rarities of England, Scotland and Wales*, which seems to be the unacknowledged source of a large amount of the material gathered into the geological and geographical sections of the *Pinax*. Childrey, born in Kent, and sent up to Magdalen College, Oxford, before the Civil War, became chaplain to Henry Somerset, Lord Herbert, at the Restoration and took occasion

1 Of Nunwell, the diarist, 1585–1655. There seems to be no allusion to this in either of the published selections from his writings.

2 See above, p. 241.

3 Cf. G. N. Clark, *Science and Social Welfare in the Age of Newton*, p. 10.

4 The title is explained in the preface, 'I have followed the precepts of my master the Lord Bacon'.

to publish next year this survey of the several counties. It is not, in the main, an original work. Camden's *Britannia*, John Speed's *Theatre of the Empire of Great Britaine* with the introductory pages to the maps of the counties, Richard Carew's *Survey of Cornwall* for his first county, Thomas Johnson's *Mercurius Botanicus* for Bath, and a few other works to which allusions are made or can be traced, are responsible for the main bulk of the contents. Some of the original material is astrological; for in 1653 he had published *Syzygiasticon Instauratum* and was still very interested in Ascendents.[1] But there are passages, very notably the long account of fossils near Badminton, and others dealing with Kent and Oxfordshire, in which he writes of nature at first hand and with evident interest. He was not a man of eminence, but his collection of material for a physical geography was a useful piece of work; and there are suggestions, for example the note appended to his record of Box (*Buxus sempervirens*) in Surrey, 'Inquiry might be made of Herbarists whether the earth be not of the same nature and composition where the same vegetables grow naturally' (p. 54), which reveal an enquiring and intelligent temper. Perhaps the most striking of these is that appended to his account of the famous invasion of mice near Southminster in Essex, recorded by Stow and Speed (see above, p. 245), and of the 'strange painted Owles'[2] (no doubt *Asio flammeus*) which followed and destroyed them (p. 100). On pp. 15–16, discussing 'the reason of several birds leaving us and returning again at set times of the year' he suggests that this is due to the fact that their special food is only seasonally present or else that the temperature hotter or colder makes them 'take their flight more northernly or southerly'; 'and if the winter prove very mild, then the winter birds (as Fieldfares, etc.) come not quite home to us; if it prove extreme sharp then they flye beyond us to the southward'. It is unnecessary to give a full and separate account of his book; for most of its data appear in Merret—without any acknowledgment. The references to *Britannia Baconica* given in our rendering of the latter part of the *Pinax* will show the extent of Merret's obligation.

Moreover, in addition to books, Merret had the encouragement of the two greatest scientists of the time. William Harvey, the first Englishman to gain worldwide fame by his discoveries, was born at Folkestone in 1578 and educated at Canterbury and at Gonville and Caius College, Cambridge, where he graduated in 1596. He took his doctor's degree at Padua in 1602, returned to England in the same year and in 1607 became a fellow of the College of Physicians. In 1615 he became lecturer in anatomy at the College and prosecuted his researches into the movements and function of the heart, with a view to testing the idea of the

1 Cf. *B.B.* pp. 122–7.
2 Stow and Speed have no mention of the Owls.

circulation of the blood which had come to him, apparently, in the previous year. His book was not published until 1628 at Frankfort. Thereafter he continued investigations and experiments, particularly into the problems of generation. In 1651 his friend George Ent got hold of his manuscripts on this subject; and his second great book was published. That the *De Generatione Animalium*[1] was not only of high value as a physiological classic but also contributed much to general zoology is clear from the use made of it by Ray in his introduction to *Willughby's Ornithology*.

For Harvey was not only a great physiologist and a famous doctor. He was an ardent though not very learned naturalist. John Aubrey[2] has recorded that when Harvey accompanied Thomas Howard, Earl of Arundel,[3] in 1636 on his embassy to the Emperor, Wenceslaus Hollar the artist, who was one of the retinue, complained that he was always making expeditions after botanical or geological specimens; and there is a long passage in his *De Generatione*, Ex. xi (*Works*, ed. R. Willis, pp. 208–10), describing a visit to the Bass rock when he was in Edinburgh in attendance upon Charles I in June 1633 (cf. J. MacMichael, *Lives of British Physicians*, 1857, pp. 41–4)[4] and discrediting the story that 'Scottish geese are produced from the fruit of trees fallen into the sea'.

Presumably it was while he was living in Oxford between 1642 and 1646 that he came to know Merret: he was Warden of Merton College for his last year in the city, and Merret had taken his M.D. and was working at Gloucester Hall. In any case, after his return to London and until his death in 1657 he must have seen very much of the young man whom he chose to superintend the books and collections for which he presented a building to the College in 1653. Indeed, there is an interesting record of the younger man's co-operation with him in Merret's *Art of Glass*, p. 300, where he records his production of 'Calamie' in powder and his having communicated it to 'the eternal glory of our nation and of Anatomy, and an excellent Chirurgian and never to be by me forgotten the incomparable Dr Harvey, a man most curious in all natural things who confessed he thought this to be the said Pompholix,[5] and with most happy success frequently used it'.

1 Detailed examination of this belongs rather to physiology than to natural history. A. W. Meyer, *An Analysis of the De Generatione* (Stanford, U.S.A. 1936), has discussed it in some detail and critically. It is inevitably somewhat disappointing; for whereas the *De Motu Cordis* is epoch-making, this in its treatment of spontaneous generation accepts current error, and in its main problem is inconclusive.

2 *Lives of Eminent Persons* (ed. 1813), II, p. 384.

3 The 'Father of vertu in England', our first great art-collector: he and Harvey had this passion in common.

4 Quoted also in D'Arcy Power, *William Harvey*, pp. 93–6.

5 A by-product of the process of the smelting of bronze: for a contemporary discussion of its precise nature and significance, cf. e.g. Mattioli, *Comment. in Diosc.* v (ed. G. Bauhin, 1598), pp. 918–20, and *Epistolae*, p. 35.

Merret's other friend, the 'father of chemistry and brother of the Earl of Cork', the admirable Robert Boyle, belonged to a younger generation.[1] Born in 1626, educated at Eton under Harrison for four years and then under a tutor at Stalbridge in Dorset, he went abroad in 1638 to Lyons and Geneva and so to Italy. Returning in 1644 he lived with his sister in London and began to attend the meetings of 'the invisible or (as they term themselves) the philosophical College' (Boyle to Francis Tallents; *Life* by Birch, p. xxxiv). These were then being held usually at the lodging of Jonathan Goddard, Professor of Physic at Gresham's College, in Wood Street—George Ent, John Wilkins, Francis Glisson, Samuel Foster, Theodore Haak and on occasion Merret being the usual members present. Boyle lived at Stalbridge from 1646 till 1654, but in 1652 visited Ireland where with the assistance of William Petty he dissected fishes, enquired into minerals and studied natural history (*Life*, p. liv). In 1654 he settled in Oxford, was joined by Robert Hooke in 1658, invented his air-pump and carried out a very large number of experiments.

He was a man of delicate health, but of singularly acute and sensitive mind, with a profound interest in religion and philosophy as well as a passion for research. As a champion of the new methods of study, his *Essay on the Usefulness of Natural Philosophy* was not only an adequate reply to Hobbes, who had attacked 'the society that is wont to meet at Gresham's College' with his customary arrogance, but a fine exposition of 'the two chief advantages which a real acquaintance with nature brings to our minds,...first by instructing our understandings and gratifying our curiosities, and next by exciting and cherishing our devotions' (*Works*, II, p. 6). During these years he was doing his best work both in observation and experiment and in the publication and exhibition of his results. Merret, who speaks of him with a reverence such as most of his contemporaries also show for him, was obviously interested in his chemical and physical studies, especially on their practical side, and had himself done some experiments. For in the earliest meetings of the Society Merret had been asked to discourse about refining (16 January 1661) and the antimony stone (22 February); to serve on a Committee to examine the making of 'ceruss' (13 March) and with Boyle and Daniel Colwall to draw up an account of the making of vitriol (3 April) (cf. Birch, *History of the R.S.* I, pp. 12–20). He had also studied the processes of glass-manufacture and the effects of frost, publishing his translation of Antonio Neri's book, *The Art of Glass*, in 1662 and *An Account of Freezing* annexed to Boyle's *New Experiments Touching Cold* in 1665. As we shall see from the *Pinax* and as is clear from his contributions on the Cornish Tin-mines and on Refining to the *Philosophical Transactions*,

1 He only met Harvey once; for this and other particulars of him, cf. L. T. More, *Life and Works of R. Boyle.*

1678, he was more deeply interested in geology and metallurgy than in botany and zoology.

Of the books written by him before his *Pinax* that on the *Art of Glass*, though a translation, yet contains a large amount of original material—Neri's work occupies pp. 1–203, Merret's pp. 205–365. The whole is dedicated to Boyle who was responsible for Merret's undertaking it and who 'will much promote by your practice the Art it's self'. The later part contains a detailed commentary upon and explanation of Neri's processes, and reveals that Merret himself had obviously performed a large amount of the actual experiments described. He pays special attention to the minerals mentioned, and notes where they can be obtained in England. There is a long and interesting section upon the making of pastes or counterfeit jewels, especially emeralds; and in this connection he gives not only an account of metallic colourings but a list of plants from which dyes may be distilled. It is obvious that Merret had a real knowledge of the materials and apparatus used in glass-making; he had, in fact, showed these to the Royal Society by request on 5 November 1662 (Birch, l.c. p. 123). Moreover, his interest extended to other forms of technology apart from mechanics and invention; and he was responsible for the list of trades which the Royal Society commissioned in 1664.

For the *Pinax* itself he had begun to prepare at least as early as 1662; for on 3 December of that year he produced for the consideration of the Royal Society 'his catalogue of the natural things of England and of the rarities thereof' (Birch, *History of the R.S.* 1, p. 138). This he was asked to complete, and to communicate to the Society. He seems thereupon to have set himself more seriously to the collection of natural curiosities in England, and during the next year to have amassed a considerable quantity. This he brought to the notice of the Society in January 1664, and was encouraged to go forward with the project. A Committee of the Society was formed 'for collecting the phaenomena of nature' and this enlisted the help of many members, although Merret's first proposal to employ a professional collector was not acted upon.

His special qualifications, though not apparently strong in actual field-work or knowledge at first hand of living plants or animals, included a good appreciation of the status of the several subjects and a fair acquaintance with the literature. Writing after the foundation of the Royal Society and during the controversy over the New Philosophy, indeed at the very time when Thomas Sprat was producing his *History* and defence, Merret was fully aware, was perhaps the first English writer on natural history to be aware, of the significance of the method of observation and experiment and of the outlook which this was producing: but to advocate this change was not his personal task. He was content to take a share in the practical work of studying and listing the natural products of England on the lines already laid down by the local floras of Gesner, Pona, Gaspard Bauhin and by 'Schoenfeldt on the plants and minerals of Silesia'.

Of British lists he recognises Johnson's works on the plants of Hampstead in 1632 and of Kent in 1634 and his *Mercurius* as the pioneers. 'Following him Dr How edited the *Phytologia Britannica* in 1650 and enlarged its scope, but was prevented from going further by his early death.' Then in 1660 'Mr Wray, fellow of Trinity College, Cambridge, published his *Cambridgeshire Catalogue*, a book of great value, planned with good judgment, and deserving a wide circulation: but in it he only dealt with plants that he and his friends had investigated and within the narrow limits of the neighbourhood of Cambridge: it could not, therefore, serve the purpose of a general catalogue'.

Merret fully recognised that the writers on natural history were almost all botanists, and assumed that this was due partly to the ease of study and the greater persistence of plants, and partly to their value to mankind. Nevertheless, in dealing with animals, birds and fishes he had recourse to the great works of Gesner and Aldrovandi and to those of John Johnstone, who had recently published the four treatises on Fishes, Birds, Quadrupeds and Serpents collectively forming his *Historia Naturalis* at Frankfort in the years 1649–53. Johnstone, whose father had migrated from Scotland and settled in Poland, had spent some years at St Andrews and in England, though his life was mainly spent in Leyden and afterwards on his estate in Silesia. He was a man of wide reading and his books, illustrated by attractive copperplates, had a large circulation and were translated into English in 1657. Ray regarded him, not unjustly, as 'a meer plagiary' (*Further Correspondence*, p. 160), but Merret found him useful.

Of English writers he makes special mention of Turner, 'the most experienced of men of his time', whose book on birds 'small in size but great in judgment' and letter to Gesner on fishes he commends. Caius, Penny and Falconer, 'lights of our College', are mentioned, and the work of Penny, Wotton and Mouffet on insects is praised as a piece of unique research which ought never to have been left so long to the moths and beetles. It would be an exaggeration, in view of his book, to say that Merret had mastered this literature. But at least he knew enough of it to see that its contributions to the various subjects were not ignored. Moreover, in the references which he makes to them in his lists he does what hardly any of his predecessors had ever done, and quotes the page on which the relevant information is to be found.

In the letter to the reader prefixed to his *Pinax*, the letter from which many of the details already given are derived, Merret has set out with candour if also with complacency the story of what he did to produce his book. This is worth a brief résumé.

First he cut out some two hundred plants from How's list as not found by himself or anyone of expert knowledge but communicated by friends or authors doubtfully competent. He also cut out those reported with insufficient localities or on hearsay evidence. He added localities. He chose the best-known names and generally used those in Gerard (Johnson's edition) or Parkinson, quoting the page for his references.

Then he gives a list of those living friends who have helped him: Dr Bowles, the only survivor apparently of How's contributors; William Browne, fellow of Magdalen, now a B.D., and already part-author of the Oxford Catalogue of 1658;[1] Gilbert Witham of Garforth, Yorkshire, who afterwards showed Ray some local plants in 1668; 'Dominus' Gunthorp, a Cornish doctor;[2] and Merret's kinsman named Jenner, whom he describes as 'theologus' and who may perhaps be the Thomas Jenner who published a collection of Emblems entitled *The Soules Solace* in 1626; and above all John Goodyer, who had already contributed so largely to Johnson's *Gerard* and the *Mercurius* and whose letters and manuscripts were lent to him by Goodyer's nephew Yalden. Finally he acknowledges that he has taken everything unfamiliar to other writers from John Ray's *Cambridgeshire Catalogue*. It is curious, considering how extensively he used them, that he does not mention Francis Willughby, from whom he quotes notes of birds and one on fishes, nor acknowledge the plant-lists which Ray must certainly have sent to him after his tour of the West country in 1662.

He then acknowledges that he has included a number of aliens like 'Hyoscyamus Peruvianus, Tobacco, cultivated in several places especially Winchcombe in Gloucester' (no doubt a sentimental reference!) so as to show what our soil can do. But he has (professedly) rejected flowers with different colours, as in his opinion these freaks are not permanent and are largely due to unsuitable soil or the age of the particular plant. Certain species inserted in the lists may not be found in fact in the localities given: they become from time to time extinct as the Beaver has done in the river Tivy in Merionethshire, and Wolves have done in England. Descriptions, even of new plants, are omitted.

With this equipment and a rather fulsome dedication to Baldwin Hamey, the wealthy doctor, benefactor of the College and patron of good causes, it is a pity that the book is not more satisfactory. The catalogue of plants upon which obviously most effort has been spent is at first sight a great improvement upon any previous list. It looks as if the removal of casual varieties, of ill-established records, and of tautologies had been

1 Merret says 'author eruditissimus' and makes no mention of Philip Stephens or Jacob Bobart.

2 I have failed to find any details: but Mary Gunthorpe of Launceston is recorded in 1667 (*Devonshire Wills*, p. 74).

fulfilled according to promise; and some at least of the new records, marked with an asterisk, are species of real interest. Moreover a beginning has been made with Fungi[1] and Mosses and several parts of the country hitherto unexplored have now been at least mentioned. John Dale, who was to have been Merret's colleague, had died and left only a couple of records, *Scirpus triqueter*,[2] p. 67, 'at the Horseferry at Westminster' and *Lathyrus nissolia*, p. 125, 'about Tyburn and Maribone Park'. But several new helpers are named: Jacob Bobart junior with varieties from 'Marlborow', p. 114, and near Oxford, p. 124; Mr Joyliff[3] from Bristol, p. 109; Mr Halilah, apothecary in Lincolnshire, p. 6; Mr Pink from Hereford, p. 3, and 'by Cumner wood in the way from Oxford to Eynsham ferry', p. 61; though their contributions are in these cases unimportant. Francis Willughby is mentioned once in connection with *Scilla verna*, p. 64; and the note is worth quoting as typical of Merret's work: it runs, 'in Barge-Island in North Wales, and at the Kings-end near Dublin Mr Heaton and in Anglesey Mr Willoughby'. The Dublin note is from *Phytologia*, p. 61; that of Anglesey refers to its discovery at Llandwyn on 24 May 1662 by Willughby and Ray.[4] The first must be a muddled version of Ray's record of it on Bardsey as 'growing there in great plenty' (*Memorials of Ray*, p. 171) and was presumably supplied by him to Merret, along with a number of his other finds. These are always without his name, though those of them taken from the *Cambridgeshire Catalogue* are usually labelled C.C. One of them, *Pinguicula lusitanica*, is entered wrongly as 'between Oakhampton and Launceston', p. 94, whereas Ray had recorded it as near Kilhampton—he did not in fact go near Oakhampton. Another, *Stratiotes aloides*, appears as 'in Lincolnshire fens [cf. *Phyt.* p. 75] and those of Ely, and at Awdery Causey in the way from London to Cambridge' which is copied from Ray, *C.C.* p. 98, 'Audrey causey abundantly' without acknowledgment and with geographical error. A number of other references may well be derived from Ray; and *C.C.* is added to the note on *Potamogeton pusillus*, although Merret gives the locality as 'in Thames near Oxford' (p. 97) instead of 'in the rivulet at Hinton Moor' (*C.C.* p. 125). Apart from these errors he uses Ray's notes

1 Among them is an interesting note on 'Mouldiness', pp. 40–1. 'These fungi the learned Mr Hook in his remarkable book the Micrographia lately published [1665] has revealed and depicted. I have seen and possess many other species of fungi; but for the present I omit them. I have put in some because I heard recently of the publication of a book at Rome in which more than three hundred were described.'

2 Gunther, *E.B.B.* p. 296, suggests that this 'juncus caule triangulari' was *Cyperus longus*.

3 The name occurs twice in How's MS. records in *E.B.B.*: pp. 284, 286. Gunther (l.c. p. 286 note) suggests that he was George Joyliffe of Wadham and Pembroke, Oxford, M.A. 1643, the eminent doctor, who lived on Garlick Hill and was visited by Evelyn (*Diary*, 19 Sept. 1657) when he was keeping two Rattlesnakes alive.

4 Cf. *Memorials*, p. 168.

and *Catalogue* very carelessly, omitting very many species, among them several new to science. Thus of the three new Clovers in Ray's *Catalogue*, *Trifolium ochroleucum*, *striatum* and *dubium*,[1] only *T. striatum* is included by Merret, unless 'Trifolium montanum maius flore albo sulphureo betwixt Northfleet and Gads-hill' (p. 121) is *T. ochroleucum*.

To Thomas Willisel no reference is made in the list: Merret evidently felt entitled to make no acknowledgment to his paid collector. But a careful comparison of the new records here with the plants ascribed to Willisel by Ray in his *English Catalogue*, 1670, makes it certain that many of the north-country plants were Willisel's. Thus on p. 111 'Sedum sive illecebra fol. oblongis on the north side of Ingleboroughhill' would seem to be *Saxifraga oppositifolia* which Ray found at the same place in 1668 (*C.A.* II, p. 269). 'Sedum minimum fl. mixto' (*Sedum anglicum*) on the same page, 'on the west of Ingleborough', *Cephalanthera ensifolia* 'in Helkwood [Helkswood, the home of *Cypripedium*] not far from Ingleborough', p. 61, *Epipactis atrorubens*, p. 89, 'in the pastures neer Setle' and *Draba incana*, pp. 90–1, 'at Clapdale in the midway betwixt Setle and Ingleborough hill on the rocks' are also probably his. *Thlaspi alpestre*, p. 118, 'above the ebbing and flowing well two miles from Giflewick [Giggleswick] in stony ground' is almost certainly his; for Ray found this also 'above the ebbing and flowing well a mile from Settle' in 1668. *Arctostaphylus uva-ursi*, p. 123, 'four miles from Heptenstall near Widdop on a great stone by the river Gorlpe [Gorple] in Lancashire' is certainly his: for Ray, *Catalogue*, p. 309, quotes this exact locality and adds, 'Th. Willisel showed me a sprig of this plant but as it had neither flowers nor fruit we cannot tell whether it is undescribed'. So, too, *Trientalis europaea*, p. 99, 'on the east end of Rumbles moor neer Helwick [? Eldwick or Kildwick, but this is south-west] in the bogs amongst the rushes above a great brook' is reprinted in *Catalogue*, p. 256, with the addition 'found and brought to us by Thomas Willisell'. Still more significant is the note on *Polemonium caeruleum*, p. 123, 'on the rocks betwixt Maw water Tarn and Mawanco where the highest rock standeth round like a castle', of which Ray wrote, *C.A.* p. 311, 'Th. Willisell also found it in the North', and in *C.A.* II, p. 299, 'more plentifully about Malham Cove' which he had visited in 1671 and of which the name printed by Merret is a barbarous version. At least two other records probably belong to Willisel: *Polygonum viviparum*, p. 16, 'at Crosby in Westmoreland and Ravenswaith; and two miles from Settle in the pastures towards Ingleborough hill in a place called Cromock', of which the first clause is a corrected version of 'in Westmerland at Crosby Mr Ravenswaith [Crosby Ravensworth Fell, near Shap]' (*Phyt.* p. 16) and the second points forward to *C.A.* II, p. 42,

1 The last-named may have been described by Johnson, *Descript.*, but cf. Clarke, *First Records*, p. 39.

'It was shown me this year, 1671, by T.W.... not far from the footway leading to Settle'; and *Polygonatum anceps*, p. 96, 'at Clapdale in the midway 'twixt Settle and Ingleborough hill', which may be from Gerard, *Herball*, p. 758, but suggests *C.A.* II, p. 238, 'This year 1671 it was shewn me by T. W. growing on the ledges of the scars near Wharfe and Settle.' What seems quite certain is that Willisel during his employment by Merret made the journey to Ingleborough which supplied the localities to which some years later he conducted Ray.

Another locality from which the records certainly belong to Willisel is Roe-hill (Rowhill) south of Dartford. From this came 'Centaurium luteum' (presumably *Blackstonia perfoliata* though this should be the next entry), p. 24, *Neottia nidus-avis*, p. 83, *Adonis autumnalis*, p. 39, and *Cephalanthera grandiflora*, p. 60. For Ray in a letter to Willughby (*Philo. Letters*, pp. 369–70) tells how he visited Rowhill on purpose to find the two last-named plants 'which Willisel declared that he had found there'.[1] There are also several other places from which the records are unsigned and may well be due to Willisel—'Bass Castle in Cornwall' (Boscastle) from which Merret states that he possesses three sorts of the 'Marish Saxifrage now commonly call'd Pearl worts', p. 109, and *Erodium maritimum*, p. 46; 'Church lench in Worcestershire' (near Evesham) where were found *Lactuca saligna*, p. 68, and *Rosa spinosissima*, p. 105, and a scarlet fungus, p. 43; and perhaps 'Hadley Castle two miles from Lee in Essex' whence came the first record of *Lathyrus hirsutus*, p. 70. Some of the excellent reports from London are almost certainly his; for he was certainly working with Robert Morison in the royal garden in St James's Park in 1662: *Matricaria inodora flore pleno*, p. 30, 'in St James's field in the upper side neer the highway'; *Hyoscyamus luteus*, p. 64, 'betwixt St James's and Hide Park'; *Lamium hybridum*, p. 69, 'in the King's new garden near Goring house'—this being on the site of Buckingham Palace and at that time the residence of Sir Henry Bennet, Baron, afterwards Earl of Arlington: it was burnt in 1674; *Rumex pulcher*, p. 69, 'in St George's fields'; *Pedicularis palustris*, p. 91, 'two miles east of Croydon below the windmill nigh the fish pond'; *Polygonum amphibium*, which he places under three different headings: 'Persicaria longissimo folio', p. 92, and 'P. foliis salicinis', p. 93, and 'Potamogeiton angustifol.', p. 97, 'in a pond in St James's Park'—presumably that called Rosamond's Pond filled up in 1770. Of two, and these perhaps the most important, *Sonchus palustris*, p. 115, 'in the meadows betwixt Woolwich and Greenwich by the banks of Thames', and *Lathyrus palustris*, p. 70, 'in a wet marsh ground on the left hand of Peckham field', we have proof that they are his; for of the first Ray, in *C.A.* p. 290, says explicitly 'Th. Willisel found it on the banks of the Thames not far from Greenwich'; and of the

1 For reasons given in my *John Ray*, p. 149, I assign this letter and visit to 1669.

second in *C.A.* p. 316, 'found and brought to us by Th. Willisel in Peckham field'. He also visited the Kentish coast between Dover and Margate (Ray, *C.A.* p. 91), and accompanied Ray, probably, on his visit to Charles Howard in May 1669 (cf. *John Ray*, pp. 149–50); for Ray records that Willisel showed him *Lathraea squamaria* 'in a shady lane, not far from Darking in Surrey, plentifully' (*C.A.* pp. 92–3).

We have elaborated the story of these records because not only are many of them plants new to our Flora, but Willisel is a man of remarkable interest and some importance. The evidence thus disclosed gives the reason for his appointment in 1668 by the Royal Society as their official collector,[1] and for Ray's commendation of him as 'the fittest man for such a purpose that I know in England both for his skill and industry'.[2] In their service he did the first of his tours for them in 1669 and we have Evelyn's account of his success in the *Diary* for 21 October: 'Our English Itinerant presented an account of his autumnal perigrination about England for which we hired him, bringing dried fowls, fish, plants, animals, etc.' From this record it is evident that Willisel, who has always been regarded as a botanist, was in fact something of a zoologist as well. It is possible that he ought to be given credit for some of Merret's observations of birds and mammals. In any case the facts of his career and the part that he played in the history of British botany are significant of the new age. He was not a herbalist nor a gardener's agent: he was a professional field-naturalist, perhaps the first in history, certainly the first in England. As such he marks an epoch. Nature has become worth studying for its own sake.

Another helper of less importance but with a place in contemporary science was Edward Morgan, who had accompanied Johnson as interpreter on the Welsh journey and now was in charge of the medical garden in Westminster (pp. 41–2). Here he had been visited by John Evelyn in 1658 (*Diary*, 10 June) and by John Ray, to whom he supplied seeds and plants for his garden in Trinity College, in 1662 and 1669. Merret records that he had found *Bellis perennis flore pleno*, p. 15, 'in Mr Seldens [probably William Sheldon of Broadway] cops neer his house in Worcestershire', *Impatiens noli-tangere*, p. 92, at 'Gwern Dhee' (but this is hard on Bowles and Johnson), *Alisma plantago-aquatica*, p. 95, 'in a small pond betwixt Clapham and South Lambeth common'; that he had received *Lavatera arborea*, p. 75, 'from the Isle of Wight'; and, most famous of his achievements, had discovered the polyanthus 'in great Woolver [Great Wolford] Wood in Warwickshire' near Moreton in the Marsh, p. 98, and had cultivated it in his garden. When the Society of

1 Cf. Birch, *History of the R.S.* II, pp. 371, 378. In this capacity he did his journey through Norfolk which yielded many first records of Breckland plants.

2 *C.A.* p. 340.

Apothecaries took over Charles Cheyne's land in Chelsea in 1676 Morgan was expected to transfer himself and his plants to it; and Gunther (*E.B.B.* p. 353) says that this was done. But disputes arose about the salary due to him 'for keeping the garden and for his plants'.[1] He was certainly still at Westminster in 1677; for Thomas Lawson, the Quaker botanist and schoolmaster, visited his garden during that summer and has left a long catalogue of its contents which is still preserved in the original manuscript notebook.[2] Probably Morgan never went to Chelsea, but returned to Wales. He was still alive in 1685 and working in the garden of the house at Bodysgallen in which he had stayed with Johnson in 1639.[3] His *Hortus Siccus* in three large volumes containing some two thousand specimens is preserved in the Bodleian.[4]

The two from whom he acknowledges help most freely were, like Willisel and Morgan, also friends of Ray. William Browne of Magdalen College, Oxford, had taken the chief part in publishing the Catalogue of the Botanical Gardens there in 1658. He had become a fellow in 1657 and stayed in the University, a pillar of its scientific interests, until his death in 1678. From Merret's references he had evidently done some field-work. The first under *Acorus calamus*, p. 2, 'found by Dr Brown neer Lyn and by Mr Brown of Oxford near Hedly in Surry' couples him with his more famous namesake Sir Thomas of Norwich. At Headley, now correctly placed in Hampshire, he also found *Chrysosplenium alternifolium*, p. 109. He is credited with two varieties of *Solanum dulcamara*, p. 34, 'at Shoram [Shoreham] in Sussex', with two of *Scutellaria minor*, p. 60, 'near Purbright in Surrey' and with three of *Colchicum autumnale*, p. 28, 'near Oxford'. But his best finds were in Berkshire, *Aristolochia clematitis*, p. 10, 'near Redding, in a place where once was a monastery' (an explanation not usually appreciated at this period), and the three Orchids, *Habenaria viridis*,[5] *Orchis militaris* and *Orchis simia*, all on p. 85, 'on several chalkey hills neer the highway from Wallingford to Redding on Barkshire side the river'. It was apparently on the other side near Caversham that Ray was sent by Browne to find *O. militaris* in 1669, though he reports both localities in *C.A.* p. 225. Some of the Oxford localities besides that of the Colchicums on p. 28 may well be his: if so they are not assigned to him.

Gilbert Witham was even more largely responsible for notes both from his native place Garforth near Methley in Yorkshire and from Stokenchurch in Oxfordshire, which he may perhaps have visited during the

1 Cf. Barrett, *History of Apoths.* pp. 96–7.
2 Kindly lent to me by Mr R. L. Hine of Hitchin.
3 Cf. Gunther, *Life of Lhwyd*, p. 74. 4 Cf. Gunther, *E.B.B.* pp. 308, 353.
5 G. C. Druce, *Flora of Berkshire*, p. 477, claims that it is this species rather than *Aceras anthropophora*.

years following his degree. He was the younger son of Cuthbert Witham, the son of Cuthbert senior who had been Rector of Garforth, and of Anne, daughter of Thomas Hemsworth of Great Purston.[1] They lived at Methley; for he reported pink *Ajuga reptans*, p. 17, 'from a close called the Wood close belonging to my father's house at Methley'. The family, which was evidently prosperous, had been in the neighbourhood for some time and must have been intimate with the Saviles, squires of Methley and patrons of the living of Darfield some twenty miles away, to which they had presented Walter Stonehouse. There can be little doubt that Stonehouse, who regularly visited Methley and has left records of plants from all this district, was responsible for infecting young Gilbert Witham with an enthusiasm for botany.

He had been to school at Westminster under the great Laurence Osbolston, had come up to Trinity College, Cambridge, with a scholarship in 1631, graduated and in 1644 became rector of Garforth, where he stayed till his death forty years later. He married Sarah Taylor of Newland in 1643 and their son Cuthbert, who was at school at Sherburn, came up to Magdalene in 1660.[2] In 1668 Witham showed Ray two local plants, *Actaea spicata* and *Pyrola minor* 'in Haselwood-woods near Sir Walter Vavasour's park-pale'[3] within half a dozen miles of Garforth. These two he had apparently not sent to Merret. Most of those sent were also from his own neighbourhood, *Stachys betonica* with pink flowers, p. 15, *Erigeron acre* with white flowers, p. 29, *Centaurea cyanus* with purple flowers, p. 32, *Viola tricolor*, p. 65, *Botrychium lunaria*, p. 74, 'plentiful in John Nuns cow-pasture adjoyning to his house at Methley' and *Antennaria dioica*, p. 47, 'on a great mountainous place on the left hand of the way as one goes from Pontfract to Wombersley and almost directly opposite to Stapleton town on the right hand'—a direction which does not lack precision. His more important contributions are the various Orchids reported from Stokenchurch, pp. 61, 88–9, though apparently the one which Ray afterwards attached to this locality (*Cephalanthera grandiflora* in *Synopsis*, II, p. 242) is not among them, unless it is the *lusus naturae* reported on p. 61. From his reports to Merret there seems reason to suppose that he was painstaking rather than expert.

It is from these contributors and particularly from Willisel that the list derives its chief value—though it must be admitted that some of their notes are inserted at the cost of violating all Merret's professions as regards varieties and duplications. If the result is not so bad as the *Phytologia*, yet there are far too many records of a trivial kind, involving repetitions and destroying any consistent idea of what constitutes a species. It seems

1 Cf. Will of Cuthbert Witham, *Yorks. Archaeol. Assoc. Records*, IX, p. 87: for pedigree, cf. Dugdale's *Visitation*, pp. 374–5.

2 Cf. Venn, *Alumni Cantab.* IV, p. 443.

3 *C.A.* pp. 71, 256.

clear that Merret did not possess enough critical knowledge or else did not take sufficient trouble to discriminate between the data sent in to him. If he knew enough to recognise the merits of Ray's arrangement of the *Salix* tribe in the *Cambridgeshire Catalogue* (as he does on p. 108), he certainly was incapable of imitating it and bringing order into his muddled aggregate of fifty supposedly different Orchids.

The plain fact is that he is not himself a botanist. Of the whole list there is almost nothing that can with certainty be ascribed to him; for most of the unassigned novelties belong to Willisel and those from Oxford are very likely from Browne. The note on *Sisymbrium irio*, p. 66, 'almost everywhere in the London suburbs on walls and ditches' is probably his; for this is the plant which appeared in such astonishing profusion after the Great Fire. But even this may be matter of hearsay—as must surely be the report of a Reed, apparently *Arundo phragmites*, 'thirty feet long, on the south of the Isle of Wight by the sea side towards the point', p. 48. But the only note in which he speaks in the first person is that recording under 'Trichomanes ramosum' (a fern in the *Phytologia*, p. 123, 'in Devonshire Mr Heaton') that 'Mr Heaton the Irish theologian enriched my hortus siccus with this plant which he found in Cornwall', p. 119. He is a collector—of notes and specimens—but not a botanist.

Nevertheless, if the editing of his botanical material is in Ray's word 'bungling', the rest of the book, which has too often been dismissed as a mere list of names, ought to make some amends for the defects of the first part. We need not spend time over the essay on the classification of plants (pp. 127–38), nor on the English list (pp. 139–54), nor upon the other lists and calendar (pp. 155–65). But the sections that begin on p. 166, though meagre and ill-arranged, represent a not wholly unsuccessful attempt to catalogue the fauna and 'fossils' of Britain.

Merret bases his faunal lists on Johnstone, Gesner and Aldrovandi, giving references to them under each species and noting whether or no there is a picture. The Quadrupeds are divided into hoofed (whole-footed, Horse, Ass and Mule, or cloven-footed, Cattle, Sheep, Goats, Deer and Pigs) and 'fingered viviparous' (untamed or domestic) and 'fingered oviparous' (Frog, Tadpole, Toad and Eft). Breeds of Dog take up nearly a page. The Seal or Sea-calf is included. 'Mus araneus the Erdshrew or Field Mouse' evidently includes both the Wood and Field Mice and the Common Shrew (p. 167). There is no mention of the Red Deer, only of Fallow and Roe; nor of the Stoat; nor any details about the habitat of the Wild Cat. The Mole is said to be sometimes white, and never found in Sheppey (p. 168).[1] Lizards and apparently the Slow-worm ('Lacertus terrestris anguiformis') are united with Newts (p. 169).

1 Cf. Childrey, *Brit. Bacon*. p. 69.

In birds, similarly, many of the entries are merely a Latin and English name and references to one or more of his authorities. Among them Turner takes a prominent place though his findings are taken not from his own book but from Gesner, and are not always followed or even understood. They are grouped into sections following exactly the order and names in Johnstone's book—'Land birds carnivorous' (including not only Hawks, Cuckoo, Owls and Crows, but the Sea-pie, the Bat, the Shell Apple—Crossbill, *Loxia curvirostra*—and the Goatsucker); 'seed-eaters not songsters' (Game-birds including Snipe, Woodcock and Hoopoe); 'dust-lovers tame' (Poultry) and 'dust-lovers that wash' (Coot, Kingfisher, many varieties of Pigeon, House-sparrow and Reed-sparrow); 'seed-eaters songsters' (Finches and Larks); 'berry-eaters' (Thrushes, Starling, 'Clotbird'); 'insect-eaters' (Woodpeckers, Swallows, Tits, Chats, Nightingale and Dotterel); 'Water fowl web-footed' (Geese, 'Capricalze', Ducks, Divers, Gulls, 'Puphin'); 'cloven-footed' (Stork, Herons, Sandpipers, Plovers, Crane, Crake); and finally three from Lincolnshire supplied by Mr Hutchinson, a London poulterer who had also supplied a Stone Curlew—these three being apparently Smew (*Mergus albellus*), Garganey (*Anas querquedula*—he calls it 'Crickaleel'), and 'Gossander'—'the flesh is yellow and when cooked turns into oil, inedible a kind of Puphin'—*Mergus merganser*.

Commenting in more detail, we note that the list contains at least two duplicates due to a misunderstanding of Turner. The Wheatear appears under that name on p. 178 with the note 'in Warwickshire Fallow Smiters': it has already been listed on p. 177 as 'Caeruleo a Clot bird, a Smatch or Arling, a Stone-check, it nests in rabbit-burrows and under stone in England Turner'. So the Dipper appears on p. 171 as the 'Water-crow, Turner saw it at Morpeth. I suspect it is the Mur of the Cornishmen'—a wildly mistaken guess—and on p. 183 as the 'Water-blackbird, Willoughby saw it flying in Cumberland'. Turner's notes, of the Siskin in Cambridgeshire, of the 'Bergander' (*Tadorna tadorna*) on the Thames and nesting in rabbit-burrows 'in insula Tenia' (Thanet), of the 'Mergus' (*Phalacrocorax carbo*) which he 'saw nesting on sea-cliffs near the mouth of the river Tyne in Norfolk [*sic*]',[1] and of the Crake (*Crex crex*) seen and heard in corn and flax in Northumberland, are duly quoted, and he is freely mentioned as the authority for the English names of birds.

Francis Willughby, as we have already seen, supplied some observations: on p. 171, where he is described as 'a most diligent and skilful observer of all nature not only in Britain but in the greatest part of Europe', on two other species of Shrike; p. 172, Crossbill (*Loxia curvirostra*) 'in Warwickshire in orchards'; p. 180, 'Colymbus' 'by Norwegians Lumme by our people Razor-bill'; Merganser (*Mergus serrator*) 'on the river

1 A conflation of two notes by Turner (*Avium Praec.* p. 90).

Tame in Warwickshire in the year 1664 when the winter was very severe';[1] p. 183, Sandpiper (*Actitis hypoleucos*) 'by fishponds and the edges of streams in Warwickshire'; 'Sea-lark' (*Charadrius hiaticula*). 'on the Welsh coast especially at Beaumaris'.

John Ray may have sent the note on the Puffin (*Fratercula arctica*) 'Anglesey and Cornwall', p. 181, and certainly sent that which precedes it on 'the Gull which the Cornishmen call a Gannet...as large as a Goose, full-webbed, with a round blue beak, a grey body which flies high and catches small pilchards ("alausas")'; for this is exactly his report as given in the itinerary of his Cornish tour in 1662 (*Memorials*, p. 188); and here as there Gannet is treated as different from Solan Goose.. Here, as before, Ray is not named.

Of other helpers mention is made of 'Mr Cole' who took the Goat-sucker in Hampshire in 1664, p. 172; of Mr Jenner his kinsman already mentioned as a botanist, who 'has bestowed on me a specimen of "Mergus" shot in Wiltshire', p. 181, and sent a small Heron ('Ardea minor J. t. 56', that is Johnstone Tab. 56)[2] also from Wiltshire, p. 182; of Mr Gunthorp another botanist who reported a mysterious Plover 'akin to the Lapwing, taller than the Snipe but smaller than a Thrush, with blue wings and a long crest from Cornwall', p. 182; of Tradescant's Museum in which he saw 'a kind of Goose called Squeed with its egg from the Bass', p. 179;[3] and of the books of Camden for the Dotterel in Lincolnshire, p. 179, and of Fuller for the arrival of Lapwings on St George's Day, p. 182.

Merret's own notes seem to be very few indeed, perhaps only two, on p. 171 of the 'Butcher-bird' (*Lanius collurio*), 'I have seen it in summer near Kingsland three or four times', and p. 183 of the Crake (*Crex crex*), 'I remember that I saw and heard it at Wheatley five miles from Oxford'. But the unsigned notes may of course be his; on p. 170 that the Eagle (*Aquila chrysaëtus*) 'migrates to us from Ireland where it abounds'; that the Lanner and Lanneret[4] are found in Sherwood and the Forest of Dean; on p. 173 that the Bustard (*Otis tarda*) occurs 'on New-market Heath and Salisbury Plain'[5] and the Hoopoe (*Upupa epops*) is

1 This is of course 1664–5 when Willughby had just returned from his continental tour.

2 Johnstone's picture thus referred to is really Tab. 50. 'Ardea minor' there appears to be the Little Bittern (*Ixobrychus minutus*) and it is not impossible that this is the species actually received by Merret.

3 In *Musaeum Tradescantianum*, p. 1, the egg is listed as 'Soland-goose Squeedes from Scotland'; on p. 4 ('whole birds') 'Solon Goose' and 'Squeede from the Basse' are consecutive items.

4 Not of course the true Lanner, which has never been found in Britain, but the Peregrine (*Falco peregrinus*): cf. Sir J. Oglander, *A Royalist's Notebook*, p. 170, 'A lannoret that was bred in the Whitecliff in Bembridge'.

5 Oglander, l.c. p. 96, reported that six Bustards came into the Isle of Wight in Oct. 1633, one being shot at Shalcombe.

'found but rarely in the New Forest and in Essex'; on p. 175 that a small kind of Reed-sparrow (surely *Acrocephalus scirpaceus*) occurs 'in the reed-beds at Kingston'; on p. 178 that the Swallow (*Hirundo rustica*) 'lives through the winter in tin-mines in Cornwall and in sea cliffs'—this being very like a quotation from Richard Carew;[1] that the 'Rough and Reev' (*Philomachus pugnax*) are found in Lincolnshire; and that the Whistling or Green Plover (apparently distinct from Lapwing, Golden or Grey) is also found on heathland there. Like Thomas Knyvet he calls the Bullfinch 'a Hoop' or 'a Nope', p. 176; he calls the male Goshawk the 'Tassel' though Tiercel is properly the male Peregrine, p. 170; quoting 'our poulterers' he lists and briefly describes the 'Gaddel' (*Anas strepera*), p. 180, but has nothing to say of the 'Boscas' or the Pochard.

From birds he goes on to fishes; and here again Johnstone supplies the headings for the list. 'Sea-fish with scales' cover the Cods and Herrings; 'without scales' the Conger, Dog-fish, Rays and Angler; the 'rock fishes' a miscellaneous lot ranging from the Lump and the Gurnards to the Sand Eel and the Halibut. Then come the 'sea and river fish', Salmon, Sturgeon, Smelt, Lamprey, Eel; the 'river fish with scales', Trout, Barbel and most of the fresh-water species except those 'without scales', Shad, Minnow, Carp, Bream, Eelpout, Perch, Pike, Bleak and Tench; and finally the Whales and Porpoises. The classification is, as will be seen, purely arbitrary—an alphabetical order like that of the plants would have been less absurd.

The species enumerated are mainly drawn from Turner's letter to Gesner. We have once more the Codlings at Bednel, p. 185, the Cook fish in the west, p. 187, the Allerfange or Brook Trout, p. 189, the Grundlin in Cumberland, the Ruff so abundant 'in the river Yare not far from Norwich in Essex [*sic*]', p. 190. A large number of the entries are bare names with references; and very few have any notes of interest attached to them. Willughby supplied one record of 'the various sorts of Wrasses in Wales near the isle of Anglesey where they are called Sea Carps', p. 186; Mr Cole of 'a relative of the Sturgeon very like it except for its head of very delicate flavour, captured in the Isle of Wight in 1664 which he painted and dried', p. 188; Mr Gunthorp of a 'Sea-pig which sailors called the Herring-eater driven ashore in the past winter in 1642 with its suckling. When it was cut up its outside was covered by fat seven or eight inches thick; the country folk extracted oil from it, salted its flesh, and ate it. The taste of its flesh was not much inferior to beef. The whole fish was full of a huge quantity of glue', p. 191. Mr Wine,[2] the king's fishmonger,

1 Cf. above, p. 246.

2 Pepys, who dined with him at the Hope Tavern, *Diary*, 9 Nov. 1660, has usually been printed as calling him Wire or Wise: but Mr F. McD. C. Turner who has kindly checked the passage for me says that the original script has Wine.

reported that a kind of Trout called a Shard[1] was brought from Lincolnshire to London, p. 189. Other fishmongers gave him the Pout and the Hake, pp. 184–5, and the 'Pril', p. 187. The description of the Halibut 'sent from Cornwall, three times as big as a Turbot, five feet long, with three rows of teeth, a long tail, recalling in colour on its lower surface a Dogfish, on its upper a Sole', p. 187, may come from the same source. To Ray he may be indebted for the Cornish notes—the Pollack, p. 184, the Pilchard, p. 185, the Tub which Ray said was 'no other than a red gurnard' (Cornish itinerary, *Memorials*, p. 190), p. 186, the Bass and its likeness to a Bream (*Mem.* p. 187), p. 190.

After fishes 'Molluscs', the Cuttle, Starfish and Turner's 'poor Cuttle'. Then 'Crustaceans', the Lobster 'newly taken, before they are cooked and turn red, their colour is blue-black and sometimes yellowish', p. 192, 'Crey fish', Prawn and Shrimp, Crabs including the 'Wrong heirs' (the Hermit Crab) and Sea Urchins. Two species in this section are recorded by 'Dr Ball honorary fellow of the College of Physicians' (Peter Balle of Leyden and Padua, whose brother William was the first treasurer of the Royal Society and according to William Coles, *Adam in Eden*, p. 308, had a garden near Syon House, admitted to the College in 1664). Then Shells headed by the Oyster, with a brief account from fishmongers of the different types 'with preference for those of Colchester and Wainfleet',[2] p. 193. 'Scollops', 'Lympets', 'Cockles a kind of Limpet', 'Muscles', 'Whilks', 'the Sheath or Razor fish', 'three or four kinds of Turbo', 'two species of Trochi', 'Wormshells on old Oysters', 'Barnacles stripped from ships returning from the East Indies'. It is no doubt a mere list of names and evidently Merret knew very little about them: but it is a first list (so far as I know) in Britain, and as such deserves to be held in respect. When Martin Lister published his great work on the subject in 1685, he had nothing else in England to help him—though by then some collecting and study had been done.

Under the heading Zoophytes there is one entry, and that obscure; then a record from Gunthorp of what Merret calls a Sea-Nettle and he Blubber—a translucent and opalescent Jelly fish which is fully described as to its appearance, habit and season; and finally two quotations, one from William Camden's *Britannia* of the 'Guiniad in Bala lake [*Coregonus pennantii*], the Torcoch in Limparey [Llyn Peris] [*Salvelinus perisii*], the Char in Winander Meer [*S. willughbeii*] which the inhabitants say is only found at a particular time of the year, and another, one-eyed, taken in the moun-

1 Charleton, *Onomasticon*, p. 155, defines Shard as Trutta minor: it is surely equivalent to the Shoat or Shote, a local name for the Common Trout, not a form of Char. Ray in his list in 1673 notes the Lincolnshire Shard, doubtless from Merret.

2 Speed, *Theatre*, p. 31, from which this is derived has 'Oysters which we call Walfleete': so Childrey, *Brit. Bacon.* p. 100: Walfleet is the salt-marsh on the south of the River Crouch, now called Wallasea Island.

tains of Carnarvon, of which I am sceptical'; and the other from Richard Carew's *Survey of Cornwall* from which he quotes a number of local names of fishes.

The insects, which occupy the next twelve pages—and of course include Spiders, Leeches, Earthworms and Slugs—are derived almost entirely from Mouffet though references are sometimes given to Aldrovandi. There is first a brief summary or classification, and then a catalogue following Mouffet's order and containing very little that is not in the *Theatrum*. References to Dr Balle who 'proved to me that a Gnat develops out of a Water-louse' (p. 197), to Hooke's *Micrographia* for its portrait of the Flea, to a number of Sea-lice at Billingsgate, to a water-flea, 'a tiny creature near the aqueduct between St Jameses and Mary-bone', p. 203, and to 'worms' in vinegar 'which my honoured colleague Dr Goddard [Jonathan Goddard of Merton College, Professor of Physic at Gresham's College] a true lover of nature always most skilfully preserves',[1] represent almost all that has any claim to being original. The list of Butterflies, pp. 198–9, which A. H. Haworth, the author of *Lepidoptera Britannica* (London, 1803–28), praised so highly is typical of the rest. Haworth, in his presidential address to the Entomological Society in 1806, published in 1812 in its *Transactions*, I, pp. 1–69, claimed to identify all Merret's species, and is followed by so recent an authority as Dr E. B. Ford, *Butterflies* (London, 1945), pp. 8, 9. Unfortunately he did not realise that all Merret's records are condensed and fragmentary extracts from *Theatrum*, pp. 98–9, 103–5. Mouffet's descriptions which Ford ridicules are, as we have seen,[2] generally recognisable: Merret's version is wholly useless. Of the 23 species listed by him every one of Haworth's identifications, with two possible exceptions, is demonstrably and often fantastically mistaken.

On p. 208 after Snake, Viper and Slow-worm have been listed with the appropriate references to Gesner, Johnstone and Aldrovandi, the last section of the book begins—without prelude or any break in the type except the insertion of the word 'Metalla'. It is in his cataloguing of the minerals and fossils of Britain that Merret breaks fresh ground (if only by rearranging existing material); and perhaps reveals to us where his own interests really lay. Hitherto, as we have seen, he has selected from the standard authorities the species which were recorded as occurring in this country, has added to them such others as his friends and correspondents had noted, and has made some attempt by the employment of Thomas Willisel and by his own reading to enlarge the resulting lists. In dealing with the geology of the country, though interest was being

1 For Goddard's interest in the 'leeches' in vinegar, cf. Birch, *History of R.S.* I, p. 231: the Vinegar Eel is a minute nematode worm, *Anguillula aceti*.

2 Cf above, p. 184.

definitely aroused,[1] he had no such number of authorities. Apart from Childrey's descriptions of the counties, from which he draws without acknowledgment, there is very little. Less than a dozen references for the more conspicuous types of organic remains to Aldrovandi, one to Robert Boyle's *On Cold*, two to Anselm de Boodt of Bruges whose *Gemmarum et Lapidum Historia* was published in 1609, one to Jean Bauhin's *Historia Balnei Bollensis*, one to Hector Boece's *History of Scotland*, one to Johnstone's *De Piscibus*, one to Fuller's *Worthies*, one to Johnson's *Gerard*, one to Schwenkfelt, one to Mattioli and four to Tradescant's *Museum*—these are his only references in the section, pp. 108–23, to any written sources. In addition he cites only six living authorities, his previous correspondent Gunthorp for a note from Cornwall, p. 221; Baldwin Hamey, to whom his book is dedicated, for a stone in a sheep's bladder, p. 210; Marmaduke Rawdon, the traveller and antiquary whom Merret describes as 'Essexiensis' (in fact he lived 1656–69 at Hoddesdon), for a 'Triorchis' from Whitby, p. 213, and an Ammonite 21 inches across which he kept in his garden, p. 215; Sir Robert Moray 'the great light of the Royal Society' for Stalactites from Wales which he presented to the Society's Museum, p. 215; William Williams 'a very learned lawyer', afterwards Speaker of the House of Commons, who told Merret about a cloud in the mountains arising from a waterfall, p. 223; Peter Balle already quoted who now appears in connection with a fossil Echinoid and gives occasion for the note on it on pp. 215–16, in which it is argued that these fossils are the actual shapes of creatures once alive impressed upon clay or soft earth and which is of some importance for the early history of geology.

The last of these must be one of the earliest statements by an Englishman on the organic nature of fossils; for though Robert Hooke's brief remarks in *Micrographia*, pp. 110–12, had been published in 1665 the great work which first defended this view with authority, Nicholaus Steno (Nils Stenson), *De Solido intra Solidum contento*, was not produced until 1669 (in Florence)[2] and was only translated into English by Henry Oldenburg in 1671. As such Merret's words deserve translation. After describing Balle's 'spatagoides' he adds: 'from this and very many others it is abundantly clear to me that many stones considered to be natural

1 The most interesting work done in England on the subject is probably the chapter on the fertilising of soil in Gabriel Plattes, *A Discovery of infinite Treasure* (London, 1639), where it is argued that 'all land hath bin sea' (pp. 27, 43) since here in England fir-trees have been buried, shells are found inland, and the hills and dales have been 'evidently graven by the water'—his theory being that the sea gradually changes its place, 'some thinke chiefly through the motion of the fixed stars', and so irrigates and refertilises the land. Plattes was a friend of Samuel Hartlib, but a man too far ahead of his time to be effective. There is no evidence that Merret knew his work.

2 His first account of his views—on glossopetrae or Sharks' teeth—was printed in his *Musculi Descriptio geometrica* at Florence in 1667.

[i.e. inorganic] are fashioned out of animals or their parts through the action of some earthen fluid (*mediante succo quodam terrestri*); that they [the animals in question] had communicated their shape to the clay or soft earth, and had then perished though their figure was preserved; as is abundantly evident in the stones of Eisleben [*Islebianis*] in which the careful observer detects manifest scales.' The Eisleben deposits had been mentioned by Georg Bauer ('Georgius Agricola') the great authority on mineralogy who had written of the smelting by faggots there in his *De Re Metallica*, Basel, 1621, p. 218; and had described the stones in the neighbouring Harz Mountains which 'express various shapes of living things, of fishes sea-plaice, pikes, perches; of birds game-cocks; sometimes salamanders, and once, as many can testify, the head of the pope of Rome, bearded and wearing the triple tiara', in his *De Natura Fossilium*, Basel, 1656, pp. 370–1. Conrad Gesner had described and figured what he calls a 'lapis Islebianus', containing a fossil fish, in his *De Figuris Lapidum*, p. 162, Zürich, 1565, and had quoted Agricola on it. Evidently such fossils were well known; for Gesner says that he had received two of them from different friends. Probably, though this of course cannot be proved, Harvey's museum had included specimens of the stones thus characterised; for Harvey had travelled in Germany when he went with the Earl of Arundel to Vienna;[1] and 'Eisleben Stones' were familiar objects to collectors.[2]

It is the more remarkable if compared with Joshua Childrey's long account of the fossils at Alderley which Merret certainly knew. Childrey, quoting John Speed's statement in *The Theatre of Britain* (ed. 1614), p. 47, that at 'Alderley are found cockles, periwinckles and oysters of solid stone', tells how when he was Chaplain to Lord Herbert at Badminton he visited and 'very diligently examined' the place, found Cockles and Scallops and 'Serpentine stones', described them at length and in detail, decided 'for many reasons that they are not Shel-fish petrified as some would have them to be', and concluded that 'natural there is no doubt they are, and such as now they are from the Creation; but how they came to put on such strange and imitating features is a secret we dare not meddle with' (*Brit. Bacon.* pp. 74–9). But that he was genuinely interested in the question and had made a collection of

1 There seems to be no evidence that he ever visited Eisleben personally: the party travelled Cologne, Coblentz, Mainz, Frankfort, Nuremberg, Regensburg, Vienna and Prague and returned by a similar route. The journey is described by William Crowne, who accompanied Arundel, in his *True Relation* (London, 1637).

2 Cf. Mercati's Catalogue of the Vatican Museum in 1574 edited by Lancisi (Rome, 1719), p. 319, where is a picture of three fishes with title Lapis Islebianus. I owe this to Dr C. E. N. Bromehead. Cf. also J. Woodward, *Catalogue of Foreign Fossils*, pp. 20–1, where six fishes from 'Isleb' are listed. They are in his collection at Cambridge (E. 26), come from the Permian Copper Slate, and are specimens of *Palaeoniscus Freieslebeni*: cf. Agassiz, *Poissons Fossiles*, II, pp. 66–78.

snake-stones is clear not only from his work in Gloucestershire but from his account of his visit to 'Hunt-cliffe' and Huntly Nab in Yorkshire, where he not only saw and described the 'Seal-fishes' (*Phoca vitulina*) that 'meet together to sleep and sun themselves' upon the rocks, but collected and measured the Ammonites whose 'outward form is just like the Gloucestershire stones with a spine and ribs' (*B.B.* pp. 160–2).

No doubt Merret's position as curator of Harvey's library and museum, of which he had made a catalogue in 1660, gave him a special interest in this part of natural history; for geological specimens and 'wrought stones' have always been favourite subjects for the collector.[1] They are noticeable, attractive and easy to preserve—'metals', 'crystals', 'pebbles', 'petrifactions' and so on—the headings of Merret's sections recall the rows of specimens under his charge. Plants in the pages of the early *Hortus Siccus*; 'cases' of birds skinned and dried; stuffed fishes and reptiles—we know from the frontispiece of Olaus Worm's *Museum* what they looked like. Specimens of the fauna needed a skilful preparation, generally beyond the power of early taxidermists; specimens of the flora were and still are unsatisfactory substitutes for living flowers. A museum naturalist like Merret would perhaps inevitably do his best work in geology; and as several of his references suggest and his later paper on tin-mines proves,[2] he had here done at least some work in the field. It is not of much importance scientifically: but at least it represents a surprisingly detailed survey of the mineral wealth and fossils of England and Wales. As such it has never, so far as can be discovered, received any attention; and it seems to deserve it.[3]

He begins with metals:

Gold extracted from tin but in small quantity.

Silver from a mine in Wales, more from tin mines, less from lead.

Tin from mines in Cornwall and West Devon is distributed all over Europe as a metal of the best quality; the tin workers distinguish it as Pyre[4] and Murdick[5] and Block Tin; and the tin-stones are called Shoad.[6]

1 Such collections had long been made. Thus in November 1565 Johann Kentmann, the famous doctor of Dresden, had sent to Gesner a picture of his cabinet of 'fossils' and a full account of its contents, cf. Gesner, *De Fossilium Genere*, Zürich, 1565. There seems however to have been no interest in the subject in England at that time.

2 Cf. Grew, *Mus. Reg. Soc.* p. 329. His paper is in *Phil. Trans.* for March 1678, no. 138, pp. 949–52.

3 For the study of this subject previous to Merret cf. E. G. R. Taylor's two valuable books, *Tudor Geography*, 1485–1583 and *Late Tudor and Early Stuart Geography*, 1583–1650, both of which contain good bibliographies of the literature of the period.

4 In his paper, *Phil. Trans.* no. 138, p. 950, Merret calls this 'pryan' tin.

5 Called Mundick in *Phil. Trans.* no. 19, p. 337 and by Grew, *Mus. Reg. Soc.* pp. 307, 325, etc.; a form of pyrites.

6 'Shoad, a fat tin-stone so called', Grew, *Mus. Reg. Soc.* p. 328.

Copper is dug at Wenloch in Shropshire [Childrey, *B.B.* p. 122];[1] in Richard the Second's time there was a rich Copper mine at Richmond in the bishopric of Durham, and also at Keswick and Newland in Cumberland [Speed, *T.B.* p. 87; *B.B.* p. 170],[2] and in smaller quantity in Cornwall in the times of Edward I and II.

Iron is abundant in many parts of England, in Sussex, in the Forest of Dean Gloucestershire [*B.B.* p. 71], in Kent, near 'Bromicham' [Birmingham] Warwickshire [*B.B.* p. 113], as also in Somerset, Hants, Durham, Salop and Sheffield in Yorkshire [*B.B.* p. 156]; the iron-workers distinguish it as 'Coldsel' and 'Redsel'; in Scotland it is called 'Slag'.

Lead in the Mendips in Somerset [*B.B.* p. 44, which tells of divining with hazel wands for ore], at Comberton in Devon [*B.B.* p. 28], at the source of the river Istwyd [Ystwyth] in Cardigan [*B.B.* p. 144], in Richmond at Fountains Abbey in Yorkshire, and especially at Wensdale [Wensleydale] and Masham ['Mask', *B.B.* p. 163]; the best and the largest bulk is dug at 'Brassenmore and Dovegang'[3] [? Brassington and Doveridge] in Derbyshire: from lead are produced lithargyrus (protoxide of lead), cerussa (white lead) and minium (red lead).[4]

Antimony is found in the lead mines of Derbyshire [Speed, *T.B.* p. 67].

There is a certain mineral earth which acquires a fiery heat when water is poured on it as the immortal Boyle records in his book on Cold.[5]

Crystals. Diamonds which they call Bristol Stones from St Vincent's rocks near the city [*B.B.* p. 37], from Cornish tin-mines, from the so-called Diamond rock in mid-Wales.

The Agate is said to come from Cornwall [*B.B.* p. 10].

Authors put on record that Crystal comes from Derbyshire: I believe it is only spar (*fluores*) [cf. Camden, *Brit.* p. 420 and *B.B.* p. 112].

Spars are found of various sorts; in Cambridgeshire six-sided: elsewhere they differ widely in colour, shape and transparency. I keep the following in my collection....

The next section is on stones found in animals:

The Toad-stone is falsely so-called: for I have demonstrated before his gracious Majesty (who deigned to attend our College for the admirable lectures of its great ornament Dr George Ent, whom he deservedly honoured with

1 Childrey says 'At Wenlock in the time of Richard the second was found a rich mine of copper'.

2 Cf. Camden, *Britannia*, ed. 1607, p. 631, 'At Newlands and elsewhere were rich veins of copper...opened afresh in our time' (E. G. R. Taylor, *Late Tudor*, p. 110).

3 Cf. Woodward, *Fossils in England*, 1729, 'Dovegang the deepest of any in the Peak...60 fathoms'.

4 These three, 'litharge, ceruss and minium', are discussed in *Art of Glass*, pp. 316–17.

5 *Works*, ed. Birch, 1772, II, pp. 462–784. These papers were presented to the Royal Society along with similar papers by Merret himself at the time of the great frost in 1664: the particular reference is on pp. 664–5, the experiment with Quicklime, discussed in *An Examen of Antiperistasis*.

Knighthood) that it is the molar teeth of a Wolf-fish [*Anarrhichas lupus*]....[1] Dr Hamey showed me stones from the bladder of a Sheep....In old people I have noted that a large part of the Aorta becomes petrified....Pearls occur frequently in some Oysters, particularly Scottish ones; larger ones are said to be extracted from the big fresh-water mussels at Kirby Lonsdale in Westmorland.

Then follows a long section on opaque stones:

Alabaster in Staffordshire, Derby and Lincoln [*B.B.* pp. 117, 111, 120] at the isle of Axholm. Sandstone everywhere familiar, of various colours and qualities; when used for buildings it is called Freestone: the name Purbeck stone is taken from the quarry where it is cut; St Paul's Cathedral here in London is built of it[2] [this of course being the old building destroyed by the Fire of London in September 1666]; Astroites, Star stone, at Shuckburgh in Warwick [Speed, *T.B.* p. 53, 'At Shugbury the pretious stone Astroites'; *B.B.* p. 113] and elsewhere: I found it yellowish and blue in a place where they dig clay of the same colour at Winchcombe in Gloucester [his native place].

Belemnites, some semi-transparent,...are commonly called Thunderbolts.

Coral or spurious coral is found on our coast especially in Cornwall [*B.B.* p. 10].

Whetstones black and yellow are brought from Ireland and are found in Derby [*B.B.* p. 111].

Calamie a constituent of bronze is found in the Mendips: from it come 'Cadmia, botrytes, placites etc.' for which see Mattioli on Dioscorides.[3]

In Staffordshire there is a place which quickly petrifies wood.

Magnes [Loadstone] of good quality in the rocks of Dartmoor [Speed, *T.B.* p. 19; *B.B.* p. 28], poorer quality elsewhere.

Manganese [*sic*] from the Mendips of which more is said in my Observations on glass-making lately published [*Art of Glass*, pp. 282, 289–90].

Marcasites from which Vitriol is made, or pyrites, Copperas stones, and Fire stones on all our East coast...[Jean not] C. Bauhin in his history of the Baths of Bollen [? pp. 33–8] and Aldrovandi have illustrated it.

Marble in Somerset, in many parts of Scotland and Ireland, in the isle of Purbeck [*B.B.* p. 31], in Durham [*B.B.* p. 164] and Hereford [*B.B.* p. 134]. A spotted kind in Yorkshire.

Grindstones from Anglesey and Flintshire [*B.B.* pp. 151, 153], in Derby, Cheshire and Staffs.

Slates or tiles are blue and of the colour of a salvia-leaf in Cornwall [*B.B.* p. 5, 'blew, sage-leaf-coloured and gray'], grey or common at Colly Weston in Northampton [*B.B.* p. 108] and in the Cotswold.

1 Ent was knighted in the Harveian Museum on 15 April 1665, after his Anatomical Lectures: presumably Merret as curator of the Museum then showed the King the teeth which he had already shown in March 1664 to the Royal Society (Birch, *Hist. of the R.S.* I, p. 392).

2 Evelyn, *Diary*, 7 September 1666, speaks of old St Paul's as of 'massie Portland stone'. It was repaired with Portland stone in 1635, cf. *Description of a Journey*, p. 72 (Camden Misc. XVI, 1936).

3 Cf. *Comm. in Diosc.* V, p. 918, 'Cadmia est e botrytis aut placitis aut ostracitis genere'.

And so on through a number of other sorts of stone: Flint, 'Emerie' in 'Guarnsey' [Speed, *T.B.* p. 94]; 'Quicksands'; Limestone; Gypsum for floors; 'Irish slate for writing memoranda'; hard blue tombstones at Pucklechurch [*B.B.* p. 83]; conglomerate at Highgate 'as you climb the hill'; 'Cauk, Black Chert, Wheatstone or Sheaf' from Derbyshire; paving-stones for the streets of London, 'Rance, Rochester-stone or Ragman'; schist or 'Warming stone because it retains heat' from Cornwall.

And so to 'objects turned into stone': Cockles, Periwinkles, Scallops, Oysters and Mussels at Badminton in Gloucestershire [*B.B.* pp. 74–9]; 'Bucardites'—'this my eldest son Robert sent me from Oxford, having found it there'; 'echinites'—'ours is white flint coated with chalk'; 'glossopetrae, tongue-stones' 'in various matrices from coal mines'; Snakestones (Ammonites) 'at Keynsham and Adderley near the surface [*B.B.* pp. 33, 76], at Farnham in Surrey but of different colour [*B.B.* p. 57], at Whitby in Yorkshire' [*B.B.* p. 159]; 'Stelechites of various shapes, some from Wales presented to the museum of the Royal Society by Sir Robert Moray';[1] 'Stalagmites, Dropping-stones' from the well at Knaresborough,[2] and in Scotland at Buchan; the Echinoid from Dr Balle already mentioned; and others from Tradescant's Museum.[3] Also 'a stone suggesting the tail of a cat or some other smaller animal consisting of many joints with a number of stripes round it at the articulations from Holy Island; the inhabitants call them St Cutbeards [Cuthbert's] beads. Mr Rawdon.' 'A shell oblong and thick of colour from Worcestershire where they call them Crow-stones and Crow-cups and from their shape Egg-stones; it is not easy to guess from which of our shells they come.'

'Burning stones' include Pit-coal from Newcastle, Tynemouth, Durham and Coquet Island [*B.B.* p. 173]; Stone-coal ('Lithantrax', so Speed, *T.B.* p. 89) in Derby, Nottingham, North Leicester, Salop, Pembroke, Caermarthen and Somerset 'where they often look the colour of bronze' [*B.B.* p. 32, 'like gold or brass']—'They say that a coal-mine set on fire some time ago is still burning at Penneths Chase in Salop' [*B.B.* p. 117, 'in Pesneth-Chase, Staffordshire, (saith Cambden)...']; Obsidian or 'Kennel-coal'; Bitumen in Salop [*B.B.* p. 122], near Edinburgh, and in Durham; Turfs in Lancashire where the folk use them for candles [*B.B.* p. 168]; and Jet 'in Norfolk and near "Mulgravii castrum" [Mulgrave Castle] in Yorkshire' [*B.B.* pp. 105, 159].

'Salts',[4] 'Alumen' in Cambridgeshire, Yorkshire and Kent 'in Tradescant's collection';[5] Vitriol; Nitre 'in wine cellars'; Sea-salt; various rocks

1 Cf. Grew, *Mus. Reg. Soc.* p. 325.

2 Cf. 'Guttulae Knasburgenses', *Mus. Trad.* p. 24.

3 An Umbilicus marinus from York presented to the Museum by Mr Man (*Mus. Trad.* p. 22); Agrifolia (or Ilex) petrificata from Lough Neagh by Mr Wybard (l.c. p. 24).

4 *Mus. Trad.* p. 21, 'Sal fossilis albus et ruber'.

5 L.c. p. 21.

from Snowdon; Red Ochre in Hereford, Hants, Lancashire and Rutland; 'Nigrica fabrilis, Black ledd this earth is peculiar to England in Europe and America, and hitherto has had no name; I gave it this name at Keswick in Cumberland' [Speed, *T.B.* p. 87; *B.B.* p. 170, 'at Keswick...likewise Black-lead is found']; Yellow Ochre in Hereford, Somerset, Oxford and Scotland; Umber from Bristol; Blue Stone called Killow in Lancashire—'we have various clays suitable for different manufactures...the best for tobacco-pipes are from Nonsuch[1] [near Epsom] in Surrey [*B.B.* p. 54, for 'crucibles'], at Poole, and in the Isle of Wight'; 'Fullers Earth from Nutley in Sussex and Woburn Berks [? Bucks], thence called Woburn Earth'; 'Rutlandshire gets its name from red earth used for dying sheeps wool [*B.B.* p. 110], the same is found in Red Horse vale Warwick [*B.B.* p. 113]'; 'Plastica, Chalk from which we get Spanish white'; 'Clay for brick-making and other uses'.

'Sea-stones', Amber and 'Ambergrise'. These complete the geological catalogue. He adds with something of an apology 'Meteora', 'so that the reader may have a synopsis of the whole circle of nature...I have extracted much of it from our authors especially Camden'. A list follows.

The first name on it is 'Ignis fatuus, the Walking-fire or Jack of the Lantern' and a note typical of the period is attached. 'This is a white and glutinous substance seen in many places which our people call "star faln"; they believe that it owes its origin to a falling star and is its stuff. But to the Royal Society I openly demonstrated that it merely arises from the intestines of frogs piled up in one place by crows;[2] and eminent men confirmed this afterwards to others of the same Society.'

Then after a list of all the 'heavenly bodies' from Lightning to Comets, Winds and Earthquakes, and a note that 'telescopes have revealed that the Galaxy is an assemblage of tiny stars', there follow 'Dew; our Manna from elm-leaves; Rime; Snow whose generation I demonstrated by experiments in a little tract on freezing; rain of blood which is certainly known to be only the excrements of insects [the detail "At Pool in the year 1653 June 20 it is reported that it rained warm blood" is given by Childrey, *Brit. Bac.* p. 30, who derives its explanation from N. C. Fabri de Peiresc: cf. W. Rand, *Life of Peireskius*, pp. 123–5]; rain of wheat which is nothing but Holly-berries eaten and excreted by Starlings' [this had been suggested to the Royal Society in a paper by Samuel Tuke on 26 June 1661];[3] Rainbows; and the five seas of Britain.

1 This clay is mentioned also in *Art of Glass*, p. 246.

2 On 2 November 1664, Merret had produced this substance and asked what it was. On 1 February 1665, he declared that it 'came out of a frog's belly that had been killed by a crow' (Birch, l.c. I, pp. 482–3, II, p. 11). Curious as his idea of phosphorescence seems, it is rational compared with some of those put forward by his contemporaries, cf. the letter of Reuben Robinson to Thomas Browne in 1659 (*Browne's Works*, ed. Wilkin, I, pp. 421–4).

3 Birch, *Hist. of R.S.* I, pp. 32–3.

After this he mentions hot springs at Bristol and Buxton [*B.B.* pp. 36, 112]; purgative waters at 'Ebbesham' (Epsom) in Surrey [*B.B.* p. 56], 'Lewsham' (Lewisham) in Kent [*B.B.* p. 70], Barnet in Middlesex [*B.B.* p. 88], 'Burntwood' (Brentwood) in Essex, near 'Stanes' in Berkshire, at 'Hailweston' (Helston) in Cornwall, 'Willinborough' in Northants, and a spring ten miles from Exeter of which Gunthorp wrote that its water lost its virtue when carried away to a distance; diuretic springs at Tunbridge, in Sussex and in Yorkshire, and at King's Newnham in Warwick [*B.B.* p. 114]; petrifying wells at Knaresborough [*B.B.* p. 157], Lenham in Kent [*B.B.* p. 61], Boxley Abbey near Maidstone [*B.B.* p. 70], Kingsmill opposite Magdalen College, Oxford [*B.B.* p. 84], Lutterworth in Leicestershire [*B.B.* p. 108], Eldenhole in Derbyshire [*B.B.* p. 112], King's Newnham [*B.B.* p. 114],[1] Boughton in Northants, Shelly near 'Chipping Unger' (Ongar) in Essex; also the so-called Dropping-well in Yorkshire (a duplicate: it being the well at Knaresborough already mentioned), Wookey Hole not far from Bristol[2] and a well at 'Tenderdon' (Tendring) in Essex.

Saline springs near Harwich in Essex and Leamington in Warwick [*B.B.* pp. 100, 113]. Salt is made at Droitwich [*B.B.* p. 115], and at Nantwich, Northwich, and Middlewich [*B.B.* p. 129], and black salt commonly called 'Bay salt'[3] in Hants [*B.B.* p. 50].

'Waters of a strange nature' like that of the Dee which does not mix with that of Bala lake [*B.B.* p. 146],[4] or the river 'Levenney'[5] (Llyfni) which passes without mixture through 'Lynsavathan' (Llyn Safaddu) in Brecknock [*B.B.* p. 136]. 'Water hot in winter and cold in summer at Luckington Hancocks in Berkshire [*B.B.* p. 50, "In Luckington is a well called Hancocks-well"—so Merret has misread him: *B.B.* correctly places Luckington in Wilts "in the edge of the shire"]. A pit near Peterborough

1 Merret quotes for this *Ger. emac.* p. 1587—a chapter on petrified wood in which Gerard after mentioning the well at Knaresborough tells how he visited King's Newnham from 'Rougby' and brought away some petrified branches of an Ash growing beside it. Childrey quotes Speed, and his own experience.

2 The association of Wookey Hole with hot springs is no doubt due to the fact that both were discussed at the same meeting of the R.S. on 25 February 1663.

3 For the mystery of this name cf. Murray, *English Dictionary*, ad loc. and J. A. Twemlow in *Eng. Hist. Review*, XXXVI, pp. 214–18.

4 Cf. Drayton, *Poly-olbion*, IX, ll. 127–33, of 'Lin-teged or Pemble-mere':

> 'That when Dee in his course fain in her lap would lie,
> Commixtion with her store his stream she doth deny,
> By his complexion prov'd, as he through her doth glide,
> Her wealth again from his she likewise doth divide:
> Those White-fish that in her do wondrously abound,
> Are never seen in him; nor are his Salmons found
> At any time in her.'

5 'Leveni'; see Giraldus Camb. *Itin. Camb.* I, ch. 2 (p. 21), for an allusion to this tale.

so deep and so cold that no diver can touch its bottom, from which bubbles up a never-frozen spring [*B.B.* p. 108]. In the water at Saltash in Cornwall peas are never cooked to the usual softness' [*B.B.* p. 23].

Subterranean rivers include 'the Mole in Surrey [*B.B.* p. 54], the Deverel [Deverill] for a mile at Wilton [*B.B.* p. 44], the Recal near Elmley in Yorks [*B.B.* p. 163], the Lid at Lidford in Devon [*B.B.* p. 28], the Alen [Alyn] in Denbigh [*B.B.* p. 152], a tributary of the Medway near Cockshead [Coxheath] in Kent [*B.B.* p. 61], and the little river Hans [?Manifold] in Stafford runs underground for three miles' [*B.B.* p. 118].

As to tides he mentions wells which rise and fall near Buxton in the Peak Forest [*B.B.* p. 113],[1] near Kilken (Cilcen) in Flint [*B.B.* p. 154], at Careg (Cwm-cerrig) in Caermarthen [*B.B.* p. 145], and in Cornwall. He notes 'the twelve foot rise four times a day' at Bristol [*B.B.* p. 33, '11 or 12 ells in height' but nothing about four times]; the ebb and flow of springs at Giggleswick [*B.B.* p. 157]; the daily change at 'Loder' (River Lowther) in Westmorland [*B.B.* p. 174]; the well on the Ogmore at Newton in Glamorgan which rises with every tide though it is a mile from the Severn [*B.B.* p. 140, 'about one hundred paces from Severn']; and finally the 'tides called Hygras [Eagres] on the Severn and Humber'.

Then after recounting a tale of the two waterfalls at 'Kan [?the river Kent] near Kendal [Speed, *T.B.* p. 85; *B.B.* p. 174, both Westmorland] in Northumberland' (?Westmorland) and the story by Williams already noted, he finishes his book with a mention of the subterranean forests in Cornwall at Land's End [*B.B.* pp. 26–7, 'at St Michael's Mount at low ebb one may see roots of mighty trees'—which leads Childrey to surmise that 'the sea has devoured much land' and that 'the Silley Islands were once all parts of the main land'], in Flint and Pembroke, in Cheshire, Cumberland and Anglesey and many places in the north-east [*B.B.* pp. 143, 131, 171, 150, quoting Hugh Lloyd]. 'The inhabitants call them Noah's Ark.'[2]

It seemed worth while to translate almost the whole of this survey of the geology and physical geography of the country and to add references to Childrey. For elementary and unoriginal as the work is, this section gives a better impression of the actual advance in scientific knowledge than we can gain from the botany or zoology of the *Pinax*. It testifies to the extent to which a serious interest in the physical features of the country had been carried since the time of Harrison's description. It points to the scientific examination of geological problems which Hooke

1 Cf. T. Hobbes, *De Mirabilibus Pecci*: he calls the 'Fons aestuans' the fifth Marvel (cf. ed. 1678, pp. 57ff.)

2 That these buried forests cannot be due to the Flood had been argued by Gabriel Plattes, *Discoverie of Hidden Treasure*, pp. 43–4.

and Ray and Woodward and Lhwyd were so soon to initiate, and gives to Speed, to Childrey and to its author places among the pioneers in the study of prehistoric life.

That even in this field Merret's work is rather that of a compiler and collector than of an investigator and student may probably be due not only to his lack of real ability and insight but to the fact that his main interest lay elsewhere. He was less concerned with science for its own sake than with the practical ends which it might serve. We have seen that he was deeply interested in technology, in glass-making and metallurgy; and the Preface to his *Pinax* makes it plain that the development of sound pharmacological and medical practice was a primary object of his energies. In this he addresses the Reader not, as we should expect, on the value of research into the flora and fauna of the country, but on the enormities of those who have taken advantage of the anarchy of the times to pose as doctors, when in fact they have no qualifications or knowledge or skill. The fervour with which he accuses them of creating diseases which they then profess to cure is justified by appeal to the concrete case of one William Trigge, a shoemaker, who, according to Baldwin Hamey in his *Bustorum aliquot Reliquiae* quoted by Munk, had been successfully exposed in the lawcourts by John Clarke,[1] President of the College of Physicians, and whose misdeeds are described at length by Charles Goodall in his *Historical Account of the College's Proceedings against Empiricks*, pp. 420–4. Merret declares that these bogus practitioners had been lately shown up by an approved 'chymist', Johnston, appointed at his suggestion by the College, and that he had himself written against Trigge, who had killed a patient suffering from arthritis by treating him for syphilis. Trigge, so he affirms, confessed that he only knew of two diseases, syphilis and scurvy, and only used one drug, sugared mercury.[2]

This delightful but wholly irrelevant discourse reveals the real outlook of its author. He was undoubtedly keen about his museum and his catalogues, and put time and money into his *Pinax*. It was published in 1666: but of this edition hardly any copies survived the Great Fire which then as in 1940 destroyed the area round St Paul's in which most of the publishing houses were situated. A reprint was produced in 1667, and a third edition was promised and in preparation.[3] But the Fire had also consumed Harvey's library and museum and the College building on Amen Corner; Merret's collection perished; he moved to Hatton Garden;

1 Of Christ's College, Cambridge, B.A. 1603, M.D. 1615, President 1645–9.

2 Goodall's record of Trigge's earlier misdeeds culminates in what appears to have been an attempt to perform a Caesarean section on a lady suffering only from dropsy (l.c. p. 421).

3 In a notice published in *Phil. Trans.* I, p. 448 (April 1667), Merret disowned the second edition of *Pinax* and announced a republication with additions and a new method within four months.

and it seems probable that his interest was not in fact great enough to survive the loss. In any case, he followed up the Preface with his *Short View of the Frauds committed by Apothecaries*;[1] this involved him in controversy with Henry Stubbe, who was already heavily engaged in attacks upon the Royal Society; and in reply to him he issued his *Self-conviction* in 1670. Soon afterwards he became embroiled with the College over his position and stipend as Curator of the Museum; this led to his formal expulsion from it; and though he lived till the last decade of the century and as is evident from Nehemiah Grew's appendix to his Catalogue of the Museum of the Royal Society[2] did not lose his interest in collections, he did nothing more of any importance as a naturalist.

There was, however, one outcome of his *Pinax* which deserves remembrance—the correspondence which it brought to him from Dr Thomas Browne of Norwich,[3] already famous as the author of *Religio Medici* in 1642 and of *Pseudodoxia Epidemica* in 1646. Browne, whose large and versatile curiosity both for information and for acquaintance with persons of importance comes out delightfully in his letters, had apparently tried to establish contact with Merret through his son Edward. But Edward, thoroughly aware of his father's weakness, had not, on this occasion, responded; and on 13 July 1668, when Edward was about to start for his second continental tour, direct approach was made by letter. In it Browne, after professing admiration for the *Pinax* and interest in the proposal of a further edition, describes how he had long 'taken notice of many animals in these parts' and three years ago had written an account of them at the request of a friend since dead. He had in fact, as is disclosed in a later letter (*Works*, ed. G. L. Keynes, VI, p. 382), made a considerable collection—'I had about fortie ["cases" or skins] hanging up in my howse': but in 1666 'the plague being at the next doores' these had been burnt by the caretaker. The account alluded to is undoubtedly the pair of papers on the Birds and Fishes of Norfolk (published by Keynes, V, pp. 317–412) afterwards sent to Ray and used in the *Ornithology* and *Historia Piscium*. These records, or a full selection from them, he now proceeded to send to Merret in a series of letters (VI, pp. 361–80) which range over a wide field and consist of jottings and enquiries set down almost at random. In the first he comments upon the mention of his own name in regard to *Acorus calamus* 'found by Dr Brown nere Lin',

1 Cf. Barrett, *History of Apoths.* pp. 82–4, for the reaction of the Apothecaries to this attack.

2 *Musaeum*, pp. 385–6 (London, 1681), a list of gums, etc. supplied by Merret.

3 Dr G. L. Keynes in his edition of Browne's *Works*, V, p. 377, note and VI, p. 360, states that the papers on Birds and Fishes in Norfolk were addressed by Browne to Merret. This seems inconsistent both with the explicit statement in his first letter to Merret (VI, p. 361) and with the dates at which these papers were written—that on Fishes in 1662, that on Birds in 1663–4.

saying 'probably there may be some mistake; for I cannot affirme nor I doubt any other, that is found thereabout. Some 25 yeares agoe I gave an account of this plant unto Mr Goodyeere and more lately to Dr How unto whome I sent some notes.'[1] He adds an account of its abundance near Norwich and the custom of strewing Heigham church with it (v, p. 362).

The chief points of interest in these letters—beyond their testimony to the extent of Browne's knowledge—are the records of the Roller (*Coracias garrulus*)[2] 'killed upon a tree about five yeares ago' (p. 364) or 'ten miles of, four yeares agoe' (p. 370); the Pelican (probably *Pelecanus onocrotalus*), 'whether you meane those at St James or others brought over, or such as have been taken or killed here I knowe not. I have one hangd up in my howse which was shott in a fen ten miles of, about four years ago;[3] and because it was so rare some conjectured it might bee one of those which belonged unto the King and flewe away' (p. 367); the Stork (*Ciconia alba*), 'I have seen two, one in a watery marsh eight miles of: another shott whose case is yet to be seen';[4] the Sword-fish (*Xiphias gladius*), one taken not long ago entangled in the herring-nets; the Sperm-whale (*Physeter macrocephalus*), 'about twelve yeares ago we had one cast up on our shoare neere Welles...and another divers yeares before at Hunstanton'—the latter presumably that of which Arthur Bacon, the apothecary at Yarmouth, 'had the cutting up and disposure' (cf. pp. 403, 312); the Golden Eagle (*Aquila chrysaëtus*), brought out of Ireland, which he kept for two years in his house and sent alive to Dr Charles Scarborough 'who told me it was kept in the Colledge' (p. 369);[5] the 'Shooing-horn or Barker from the figure of the bill and barking note; a long made bird of white and blackish colour; fin-footed; a marsh-bird; and not rare some times of the yeare in Marshland. It may upon view be called recurvirostra nostras or avoseta' (p. 369); the pictures of the Shearwater (*Puffinus puffinus*) and the Little Auk (*Alle alle*) which appeared later in *Willughby's Ornithology*; and the note 'Burganders [*Tadorna tadorna*] not so rare as Turner[6] makes them, common in Norfolk, so abounding in vast and spatious warrens' (p. 372).

1 See above, p. 304.

2 'On the thirteenth of May, 1664, kild about Crestwick', Browne, *Birds in Norfolk* (v, p. 387).

3 'Upon Horsey fenne 1663, May 22 wch stuffed and cleansed I yet retain', *Birds in Norfolk* (v, p. 381).

4 Is this the Stork 'shott in the wing by the sea neere Hasburrowe and brought alive unto mee', cf. *Birds in Norfolk* (v, p. 394)?

5 'There it perished in the common fire', *Birds in Norfolk* (v, p. 389).

6 Both Wilkin and Keynes read 'Turn'. The reference is to *Avium Praec.* p. 23; but Browne got it from Merret, *Pinax*, p. 179, and copied the abbrevation, presumably without understanding it.

To these voluminous contributions only two of Merret's answers have been preserved.[1] The first, in answer to Browne's second, is little but an expansion of thanks and of assurance that the notes are 'either confirmative or additional'. As to the fishes he quotes 'the king's fishmonger' (presumably the Mr Wine of the *Pinax*) for the 'lupus marinus' (*Anarrhichas lupus*), and 'fishermen' for the story that Salmon 'wear off on the banks' some part of the hook on the lower jaw on which Browne had commented (VI, p. 366). He mistakes Browne's May chit (?Sanderling, *Crocethia alba*) for a duck; and fails to recognise his Dorhawke as the Goatsucker which he had included in the *Pinax*. His second letter, dated 8 May 1669, refers to the last but one of Browne's letters, one of those printed by Keynes but not by Wilkin—the 'Norwich aspredo' (presumably the Ruff); 'Elk's bill' (Whooper Swan, *Cygnus cygnus*); the Crackling Teal (Garganey, *Anas querquedula*); the Clangula (presumably Goldeneye, *Bucephala clangula*) and the Willock, 'which I have seen brought from Greenland but never thought was found in England' (?Guillemot, *Uria aalge*). He then reports on Boyle's exhibit of 'black amber' to the Royal Society; and asks that 'if it would please you let it be done without your charge and trouble' Browne should set a collector to obtain specimens 'to furnish our colledge with them as curiosities all beeing lost by the fire' (p. 382).

At this time, therefore, he was evidently anxious to reform the Museum and to re-issue the *Pinax*—for which he claims to have got already at least a hundred new items. If only he had not fallen foul of the authorities of the College, his plans might have been fulfilled. But it may well be doubted whether he was in fact the right man for such a task. Probably justice is done to his memory if he is credited with having prepared the way (as much by his 'bungling' as by his virtues) for Grew and the Museum of the Royal Society and for Ray and the *Synopses*.

1 In Wilkin, *Browne's Works*, I, pp. 442–5; Keynes, l.c. VI, pp. 365–6, 380–2.

E. EPILOGUE

CHAPTER XVIII. THE COMING OF MODERN MAN: *RELIGIO MEDICI*

To chronicle in all its details the story of the coming of the modern scientific outlook is inevitably to be concerned with a succession of specialised workers by whose contributions a knowledge of natural phenomena based upon observation and tested by experiment has been gradually achieved. Such concern in itself gives very little idea of the scope and character of the change as a whole. If this is to be appreciated it must be by a more general treatment of the effects of this knowledge in producing a radically new 'climate of opinion'. We will bring our study to an end by attempting such an exposition of what is in fact one of the most momentous developments that has ever taken place in human history.

We have seen that the medieval concept of nature, deduced in principle from the axioms of scholastic theology and expounded in terms of symbolism and fable, existed alongside of the artistry, appreciation and skill of men who loved and lived by the world of plants and animals. It is manifestly untrue to say, as is so often said, that the Middle Ages possessed an integrated and consistent Weltanschauung: their world-view was in fact divided theologically into the two spheres of reason (or nature) and revelation (or supernature). It would be an exaggeration to say that this division was identical with that between the cities of earth and the City of God as Augustine described them, but it certainly expressed a wide difference between the world of religion and the world of common life. This difference was accepted by all parties and persisted with an astonishing tenacity. Indeed, as restated in an altered form by Descartes, it has not yet wholly disappeared.

The first threat to this concept came when the claim of the theological tradition to represent the wisdom of the classical and early Christian sages was shattered by the discovery of their works. What the Greek New Testament did to medieval religion, that the Greek of Aristotle and Galen did to medieval science: it revealed the vast size of the accretions, and the startling contrast between the originals and their contemporary representatives. It was not that the fables were false to the facts but that they were false also to their own sources.

The contrast and resulting conflict gave for a time to the ancient writings an authority, indeed an infallibility, very damaging to progress.

So long as Aristotle and Galen, or for that matter Genesis and the Apocalypse, were accepted and imposed as inerrant, it was difficult to encourage the investigation of the facts with which any real advance must begin. This delayed, though it could not prevent, the coming of the new outlook.

Nevertheless, when once traditional beliefs were questioned—and the questioning though less severe in science than in religion was definite and searching—the opportunity of the artists had arrived. There must plainly be an appeal to the facts, to the actual data of religion and science, that is, in the case of science, to the natural order, to heaven and earth, to flora and fauna. Exact observation and sensitive interpretation were needed; and above all the delight in the beauty and strangeness and coherence of the world which is probably the mainspring of scientific as it certainly is of artistic attainment. Wonder, curiosity, understanding—these now found outlet in the study, exploration and interpretation of nature.

Finally there was the discovery that this appeal to the facts furnished the starting-point for a 'new philosophy'; and suggested a method by which its pursuit might be undertaken. We have said little of Francis Bacon; for his study of nature though wide and often ingenious was neither accurate nor profound; he built an aviary at York House and kept a bird-sanctuary at Gorhambury, he had his dinner-table decked with flowers and laid out elegant gardens and boscages: but, as his posthumous *Sylva Sylvarum or Natural History* proves, he was much more skilful in propounding questions than in answering them, and accepted large elements of the fabulous without scruple or examination. His merit was that he formulated, illustrated and publicised the inductive method; and gave the dignity and strength of a conscious movement to what had hitherto been largely tentative and casual.

The result was as we have seen not only the creation of a specific Society pledged to be the instrument of the 'New Philosophy', but a widespread, indeed almost general change in outlook among educated men. John Aubrey, who though a gossip and a dilettante had a shrewd knowledge of the men of his time, claimed that the date of this change could be fixed with precision; and gave his reasons for so fixing it. There was of course a large overlap: Joseph Mead, the great tutor of Christ's, who used to ask his pupils daily 'Quid dubitas?' and who may well be regarded as the first of the Cambridge Platonists, belongs as obviously to the modern age as John Duport his junior, the equally great tutor of Trinity who wrote Latin epigrams on every conceivable subject, but never quite approved of the world which his pupils Barrow and Ray were shaping, belongs to the old. Nevertheless the issue though still contested was in fact decided. The child had come to the birth; and though neither Puritan austerity nor Restoration frivolity was an entirely appropriate climate for it, its vitality was strong enough to survive the storms of its infancy and to convince even the most sceptical that it had come to stay.

To demonstrate the extent and significance of the change it should only be necessary to look at the galaxy of men of genius who were the effective leaders of the Royal Society in its first generation of life. John Wilkins with his vivid imagination and prophetic wisdom and above all his brilliant power of drawing together and inspiring men of widely different tempers and beliefs—Wilkins who foreshadowed the telephone with his 'secret messenger' and anticipated Esperanto with his 'Universal Character'. William Petty, statistician and economist as well as doctor, musician, inventor and Latin poet,[1] who had in rare combination exceptional intellectual and exceptional practical abilities, and who might have been among the greatest if he had not preferred money-making and a career to research. Christopher Wren, known to all the world as the greatest of British architects, but known to his contemporaries as a man of many-sided competence, capable of leadership in almost every field of scholarship or technology. Robert Boyle, John Ray, Isaac Barrow, Nehemiah Grew, Isaac Newton—each with a strong claim to be the founder of a new department of science. Robert Hooke, in some respects the most versatile and certainly the most inventive of them all—little Hooke so sensitive and crabbed, so fertile of ideas and so ingenious in putting them into practical form—little Hooke of whom Pepys wrote when he first met him that he 'is the most and promises the least of any man in the world that ever I saw' (*Diary*, 15 February 1665). In any other period the following pioneers would also deserve special mention: John Wallis in mathematics, Walter Needham in embryology, Francis Willughby in entomology, Martin Lister in conchology, Thomas Sydenham in clinical medicine, Richard Lower and John Mayow in surgery, Edmund Halley and John Flamsteed in astronomy, Edward Lhwyd and John Woodward in geology, John Evelyn the bibliophile, and Samuel Pepys, civil servant, diarist and sometime President of the Royal Society. There has seldom been in all history—never been in our own—so remarkable a group.

But it is not by the study of genius that the extent of a movement of thought is best appreciated; for genius is distinguished not only by its originality, but usually by a measure of specialisation. The 'climate of opinion', the spirit of the age, the characteristic changes which mark a new epoch—these are better studied in men of more general and average ability who can claim to represent rather than to surpass their fellows. Among the typical figures whose range of interests is wide enough to

1 Cf. the eulogy of him in Evelyn, *Diary*, 22 March 1675. He wrote a Latin poem, *Scholaris Scholifuga*, on his own forsaking of medicine for land-surveying; cf. *The Petty Papers*, II, p. 261.

illustrate the intellectual outlook of the time and whose literary remains are sufficiently numerous to interpret it to us, two are particularly suitable for our purpose, Henry More and Sir Thomas Browne; and of the two Browne, for the unflattering reason that he is the more ordinary, is the better. We may well follow Edmund Gosse in regarding him as a mirror of his age.[1]

The son of a London merchant—but, it is duly added, sprung from an old Cheshire family; educated at Winchester and at Broadgate Hall, later Pembroke College, Oxford; after graduation in 1626–7 and a further period in Oxford accompanying his step-father to Ireland, and thence travelling to Montpellier and Padua and Leyden, he settled in Norwich in 1636 and lived in practice there for the rest of his days, 'his whole house and garden being', according to John Evelyn,[2] who visited him in 1671, 'a paradise and cabinet of rarities'. He married a Norfolk lady in 1641, made a great reputation by publishing his *Religio Medici* in 1642, followed this with *Pseudodoxia Epidemica* in 1646, with *Hydriotaphia* and *The Garden of Cyrus* in 1658, and wrote a large number of other papers unpublished during his life. In 1671 when Charles II visited Norwich he was knighted. He died in 1682, and was buried, 'pientissimus, integerrimus, doctissimus', in the Church of St Peter Mancroft.

Seventeenth-century epitaphs, like other obituary notices, are not usually reliable guides to character: but, superlatives excepted, the three adjectives on his memorial tablet are not in Browne's case undeserved or ill-chosen.

Of his religion his books, his letters, the attitude of his contemporaries, and the quality of his conduct are sufficient evidence. He was not a man of profound experience or deep and searching insight. Despite his claim that his life has been 'a miracle of thirty years' he does not seem to have known the sense of dereliction or of communion, of guilt or of redemption in any very poignant form—indeed he had no close acquaintance with the educative facts of personal failure or pain. His temperament is that of the 'once-born', and if this gives him a real calm and joy in accepting the simple happenings of daily life as evidences of God's care, it leads him on occasion into a superficiality which almost sinks into complacency. When he thanks God that he has escaped the sin of pride, and then proceeds with the naïveté of a child to display his accomplishments—his six languages 'besides the jargon and patois of several provinces', his knowledge of the 'names and something more' of all the constellations

1 *Sir Thomas Browne*, p. 37. In view of the large literature dealing with him and especially of T. Southwell's edition of his 'Notes and Letters on the Natural History of Norfolk' (London, 1902) detailed treatment of his work as a naturalist seems unnecessary.

2 *Diary*, 17 October.

and of 'most of the plants of my country'—it is difficult not to feel that the charge of egotism is well-deserved.[1] But, even so, no one but a German pietist would call him an atheist, or suggest that his religion even if immature was unreal or unchristian.

Indeed, even if his consciousness of sin is defective, this is due more to the strength of his sense of wonder and of the reality and presence of God in the world of nature than to any grave defect in his faith or morals. Like most of his contemporaries he has an intense *joie de vivre*, a delight in being alive in a day of new beginnings and of boundless possibilities. In the recovery of a knowledge of 'the wisdom of God in the works of creation' as Ray called it, Browne and his friends found a thrill and a fascination that made every day a 'May morning'. Withdrawing as they did from the quarrels of the sectaries and the self-seeking of the politicians, hating the cant of the Puritans only less than the vices of the Court, these men found in the world that the new philosophy was opening to them a complete satisfaction, an object for their contemplation, a stimulus for their intelligence, a scope for their endeavours. It is no accident that John Milton, whose temper was much nearer to theirs than is commonly recognised, not only found his great theme in Paradise Lost and Regained, but rose to his fullest splendour of phrase and imagery, of insight and artistry, in his descriptions of nature before the Fall.

In Browne as more explicitly in Henry More, and in Benjamin Whichcote, John Smith and Ralph Cudworth, this conviction of the worth of nature and the delight and duty of its exploration finds expression in a definitely Platonic theology. 'The severe schools', he writes in *Religio Medici* (*Works*, ed. G. L. Keynes, I, p. 17), 'shall never laugh me out of the Philosophy of Hermes, that this visible World is but a Picture of the invisible, wherein as in a Pourtraict, things are not truely but in equivocal shapes, and as they counterfeit some more real substance in that invisible fabrick'; and again on p. 20: 'Therefore sometimes, and in some things, there appears to me as much Divinity in Galen his books *De Usu Partium* as in Suarez Metaphysicks'; and again on p. 21: 'Thus there are two Books from whence I collect my Divinity; besides that written one of God, another of His servant Nature, that universal and publick Manuscript that lies expans'd unto the Eyes of all: those that never saw Him in the one, have discover'd Him in the other.' Thus he approximates to the ideas and language of the symbolists and the emblem books—but with the basic difference that for them the meaning of the symbols is determined by a traditional lore which has little relation to the facts of nature, whereas to Browne the unbiassed and realistic study of the natural order is the inescapable preliminary to the interpretation of its significance.

1 Cf. A. Whyte, *Sir T. Browne*, p. 22; and S. Johnson, *Life of Sir T. Browne* (1761), pp. x–xiii.

He is not, it must be admitted, consistent in his appeal for objective study, or free from the prejudices and superstitions of an earlier time. Like Milton he refused, at least at this time in his life, to accept the Copernican astronomy and insisted that 'to make a revolution everyday is the Nature of the Sun' (p. 21): like More he could say, 'I have ever believed, and do now know, that there are Witches: they that doubt of these do not onely deny *them* but Spirits, and are obliquely and upon consequence a sort not of Infidels, but Atheists' (p. 38)—which is exactly the contention of More's *Antidote against Atheism*, Book III and of *Saducismus Triumphatus*: like Boyle and Newton he believed in the creation-stories in Genesis and in the universality of Noah's flood (pp. 51, 30), but he is prepared to allow allegorical interpretations and to argue that many such matters are not worth arguing nor all miracles equally credible.

But it is not in his acceptance of this or that part of the tradition but in his general insistence upon the right and duty of enquiry, upon the competence and responsibility of human reason, and upon the method of verification and induction applied to all phenomena that the character of his work is plainly seen. Whatever the particular findings, a theology thus attained differs radically from the traditional and authoritarian systems whether of the papal church or of the Continental reformers. It belongs in essence to the modern world even if no modern of to-day would accept it as his own.

This leads us directly to the second epithet of his tombstone. His desire is to see life as a whole, to find an intimate correspondence between the two worlds of man's experience and of the tradition, and to interpret them both so coherently that thought and conduct may be integrated into a consistent unity of achievement. He wants to face the problems open-eyed and to attempt their solution by honest effort of mind and will.

That he accepted a duality of 'divided and distinguished worlds'—'for though there be but one to sense, there are two to reason, the one visible the other invisible'—and has coined the phrase 'Man that great and true Amphibium' to express our double status (p. 43) is of course true. The time had not come, has not yet come, when the distinction between outward and inward, temporal and eternal, can be resolved. But he is very sure that the worlds of our sojourn are phases in the experience of a single organism; and compares the passage from one to another to 'those strange and mystical transmigrations that I have observed in silk-worms' (p. 50). It is presumably on this ground that his German critics labelled him 'the patron of syncretism'.[1] Rather is it to his credit that he refuses

1 Cf. the passages quoted from eighteenth-century German critics by Wilkin, 'Supplementary Memoir' (*Browne's Works*, I, pp. lxvi–lxviii). Similarly B. Willey, *The Seventeenth Century Background*, in a sympathetic account of him speaks of 'the inter-availability of all his worlds of experience' (l.c. p. 44).

either to solve the problem by denying one or the other world or to rest content with their traditional contrast and juxtaposition.

It is in his second book that his 'Enquiry into Vulgar and Common Errors' discloses the range and honesty of his attempt to see life steadily and see it whole. *Pseudodoxia Epidemica or Enquiries into very many received Tenents and commonly presumed Truths* was first published in London, 'printed by T.H. for Edward Dod and are to be sold in Ivie Lane', 1646. A second edition 'corrected and much enlarged by the author together with some marginall observations and a table alphabeticall at the end' was 'printed by A. Miller for Edw. Dod and Nath. Ekins at the Gunne in Ivie Lane 1650'. A third 'corrected and enlarged' was 'printed by R.W. for Nath. Ekins at the Gun in Pauls Church-Yard 1658'. A fourth bound up with his *Urn-burial* and *Garden of Cyrus* appeared in the same year, 'printed for Edward Dod and are to be sould by Andrew Crook at the Green Dragon in Pauls Churchyard'. A fifth was published in 1669 'printed for the Assigns of Edward Dod'. A sixth which he declared to be 'compleat and perfect' was 'printed by J.R. for Nath. Ekins 1672'.[1]

Being in each case freshly set up there are in all the editions differences of pagination and spelling; and as G. L. Keynes has pointed out the printing deteriorates until in the fifth edition it is thoroughly bad. But major differences only occur in the second, third and sixth editions, the fourth and fifth being virtually reprints of the third. The largest changes were made in the second edition, where fresh chapters and paragraphs were freely inserted and there was a considerable amount of excision and revision. Smaller additions and a new chapter in Book III on 'Sperma-Ceti', based upon the two Whales stranded in Norfolk 'near Hunstanton above twenty years ago' and 'near Wells not many years since'[2] to which we have already referred, were inserted in 1658. A final revision and new details, but not such as to alter the substance of the book, are evident in 1672, one of the most interesting being the statement in Book III, chapter 13, that 'Toadstones are at last found to be taken out of a Fishe's mouth... the Lupus Marinus a Fish often taken in our Northern seas, as was publickly declared by an eminent and learned Physitian' (ed. Keynes, p. 229): in the margin is added a note, 'Sir George Ent', an error on Browne's part due to a mis-reading of the *Pinax*;[3] for this theory of Toadstones was put forward by Christopher Merret at the meeting of the Royal Society on 9 March 1664 (cf. Birch, *History of the R.S.* I, p. 392).

1 Full bibliographical details of these editions are in G. L. Keynes, *A Bibliography of Sir T. Browne*, pp. 48–56.

2 Apparently in 1656, cf. Browne's letter to Merret (*Works*, VI, p. 367); E. Gosse, *Sir T. Browne*, pp. 78–80, reverses the dates of these and represents Browne as visiting that at Hunstanton—which is a mistake: he visited that at Wells.

3 Cf. above, p. 329 for Merret's account of this. Ray, *Hist. Pisc.* p. 131, correctly ascribes the theory to Merret.

The intention of the book is admirably stated in the aphorism from Lactantius, inserted at its end, in the second edition, 'Primus sapientiae gradus est falsa intelligere'. Browne in his *Religio Medici* had set out his convictions and philosophy. Here he applies them by a long and detailed examination of popular errors, classifying these into six groups after an introductory book on their sources and the authors chiefly responsible for their propagation. Among these he mentions first Herodotus in spite of acknowledgment of his defence by Camerarius and Stephanus, but is most critical of Dioscorides and Pliny of whom he remarks 'what is very strange, there is scarce a popular error passant in our dayes which is not either directly expressed or diductively contained in this work': Solinus, Athenaeus, St Basil and St Ambrose, Isidore, Albertus Magnus, Hieronymus Cardanus and several others are strongly criticised: Oppian he highly praises—though picking out certain mistakes: of Sir John Mandeville's work he says that it 'may afforde commendable mythologie but in a naturall and proper Exposition it containeth impossibilities and things inconsistent with truth' (ed. Keynes, II, pp. 56–64).

The errors are: Book II, 'concerning minerall and vegetable bodies'; III, 'Animals'; IV, 'Man'; V, 'Pictures'; VI, 'Tenents geographicall and historicall'; VII, 'Tenents generally received and some deduced from the history of holy Scripture'. The range of reading is not unworthy of the subjects and shows a remarkable acquaintance with almost the whole of the relevant literature, real scholarship in the treatment of its text, and a high degree of insight and ingenuity in its interpretation. Even in the case of Scriptural records he encourages allegorical and symbolic explanations as in his discussion of the Fall in Book I, chapters 1 and 2, and criticises his contemporaries because 'their apprehensions are commonly confined unto the literal sense of the text; from whence have ensued the gross and duller sort of Heresies' (II, p. 26). It is not therefore surprising that he lays down as an axiom, 'Nor is onely a resolved prostration unto Antiquity a powerful enemy unto Knowledge, but any confident adherence unto authority, or resignation of our judgements upon the testimony of any Age or Author whatsoever' (II, p. 50). Such testimony depends for its credibility upon 'the solid reason or confirmed experience' of him who utters it. As examples of the right kind of evidence he quotes 'Raymund Sebund [died 1432 at Toulouse] a Physitian of Tholouze' who 'hath written a natural Theology'—abridged by Jan Amos Comenius with title *Oculus Fidei*, Amsterdam, 1661; and 'Hugo Grotius, a Civilian' and his 'excellent Tract of the verity of the Christian Religion'—translated into English as early as 1632 (II, p. 53).

A good example of his handling of a subject, which will illustrate the principles that he enunciates, is to be found in his exposition of magnetism in the chapters 'concerning the Loadstone', Book II, chapters 2 and 3. He begins: 'And first we conceive the earth to be a magnetical body'; and so though accepting it as 'the center of the universe' he recognises the 'effluviums' or attractions of all bodies as both astronomically and philosophically important (II, pp. 100–1), quoting the different theories of René Descartes[1] and of Sir Kenelm Digby as to their nature. In the discussion he makes full use of the work of 'Dr Gilbert not many years past a Physitian in London...the Father Philosopher who discovered more in it than Columbus or Americus did ever by it [William Gilbert, 1540–1603, physician to James I, author of *De Magnete*, 1600]' (p. 110), but refers freely to the 'Tables of declination which Ridley [Mark Ridley, 1560–1624, physician to Czar of Russia, author of *Short Treatise of Magneticall Bodies*, 1613] received from Mr Brigs in our time Geometry Professor in Oxford' (Henry Briggs, 1561–1630; this Table was published in 1602), 'Renatus des Cartes in his *Principles of Philosophy*' (René Descartes, 1596–1650), 'Cabeus' (Niccolo Cabeo, 1585–1650, wrote *Philosophia Magnetica*, 1629), 'Helmontius' (Jan Baptista van Helmont, 1577–1644), 'Kircherus' (Athanasius Kircher, 1602–80) and 'Licetus' (Fortunio Liceti, 1577–1657). Having stated his view of the nature of Magnetism he deals with the 'errors' connected with it—that Garlic destroys attraction as stated by Pliny, Solinus and many others including Mattioli and Johann Lange; that 'mercurial oil' has the same effect, by Paracelsus; that the nails fly out of a ship approaching a loadstone mine, by Serapion the Moor; that Mahomet's tomb is suspended in air between two magnets as generally believed; and a number of other medical and magical beliefs.

Of plants which should surely have attracted his special attention as doctor and pharmacist he has very little to say in this book,[2] and that only in a couple of chapters. At this very time Oswald Croll, a follower of Paracelsus, 'a great philosopher and hermeticall physitian',[3] who in 1609 had delivered himself in his book *De Signaturis* of the doctrine that 'God has stamped upon each plant legible characters to disclose the uses', was being popularised by William Coles,[4] fellow of New College, who in his *Adam in Eden* was classifying all plants 'by appropriating to every part of the body (from the Crown of the Head to the Soal of the Foot) such herbs whose grand uses do most specifically and by signatures thereunto belong'—so that he begins with the walnut 'the perfect Signature of the

1 Gosse, l.c. p. 37, states that Browne 'never once makes mention [of Descartes] in any portion of his writings'—an error pointed out by R. Sencourt, *Outflying Philosophy*, p. 89.

2 He left a large number of miscellaneous notes printed by Keynes, v, pp. 347–76: but these are not of great importance or interest.

3 So Coles, *Adam in Eden*, p. 247. Croll's chief book was translated into English in 1657 by H. Pinnell, *Philosophy Reformed*.

4 Cf. above, p. 234.

Head' and ends six hundred pages later with the Ladies' Bedstraw 'to bath the Feet of Travellers'. Here should be abundant material for our critic—and in his particular province. Yet the doctrine of 'signatures', the equally absurd doctrine that every country bore the plants specially suited to its diseases, and the vast mass of legendary 'cures' are ignored by him—which suggests that he did not carry his principles into his practice! He condemns the 'false conceptions of Mandrakes' as Turner had done a century before; suggests that 'the Thorn at Glassenbury' flowering at Christmas is not a miracle since there are others including one at Parham Park in Suffolk; and denies that 'ferrum equinum' draws iron, or bay-leaves protect from lightning, or almonds prevent intoxication. But almost his only other point, that Mistletoe (*Viscum album*) is not, as is commonly supposed, generated from seeds let fall by birds, but is 'an arboreous excrescence bred of a superfluous sap which the tree itself cannot assimilate' (II, p. 167), is a grave blunder.

Nevertheless there is in chapter 7 a short paragraph that raises a point of interest. It deals with the familiar prodigy of the shower of wheat. 'Only this much', he says (II, p. 175), 'we shall not omit to inform, that what was this year found in many places and almost preached for wheat rained from the clouds was but the seed of Ivy-berries which somewhat represent it; and though it were found in steeples and high places, might be conveyed thither and muted out by birds: for many feed thereon and in the crops of some we have found no less than three ounces.' This paragraph was added in the second edition and the year in question is therefore 1649–50. Browne plainly claims the explanation as his own. Yet in June 1661 Colonel afterwards Sir Samuel Tuke, Evelyn's cousin, brought to the Royal Society the history of the seeds rained down in Warwickshire—that on 30 May, Henry Puckering had collected these seeds at Tuchbrook two miles from Warwick and sent them to the King, and that Arthur Mason had reported the same from Shropshire. Tuke had then declared that these seeds were not wheat but ivy and had been excreted by Starlings—all this without any sort of reference to Browne.[1] Merret in his *Pinax*, as we have seen, gave this same explanation of the prodigy.

In the book on animals he is much more voluminous and correct. He deals at length with some of the best-loved traditions of the Elephant and the Bear, the Wolf and the Deer, the Kingfisher and the Salamander, the Viper and the Chamaeleon, the Ostrich digesting iron and the Stork that lives only in republics.[2] Many of these he condemns by appeal to his own experiments; others he shows to be ridiculous; others he refutes by evi-

1 Cf. Birch, *History of the R.S.* I, p. 29.

2 For this belief, cf. Fynes Moryson in *Shakespeare's England*, ed. C. Hughes, p. 349.

dence. He utterly rejects the Gryphon and the Phoenix, is very sceptical of the Amphisbaena, and shows how contradictory are the witnesses to the Unicorn. Perhaps the most interesting sections are those in chapter 27 where he discusses the singing of the Swan, citing Aldrovandi's excellent account and picture of the windpipe of the 'Elk' (or Whooper, *Cygnus cygnus*) in *Ornithologia*, III, p. 5, though he mocks at his report of the music of the Swans on the Thames (*Ornith.* III, p. 7, quoting the evidence of George Brown!); where he expounds the 'mugient noise' of the Bittern (*Botaurus stellaris*); where in regard to the traditional enmity of Toad and Spider he remarks, 'what we have observed herein we cannot in reason conceal; who having in a glass included a Toad with several Spiders, we beheld the Spiders without resistance to sit upon his head and pass over all his body; which at last upon advantage he swallowed down and that in few hours unto the number of seven' (II, p. 296); and where he pays a warm tribute to Thomas Penny[1] for his proof that an Earwig could fly.

To comment upon the rest of the book is unnecessary, though as proof of the vast difference between the medieval and the modern world and evidence of the wide extent of the changes necessitated by the New Philosophy—as well as an occasional reminder of the persistence of superstitions even till to-day—such an examination is of interest. In Book V there are a number of other points bearing upon our subject—a good discussion of the traditional 'Pelican in its piety' as compared with the actual bird which Browne knew; an excellent protest against the painting of Adam with a navel, as by Raffaelle and Michelangelo, since this would imply that 'the Creator affected superfluities, or ordained parts without use or office' (III, p. 97); and an admirable dismissal of palmistry on the ground that if true 'it seems not confinable unto man, but other creatures are also considerable; as is the forefoot of the Moll [Mole, *Talpa europaea*] and especially of the Monkey wherein we have observed the table line, that of life, and of the liver' (III, p. 152).

That Browne stood between two epochs, that he carried over from the medieval a belief in the devil similar to that of Milton, that he did not fully accept the Copernican astronomy or approve the mechanical view of nature as sufficient, and that there are frequent inconsistencies and contradictions in his work, is no doubt true. But it is difficult to feel that Professor Willey does justice to him when he insists that he is typical 'of the double-faced age in which he lived, an age half scientific and half magical, half sceptical and half credulous' (l.c. p. 41). The age, as Browne reflected it, is no longer facing both ways; it has made the turn; and in

1 That he knew the *Theatrum* is shown by his repetition in the *Garden of Cyrus*, *Works*, IV, p. 103, of the statement that the head of the caterpillar becomes the tail of the butterfly: cf. G. Taylor, *Insect Life in Britain*, p. 34, and above, p. 187.

intention Browne is definitely modern. He desired to reject all infallibilities, to test all propositions, to expose all superstitions. But, unlike too many of his successors, he refuses to limit his view of the universe to the realm of the mathematical and mechanistic. Being a man of well-developed and wide interests with a common sense which prevents him being blinkered by logic or swept off his feet by passion, he insists upon treating the whole content of his experience as somehow congruous and explicable. He intends, so far as he can, 'to see life steadily and see it whole', and to tell what he has seen. He will not easily call anything 'common or unclean' or say that it is no concern of his. He is not a great thinker, a great theologian, or a great scientist: and in consequence his attempts are not as satisfactory as those of some of his contemporaries. But at least he represents the New Philosophy at its best, before it had been narrowed and desiccated by the pre-eminence of Descartes and Newton; and shows us the quality of those first modern men, the founders of the Royal Society, before Walter Moyle in 1719 was compelled to lament that 'I find there is no room in Gresham College for Natural History: Mathematics have engrossed all' (*Works*, ed. Sargeant, I, p. 422).

For he combines, however incompletely, the two worlds of observed fact and of inferred meaning, of outward sign and inward significance, which had been sundered by the medieval failure to relate its symbolism to actuality and which were even then being driven into conflict. In the *Pseudodoxia*, as we have seen, it is to the test of experiment that he continually refers: where possible, he applies the test himself; if not, he quotes such evidence as he deems reliable. But he does not confine himself to the bare fact or refuse to recognise the general principles which it illustrates or the hypotheses which it suggests. This involves for him constant references to the emblematists: he knows the work of Andrea Alciati (cf. III, p. 317) and repeatedly refers to the huge *Hieroglyphica* of Giovanni Pietro Valeriano: nor does he treat them with contempt. However strongly he may criticise their errors, they were standing for a belief in the unity, the coherence, and the meaning of the universe; and though their data must always be subject to correction, their interpretations had an intention which must not be despised. In Browne's world they too have their place.

And so to his learning—the learning which for his whole philosophy was the indispensable condition of intellectual, moral and religious well-being. That he had special aptitudes for it, a zest for knowledge, a capacious and retentive memory, a lively brain, and an excellent education, gave him great advantages. But it is not so much the mass and variety of his intellectual interests as their peculiar quality that deserves our attention. For this quality which he shares with very many of his contemporaries is a new and modern thing.

We can see it wherever we study anything of his output. Like all educated men until recent times he knew and revered, could quote and expound, the Greek and Latin Classics. But his attitude to them is one of interest and admiration, not of subservience or reverence. Unlike the men of the Renaissance he has no sort of sense that when Aristotle has spoken the case is ended, no sort of feeling that Galen or Dioscorides any more than Augustine or Luther carries a final authority. Even in the case of Scripture though he accepts its utterances by faith and even asserts his agreement with Tertullian's 'Certum est quia impossibile est' (I, p. 13) he does not hesitate to admit the attractiveness of rationalistic interpretations or the right of reason and scholarship to scrutinise and comment. Unlike earlier generations he and his were not content to accept or even to find formulae, and while they did not venture to renounce all traditional authorities they at least maintained axioms which made a drastic criticism of them inevitable.

So, too, and this comes out plainly in his *Hydriotaphia*, his antiquarian interests, very strong and constant in a man temperamentally conservative, were those of a genuine student seeking to find out the significance of the past for its own sake and not in order to buttress up a theory or to reinforce a prejudice. Thus when his son Edward is travelling to Hungary and the Balkans, he is as keen as any member of the Royal Society that he should make the most of his opportunities, and do his best to answer the questions which they or Oldenburg so searchingly supplied (cf. VI, p. 46)—not, be it noted, for any political or practical advantage but simply because so men's knowledge could be increased.[1]

Characteristic of this desire for knowledge is his very evident interest in geography and map-making. From many allusions in his letters he had evidently made a considerable collection of maps both British and European and had also a number of books descriptive of foreign countries in his library, such as those named in his letter of 28 April 1669 (VI, pp. 50–1). It was of course the age in which cartography first began to be taken seriously: Christopher Saxton and John Speed had mapped the counties of England; John Taylor had explored its waterways; John Ogilby was making his admirable survey of its roads; since Abraham Ortel had produced his *Theatrum Orbis Terrarum* in 1570, atlases had been steadily improved; Browne's knowledge, though much more evident than in most of his contemporaries, is nevertheless typical of their outlook.

So, too, in his own profession though he is wholly honest in claiming the work of Galen as a contribution to Divinity, he has no such veneration for it or for any other authority as the physicians of the generations from Linacre to Culpeper had felt and shown. ''Εκ βιβλίου κυβερνῆτα [a statesman by the book] is grown into a proverb; and no less ridiculous are they

1 Cf. Browne's letters to Oldenburg in *Works*, VI, pp. 387–91.

who think out of book to become physicians' is the sentence with which he begins his long letter of advice to the young medical student, Henry Power (VI, p. 277). He is insistent that the foundation must be laid in anatomy; and very evidently practised what he preached. Harvey's work he accepted with enthusiasm ('which discovery I prefer to that of Columbus', l.c.); and later he wrote warmly of the researches of John Mayow on Respiration. For drugs, though he lays first stress on herbs, yet he does not ignore 'the useful part of chymistry'. Paré, Hartmann, Fernel and Rivière, representatives of very different types of practice, are all recommended by him: but they are to be used only as guides, and the doctor's own observations and clinical experience are the essential qualification. For himself he seems to be more concerned with the right use and dosage of Peruvian bark (quinine, then just coming into regular employment for agues) than with any other single problem; and this for one living near the fenland is a very proper concern.

In natural history his interest was all-inclusive: but zoology attracted him more than botany. Of the latter he seems to think that there is now not much more to be learned; that the physic gardens and books of dried specimens supplementing the relatively large literature of catalogues, herbals and pharmacologies have almost exhausted the subject. Thus in June 1681 he wrote to his son 'I shall not persuade you to buy Dr Morison's [Robert Morison, 1620–83] herball of five pound price [the *Plantàrum Historiae Universalis Oxoniensis pars secunda* issued 1680]. It was ill contrived to print it first in small volumes [presumably his *Praeludia* and *Umbelliferarum Distributio*] and then afterwards other peeces in large volumes: and fewer than ever are like to bee so criticall as formerly in botanicks especially in the nomenclature and distinction of vegetables' (VI, p. 222). Indeed, though in 1648 Power when at Cambridge had echoed Turner's complaint, which Ray repeated not less vehemently in 1660, that there was no competent teaching of the subject, the practice of issuing and circulating collections or 'dry gardens' from Padua, which was well established by the beginning of the century and to which Browne alludes in a letter of 28 June 1679,[1] must have made the students' path much less arduous. This and the gardens of Morgan at Westminster and of the University at Oxford gave opportunities for the identification and study of plants which put that side of botany within the reach of any diligent student. Nevertheless, even if the subject never greatly attracted him, Browne welcomed eagerly the proposals of Nehemiah Grew in May 1682 for subscriptions to his forthcoming book, *The Anatomy of Vegetables*, and collected contributions for copies from three of his neighbours.

In zoology his keenness is much more evident. As we have seen he had made a substantial collection of stuffed or preserved specimens, and his

1 VI, p. 137 and see above, p. 265.

must have been one of the earliest of private museums. No doubt such 'cases' were not very savoury; and we can sympathise with the domestic who with plague next door thought it safer to burn them; but if afterwards he confined his collection to 'the eggs of all the foule and birds he could procure'[1] ('the Cassowary[2] or Emeu whose fine green channelled egge I have', VI, p. 179), to bones and shells and minerals, he replaced the skins by pictures drawn either by his wife and daughter or by professionals. There is an attractive reference to this in a letter of 6 September 1680 to his son (VI, p. 182): 'I thought to have sent a spider by him, which was brought mee out of the feilds, large and round, and finely marked green, and even allmost as bigge as the figures inclosed, drawne by your mother, for your sisters dared not doe it. It may bee seen in Moufetus, and I have had of them before, and one drawne out in oyled colours upon an oyled paper. I do not find it in Dr Lister's [Martin Lister, 1639–1712] table of spiders though hee hath writt well *De Araneis* [*Tres Tractatus*, 1678].' His pictures of birds—the admirable Little Auk and others which he sent to Ray—were used in the *Ornithology* and though not great works of art are easily identifiable. He seems also to have collected pictures of Fishes; and here perhaps his most original work in natural history was done. The document on Fishes found in Norfolk is a serious and useful contribution—far more complete than Turner or Merret and containing not only a good survey of the species and distribution of the commoner fishes, but several records of real importance. His identifications are based upon Rondelet's great book, and are in the main accurate. Southwell's notes in his edition of Browne's *Natural History of Norfolk* draw attention to the report of young Whales (probably not Cachelots), of a Saw-fish (*Pristis antiquorum*) at Lynn, of the 'Moonefish' (*Orthagoriscus mola*) of 200 pounds weight in 1667 at Mundesley (a note added later to the script), and many others of special interest.

Moreover, he was not content with dead specimens but early in his career began to keep them alive. Besides the Eagle out of Ireland (*Aquila chrysaëtus*) which he had kept for two years and the 'Fenne Agle' (*Haliaëtus albicilla*) which 'grewe so tame that it went about the yard feeding on fish' (V, p. 378), he had managed to keep two Shearwaters whose 'draught' he sent to Merret—it was afterwards published by Ray—'cramming them

1 So Evelyn, *Diary*, 17 October 1671, describing his visit to Browne at Norwich.

2 A Cassowary had been presented to James I by Prince Maurice of Orange and laid eggs as described by William Harvey (*Works*, p. 188). The first to be brought to Europe came to Holland from 'Banda one of the Moluccas [Ceram]' in 1596 (so Aldrovandi, *Ornith.* III, p. 54: he calls it Eme) and lived for some time in various noble quarters: Christiaan Porret of Leyden apparently obtained a skin and an egg in 1603: cf. De l'Ecluse, *Exoticarum*, V, pp. 97–9. Ray in *Ornithologia*, p. 105, says: 'We have seen four in London, one in St James's [it died there in 1656, *Mus. Trad.* p. 3], and three belonging to Mr Maydston a merchant, brought from the East Indies.'

with fish' for five weeks though they refused to feed themselves and when he 'wearied with cramming them' lived for seventeen days without food (VI, p. 372)—an unpleasant observation which is yet not without interest in view of their survival through weeks of stormy weather when their habits must make feeding almost impossible. He also kept in his garden a Bittern (*Botaurus stellaris*) for two years 'feeding it with fish, mice and frogges' (*Birds in Norfolk*, v, p. 382), and Ruffs (*Philomachus pugnax*) 'from May till the next spring' (p. 383) noting the loss of their ruffs in autumn. His eagerness to obtain and study new species is clear from the flood of letters which he sent to his son in the last winter of his life when Edward had obtained one of the Ostriches[1] which the Ambassador of the King of Fez and Morocco had brought to England (VI, pp. 235–40); and his advice is not only as to its structure and the accuracy of the pictures in Aldrovandi, Johnstone and Willughby, but as to its diet and warmth. Even though the bird died in a day or two, his interest continued; and the records of its dissection sent by Edward provoked two further letters of praise, comment and suggestion (pp. 240–3).

The result of this interest in the observation of anatomical and physiological details is to introduce a new and relatively modern outlook in biological studies. Browne unlike Turner or Johnstone is as we have seen wholly scornful of the phoenix and other traditional creatures, and highly critical of Pliny and his successors. He retails indeed the story of the 'Gulo' (or Wolverine) 'a devouring ravenous quadruped frequent in Lithuania and Moscovie....which filleth itself with any caryon and then when it can eat no more compresseth itself between two trees standing neere together and so squeezeth out through the gutts what it hath devoured, and then filleth itself agayne' (pp. 85–6), a story which goes back to Olaus Magnus and had been repeated by Gesner and Topsell and Robert Lovell. But he does so after raising the difficulty of 'the division of the gutts, their complications, foulds and caecum' and on the assurance that Pieter Pauw (1564–1617), the famous anatomist of Leyden, had dissected a 'gulo' and found that it had only one 'intestinum rectum', eminently suited to the treatment stated in the legend.

It is this appeal to factual evidence, to observation and experiment, that characterises Browne and his generation, and proves how completely the method of philosophy had been revolutionised. We say method, rather than goal; for the attempt of Macaulay, in his essay on Francis Bacon, and of many others including some of the modern historians of science to suggest that the New Philosophy aimed at utility rather than truth, thus setting the new and the old at variance in respect to the end that

1 Ostriches were kept in his park by James I in Harvey's time (*De Generatione*, *Works*, p. 188), and to these Browne perhaps refers when he says that he saw an Ostrich at Greenwich when he was a boy (*Works*, VI, p. 179).

they had in view, is at once exaggerated and superficial. The fact is not that Plato or Thomas Aquinas was indifferent to the utility or moral effect of their teaching (few suggestions could be more unjust to them), but that Plato regarded the phenomenal and Thomas the natural as unreliable guides to the knowledge of the real or the divine. They did not and could not stress the study of nature as a necessary discipline for and element in philosophy: indeed each of them tended to regard such study as a waste of time and on occasion a deadly illusion. In consequence they promoted the formation of a cosmology in which metaphysical theories formed a bed of Procrustes for physical facts—with the result that 'truth' (thus conceived) soon ceased to have any exact relation with utility or even with reality.

It is an exaggeration to say that Bacon and the Christian thinkers, who followed him and formed the Royal Society, were primarily concerned with the benefit of their discoveries to mankind—though being Christian they were not and could not be indifferent to the fruit of their thought. Their basic conviction was that from the accurate understanding of the creation they could gain a true and effective knowledge of the Creator and so could bring the life of mankind into a fuller and more intelligent correspondence with His will. Thus philosophy was to them not the exposition of something known or, rather, supernaturally taught, but the exploration of something knowable, to be sought along every road opened by human research and divine revelation. If they were still disposed to separate revelation and research, at least they were convinced that their lessons were similar and mutually corroborative. Because there was no conflict between them, there was no antithesis between the true and the beneficial. God's will is man's welfare; and the search for knowledge is therefore a service of the highest value and utility. Speculation without reference to verifiable data was to them as unprofitable as observation without explanatory hypothesis. They were neither theorists nor technicians; for the plain reason that they regarded theory and practice as inevitably interdependent.

That despite his affectations and inconsistencies this was Browne's belief is clear from the passage which Professor Willey[1] quotes from his last work, *Christian Morals*. There at the end of his life, set in a huge collection of slightly laboured aphorisms, is a declaration of his faith in the principles of the New Age which might stand alongside Ray's similar apostrophe in the preface to his *Synopsis Britannica*. 'Let thy Studies be free as thy Thoughts and Contemplations, but fly not only upon the wings of Imagination; joyn Sense unto Reason, and Experiment unto Speculation, and so give life unto Embryon Truths and Verities yet in their Chaos. There is nothing more acceptable unto the Ingenious World

1 *Seventeenth Century Background*, pp. 48–9.

than this noble Eluctation of Truth; wherein, against the tenacity of Prejudice and Prescription, this Century now prevaileth. What Libraries of new Volumes aftertimes will behold, and in what a new World of Knowledge the eyes of our Posterity may be happy, a few Ages may joyfully declare' (I, p. 123).

In the New Philosophy thus conceived there was indeed attained a synthesis of the two movements which as we have seen had been responsible for the breakdown of medievalism. The scholars, who appealed to antiquity, to the Greek and Latin Classics, to the Greek New Testament, and to the example and teaching of the earliest Church, actually found in the writings of the Greek Apologists and the Christian Platonists of Alexandria an attitude towards nature, a concept of progressive revelation, and an insistence upon education and intellectual effort wholly appropriate to the new insistence upon observation and experiment. The naturalists, striving to develop hypotheses consistent with the fresh data disclosed in astronomy and geology, botany and zoology, found themselves anticipated by thinkers, who had drawn their conclusions not from the study of the physical world, but from the ancient Logos-theology of Justin, Clement and Origen.

Joseph Mead, the forerunner of the School, an 'accurate herbalist and excellent anatomist';[1] Henry More, his pupil, the friend of Descartes; Benjamin Whichcote the critic of orthodox Calvinism; Ralph Cudworth in his *True Intellectual System of the Universe*; these and others developed into a great Christian philosophy the faith to which Browne's *Religio Medici* had given popular expression. Robert Boyle in his *Disquisition about the Final Causes of Natural Things*; John Ray in his *Wisdom of God in the Works of Creation*; Richard Bentley in his Boyle Lectures; Nehemiah Grew in his *Cosmologia sacra*; William Derham in his *Physico-Theology*; these and others expounded the findings of their scientific researches in terms of a similar philosophy. By the close of the century it seemed as if the lines for the development of a coherent, integrative and verifiable outlook had been securely laid. If these were followed, difficulties, however large and whether raised by tenacious defenders of the old or narrowly arrogant champions of the new, could in time be happily surmounted.

That these hopes were not fulfilled and that we have still failed to reach a satisfying and integrated *Weltanschauung* is due to a wide variety of causes, some of which were unforeseen when our period closed. But already the obstacles which have proved chiefly destructive had made their appearance. When Browne put the book of nature and the book of revelation side by side such 'syncretism' horrified the reformed churches of the Continent although his concessions to tradition were in fact large enough to foreshadow the conflict between Genesis and geology. Protes-

1 So J. Worthington, *Life*, pp. iv, viii.

tant bibliolatry, springing out of the recovery of knowledge of the Bible through the work of scholars and translators, had been so valuable an instrument in effecting the Reformation that its advocates could not but erect it into an irreformable dogma—with results that were already becoming unfortunate. Evidently it would take time, a long time, to convince the religious that the same methods of exact study and constant verification must be applied to the origins of their faith as were given to the origins of life or of the solar system.

In the countries of the Reformation this process has gone forward, more slowly than might have been expected, but in the main without serious set-backs. But in the countries of the Roman Catholic Church far more deeply-seated hostility had to be faced; for the medieval tradition which had there succeeded in resisting the Reformation was thereby hardened against change. A compromise, segregating science from religion, and defining their respective spheres, was perhaps inevitable: and already, before the scientific movement in England had begun to manifest itself, an attitude in philosophy definitely antagonistic to any synthetic outlook had made its appearance. This arose largely out of the necessary development of scientific studies, that is out of the fact that any advance in biology must wait upon and be conditioned by a thorough investigation of physics and chemistry. Physiological work like that of Mayow or Grew revealed how important was a true understanding of the mechanisms of life and behaviour. The achievements of technicians and engineers supplied appropriate analogies and suggested methods which anatomists and vivisectors were not slow to follow. The consequent emphasis upon quantitative and materialistic interests was speedily made the basis for an accommodation with the representatives of the medieval tradition. Descartes, protecting himself by the dogma that all animals were mere automata, found the Roman Church, which denies that animals have 'souls' and allows their use as machines, in agreement with him on this issue; and it was easily admitted that, provided he left mankind and the supernatural unassailed, he might legitimately apply mechanistic analogies to all the natural order and even suggest that for its interpretation no other categories were required.

Both as a philosopher and as a scientist Descartes hardly seems to deserve the reputation and influence which he has certainly attained.[1] But the defects of his thought and work were in practice outweighed by

1 Professor C. D. Broad, *Camb. Hist. Journal*, VIII, p. 54, has recently summed up the value and defects of his work in an article which ends with the two sentences: 'At certain periods in the development of human knowledge it may be profitable and even essential for generations of scientists to act on a theory which is philosophically quite ridiculous. And the success of this procedure may blind people for centuries to the fact that its assumptions are quite incredible if taken to be the whole truth and nothing but the truth.'

its value in providing an escape for scientists from the dilemma with which Catholicism presented them. This dilemma had been familiar ever since the establishment of the Inquisition: either matters of general human importance must be left out of account, or discussion of them must conform to the rigid definitions of the tradition. The fate of Galileo illustrated once for all the grave effects of such an alternative. Descartes succeeded in concealing the direct challenge to the Church which his theories in fact involved, by asserting the traditional duality of matter and spirit, and insisting that in his mechanistic explanations he was dealing only with the soulless animals and with the material elements in mankind.[1] That in consequence science would confine itself to the sphere of weight and measurement, that biological, psychological and religious studies would be distorted, and that a radical dichotomy of such a kind would have disastrous effects upon human welfare, was a result which Descartes, Borelli and the Cartesians neither intended nor foresaw. That it has occurred and been widely accepted not only by those who are unaware of its difficulties, but by scientists and Christians, both of whom ought to stand for an all-embracing interpretation of the universe, shows how far the hopes of the early advocates of the New Philosophy have been disappointed. It is something of an irony that the scientific interpretation of nature should after four centuries of effort have replaced the hieroglyphs of medieval Catholicism by the robots of modern Behaviourism. For it may well be doubted whether the attempt to equate fable with fact or mechanism with meaninglessness is the more absurd. But that is not the end of the story. There are abundant signs that the quest for a new Religio Medici is beginning again; and that both of Browne's 'two books' will be employed in it.

1 For the effect of this compromise, cf. a pleasant account of Stephen Hales' Sunday, in J. Gilmour, *British Botanists*; pp. 22–3.

INDEX OF SUBJECTS

INDEX OF FLORA

INDEX OF FAUNA

INDEX OF PERSONS

Printed in the United States
By Bookmasters